Undergraduate Texts in Mathematics

Undergraduate Texts in Mathematics

Undergraduate Texts in Mathematics are generally aimed at third- and fourth-year undergraduate mathematics students at North American universities. These texts strive to provide students and teachers with new perspectives and novel approaches. The books include motivation that guides the reader to an appreciation of interrelations among different aspects of the subject. They feature examples that illustrate key concepts as well as exercises that strengthen understanding.

More information about this series at http://www.springer.com/series/666

Ramin Takloo-Bighash

A Pythagorean Introduction to Number Theory

Right Triangles, Sums of Squares, and Arithmetic

 Springer

Ramin Takloo-Bighash
Department of Mathematics, Statistics,
   and Computer Science
University of Illinois at Chicago
Chicago, IL, USA

ISSN 0172-6056   ISSN 2197-5604   (electronic)
Undergraduate Texts in Mathematics
ISBN 978-3-030-80529-6   ISBN 978-3-030-02604-2   (eBook)
https://doi.org/10.1007/978-3-030-02604-2

Library of Congress Control Number: 2018958346

Mathematics Subject Classification (2010): 11-01, 11A25, 11H06, 11H55, 11D85

This Springer imprint is published by the registered company Springer Nature Switzerland AG
The registered company address is: Gewerbestrasse 11, 6330 Cham, Switzerland

To Paria, Shalizeh, and Arad.
In the memory of my father.

Preface

This book came out of an attempt to explain to a class of motivated students at the University of Illinois at Chicago what sorts of problems I thought about in my research. In the course, we had just talked about the integral solutions to the Pythagorean Equation and it seemed only natural to use the Pythagorean Equation as the context to motivate the answer. Basically, I motivated my own research, the study of rational points of bounded height on algebraic varieties, by posing the following question: What can you say about the number of right triangles with integral sides whose hypotenuses are bounded by a large number X? How does this number depend on X? In attempting to give a truly elementary explanation of the solution, I ended up having to introduce a fair bit of number theory, the Gauss circle problem, the Möbius function, partial summation, and other topics. These topics formed the material in Chapter 13 of the present text.

Mathematicians never develop theories in the abstract. Despite the impression given by textbooks, mathematics is a messy subject, driven by concrete problems that are unruly. Theories never present themselves in little bite-size packages with bowties on top. Theories are the afterthought. In most textbooks, theories are presented in beautiful well-defined forms, and there is in most cases no motivation to justify the development of the theory in the particular way and what example or application that is given is to a large extent artificial and just "too perfect." Perhaps students are more aware of this fact than what professional mathematicians tend to give them credit for—and in fact, in the case of the class I was teaching, even though the material of Chapter 13 was fairly technical, my students responded quite well to the lectures and followed the technical details enthusiastically. Apparently, a bit of motivation helps.

What I have tried to do in this book is to begin with the experience of that class and take it a bit further. The idea is to ask natural number theoretic questions about right triangles and develop the necessary theory to answer those questions. For example, we show in Chapter 5 that in order for a number to be the length of the hypotenuse of a right triangle with coprime sides, it is necessary and sufficient that all prime factors of that number be of the form $4k + 1$. This result requires determining all numbers that are sums of squares. We present three proofs of this fact:

using elementary methods in Chapter 5, using geometric methods in Chapter 10, and using linear algebra methods in Chapter 12. Since primes of the form $4k + 1$ are relevant to this discussion, we take up the study of such primes in Chapter 6. This study further motivates the Law of Quadratic Reciprocity which we state in Chapter 6 and prove in Chapter 7. We also determine which numbers are sums of three or more squares in Chapters 9, 10, 11, and 12.

When I was in high school, I used to think of number theory as a kind of *algebra*. Essentially everything I learned involved doing algebraic operations with variables, and it did not look like that number theory would have anything to do with areas of mathematics other than algebra. In reality, number theory as a field of study sits at the crossroads of many branches of mathematics, and that fact already makes a prominent appearance in this modest book. Throughout the book, there are many places where geometric, topological, and analytic considerations play a role. For example, we need to use some fairly sophisticated theorems from analysis in Chapter 14. If you have not learned analysis before reading this book, you should not be disheartened. If anything, you should take delight in the fact that now you have a real reason to learn whatever theorem from analysis that you may not otherwise have fully appreciated.

Each chapter of the book has a few exercises. I recommend that the reader tries all of these exercises, even though a few of them are quite difficult. Because of the nature of this book, many of the ideas are not fully developed in the text, and the exercises are included to augment the material. For example, even though the Möbius function is introduced in Chapter 13, nowhere in the text is the standard Möbius Inversion Formula presented, though a version of it is derived as Lemma 13.3. We have, however, presented the Möbius Inversion Formula and some applicants in the exercises to Chapter 13. Many of these exercises are problems that I have seen over the years in various texts, jotted down in my notebooks or assigned in exams, but do not remember the source. The classical textbooks by Landau [L], Carmichael [Car], and Mossaheb [M] are certainly the sources for a few of the exercises throughout the text. A few of the exercises in the book are fairly non-trivial problems. I have posted some hints for a number of the exercises on the book's website at

http://www.math.uic.edu/~rtakloo

In addition to exercises, each chapter has a Notes section. The contents of these sections vary from chapter to chapter. Some of them are concerned with the history of the subject, some others give references to more advanced topics, and a few describe connections to current research.

Numerical experiments and hands-on computations have always been a cornerstone of mathematical discovery. Before computers were invented, or were so commonplace, mathematicians had to do their numerical computations by hand. Even today, it is hard to exaggerate the importance of doing computations by hand —the most efficient way to understand a theorem is to work out a couple of small examples with pen and paper. It is of course also extremely important to take advantage of the abundant computational power provided by machines to do numerical computations, run experiments, formulate conjectures, and test strategies to prove these conjectures. I have included a number of computer-based exercises in each chapter. These exercises are marked by (✠). These exercises are not written with any particular computer programming language or computational package in mind. Many of the standard computational packages available on the market can do basic number theory; I highly recommend SageMath—a powerful computer algebra system whose development is spearheaded by William Stein in collaboration with a large group of mathematicians. Beyond its technical merits, SageMath is also freely available both as a Web-based program and as a package that can be installed on a personal computer. Appendix C provides a brief introduction to SageMath as a means to get the reader started. What is in this appendix is enough for most of the computational exercises in the book, but not all. Once the reader is familiar with SageMath as presented in the appendix, he or she should be able to consult the references to acquire the necessary skills for these more advanced exercises.

This is how the book is organized:

- We present a couple of different proofs of the Pythagorean Theorem in Chapter 1 and describe the types of number theoretic problems regarding right triangles we will be discussing in this book.
- Chapter 2 contains the basic theorems of elementary number theory, the theory of divisibility, congruences, the Euler ϕ-function, and primitive roots.
- We find the solutions of the Pythagorean Equation in integers in Chapter 3 using two different methods, one algebraic and the other geometric. We then apply the geometric method to find solutions to some other equations. We also discuss a special case of *Fermat's Last Theorem*.
- In Chapter 4, we study the areas of right triangles with integer sides.
- Chapter 5 is devoted to the study of numbers that are side lengths of right triangles. Our analysis in this section is based on Gaussian integers which we briefly review. We also discover the relevance of prime numbers of the form $4k + 1$ to our problem.
- Chapter 6 contains a number of theorems about the infinitude of primes of various special forms, including primes of the form $4k + 1$. This chapter also makes a case for a study of squares modulo primes, leading to the statement of the *Law of Quadratic Reciprocity*.
- We present a proof of the Law of Quadratic Reciprocity in Chapter 7 using *quadratic Gauss sums*.
- Gauss sums are used in Chapter 8 to study the solutions of the Pythagorean Equation modulo various integers.
- In Chapter 9, we extend the scope of our study to include analogues of the Pythagorean Equation in higher dimensions and prove several results about the distribution of integral points on circles and spheres in various dimensions. In this chapter, we state a theorem about numbers which are sums of two, three, or more squares.
- Chapter 10 contains a geometric result due to Minkowski. We use this theorem to prove the theorem on sums of squares.

- Chapter 11 presents the theory of quaternions and uses these objects to give another proof of the theorem on sums of four squares.
- Chapter 12 deals with the theory of quadratic forms. We use this theory to give a second proof of the theorem on three squares.
- Chapters 13 and 14 are more analytic in nature than the chapters that precede them. In Chapter 13, we prove a classical theorem of Lehmer from 1900 that counts the number of primitive right triangles with bounded hypotenuse. This requires developing some basic analytic number theory.
- In Chapter 14, we introduce the notion of height and prove that rational points of bounded height are equidistributed on the unit circle with respect to a natural measure.
- Appendix A contains some basic material we often refer to in the book.
- Appendix B reviews the basic properties of algebraic integers. We use these basic properties in our proof of the Law of Quadratic Reciprocity.
- Finally, Appendix C is a minimal introduction to SageMath.

How to use this book. The topics in Chapters 2 through 7 are completely appropriate for a first course in elementary number theory. Depending on the level of the students enrolled in the course, one might consider covering the proof of the Four Squares Theorem from either Chapter 10 or Chapter 11. In some institutions, students take number theory as a junior or senior by which time they have, often, already learned basic analysis and algebra. In such instances, the materials in either Chapter 13 or Chapter 14 might be a good end-of-semester topic. When I taught from this book last year, in a semester-long course, I taught Chapters 1, 2, Example 8.6, 3, Chapters 6 and 7, the proofs of the Two Squares and Four Squares Theorems from Chapter 10, Theorem 9.4, and Chapter 13.

The book may also be used as the textbook for a second-semester undergraduate course, or an honors course, or a first-year master's level course. In these cases, I would concentrate on the topics covered in Chapters 8 through 14, though Chapter 4 might also be a good starting point as what is discussed in that chapter is not usually covered in undergraduate classes. Except for the first two sections of Chapter 9 that are referred to throughout the second part of the book, the other chapters are independent of each other and they can be taught in pretty much any order. Many of the major theorems in this book are proved in more than one way. This is aimed to give instructors flexibility in designing their courses based on their own interests, or who is attending the course.

I wish to thank the students of my Foundations of Number Theory class at UIC in the fall term of 2016 for their patience and dedication. These students were Samuel Coburn, William d'Alessandro, Victor Flores, Fayyazul Hassan, Ryan Henry, Robert Hull, Ayman Hussein, McKinley Meyer, Natawut Monaikul, Samantha Montiague, Shayne Officer, George Sullivan, and Marshal Thrasher. They took notes, asked questions, and, in a lot of ways, led the project. Without them, this book would have never materialized.

I also wish to thank Jeffery Breeding-Allison, Antoine Chambert-Loir, Samit Dasgupta, Harald Helfgott, Hadi Jorati, Lillian Pierce, Lior Silberman, William Stein, Sho Tanimoto, Frank Thorne, and Felipe Voloch, as well as the anonymous readers for many helpful suggestions. This book would have never seen the light of the day had it not been for the support and encouragement of my editor Loretta Bartolini.

My work on this project is partially supported by a Collaboration Grant from the Simons Foundation.

This book was written at the Brothers K Coffeehouse in Evanston, IL. The baristas at Brothers K serve a lot more than just *earl gray*. I thank Yelena Dligach who suggested that I write this book and Dr. Joshua Nathan for his care and support during the past few years.

Finally I thank my wife, Paria, and my children, Shalizeh and Arad, for their patience and encouragement. It is to them that this book is humbly dedicated.

Chicago, IL, USA   Ramin Takloo-Bighash
July 2018

Contents

Notation

The following notations are frequently used in the rest of the text:

- $\mathbb{R}$: The field of real numbers.
- $\mathbb{C}$: The field of complex numbers.
- $\mathbb{Q}$: The field of rational numbers.
- $\mathbb{Z}$: The ring of all integers.
- $\mathbb{N}$: The set of all natural numbers, i.e., all positive integers.
- $R[x]$: For a ring R, this is the ring of all polynomials in the variable x with coefficients in R.
- $[x]$: The integer part of a real number x, i.e., the largest integer m with the property that $m \leq x$.
- $\{x\}$: The fractional part of x, i.e., $x - [x]$.
- $|||x|||$: The distance of x to the closest integer, i.e., $\min(\{x\}, 1 - \{x\})$.
- $a \mid b$ for integers a, b: a divides b, i.e., there is an integer c such that $b = ac$.
- $a \nmid b$ for integers a, b: b is not divisible by a.
- $a \equiv b \bmod c$, with a, b, c integers such that $c \neq 0$: $c \mid a - b$.
- $M_n(R)$: The ring of $n \times n$ matrices with entries in the set R.
- $\mathrm{GL}_n(\mathbb{Z})$: The group of $n \times n$ integral matrices with determinant equal to ± 1.
- $\mathrm{SL}_n(\mathbb{Z})$: The group of $n \times n$ integral matrices with determinant equal to $+1$.
- $f(x) = O(g(x))$ for real functions f, g: If there is a constant $C > 0$ such that for all x large enough, $|f(x)| \leq C|g(x)|$.
- $f(x) = o(g(x))$ for real functions f, g. If

$$\lim_{x \to \infty} \frac{f(x)}{g(x)} = 0.$$

- $\phi(n)$ for a natural number n: Euler totient function.
- $\sigma(n)$ for a natural number n: The sum of the divisors of n.

- $d(n)$ for a natural number n: The number of divisors of n.
- $sqf(n)$ for a natural number n: The square-free part of n, i.e., the smallest natural number m such that $n = k^2 \cdot m$ for some natural number k.
- δ_{kl}: Kronecker's delta function, equal to 1 if $k = l$, 0 otherwise.
- χ_S for the subset S of a set X: The characteristic function of S, i.e., $\chi_S(x) = 1$ if $x \in S$, $\chi_S(x) = 0$ if $x \in X - S$.
- $\#A$ for a finite set A: The number of elements of the set A.

Part I
Foundational material

Chapter 1
Introduction

In the first section of this opening chapter we review two different proofs of the Pythagorean Theorem, one due to Euclid and the other one due to a former president of the United States, James Garfield. In the same section we also review some higher dimensional analogues of the Pythagorean Theorem. Later in the chapter we define Pythagorean triples; explain what it means for a Pythagorean triple to be primitive; and clarify the relationship between Pythagorean triples and points with rational coordinates on the unit circle. At the end we list the problems that we will be interested in studying in the book. In the notes at the end of the chapter we talk about Pythagoreans and their, sometimes strange, beliefs. We will also briefly review the history of Pythagorean triples.

1.1 The Pythagorean Theorem

Proposition XLVII of Book II of Euclid's *Elements* [20] is the following theorem:

Theorem 1.1. *In a right triangle ABC the square on the hypotenuse AB is equal to the sum of the squares on the other sides AC and BC, that is,*

$$AB^2 = AC^2 + BC^2.$$

Theorem 1.1 is usually attributed to Pythagoras (580 BCE-500 BCE) or at least to the Pythagorean school, and for that reason the equation

$$x^2 + y^2 = z^2, \tag{1.1}$$

satisfied by the side lengths of a right triangle, is referred to as the *Pythagorean Equation*.

There are hundreds of proofs for the Pythagorean Theorem. We will momentarily give the proof contained in Euclid's *Elements*. The proof is truly geometric and very

© Springer Nature Switzerland AG 2018
R. Takloo-Bighash, *A Pythagorean Introduction to Number Theory*,
Undergraduate Texts in Mathematics, https://doi.org/10.1007/978-3-030-02604-2_1

Fig. 1.1 Euclid's proof of
Theorem 1.1. The triangle
ABC is a right angle triangle
with *C* being the right angle

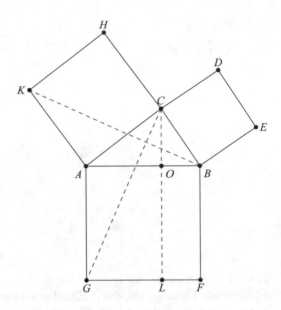

much in the Pythagorean tradition. In the argument, AB^2 is interpreted as the area of
the square built on the edge *AB*, and the theorem is proved by showing that the area
of the square built on *AB* is equal to the sum of the areas of the squares built on *AC*
and *BC*.

Proof (Euclid). Draw squares *ACHK*, *CBED*, and *ABFG* as in Figure 1.1. Pick a
point *O* on *AB* such that $CO \perp AB$. Draw the altitude *CO* from *C* and extend it to
intersect *GF* at *L*. Draw *CG* and *KB*.

   Since *ABFG* is a square, $AG = AB$. Similarly, $AC = AK$. Since $\angle GAB$ and
$\angle CAK$ are right angles, $\angle GAC = \angle BAK$. Putting these facts together, we conclude
$\triangle KAB \simeq \triangle CAG$. In particular the areas of these triangles are equal.

   Since *ACB* and *HCA* are both right angles, the line segment *HB* passes through *C*.
Consequently, the area of *KAB* is half the area of the square *ACHK*. Next, the area
of *CAG* is half the area of the rectangle *OLGA* as the shapes share the same base *AG*
and have equal heights. Hence, the area of *ACHK* is equal to the area of *OLGA*. A
similar argument shows that the area of the square *CBED* is equal to the area of the
rectangle *OLFB*. Finally, the sum of the areas of *OLGA* and *OLFB* is the area of the
square *ABFG*.   □

   This is by no means the easiest proof of the Pythagorean Theorem. Here we record
a famous proof published by James Garfield, the 20th president of the United States,
five years before he took office. This proof appeared in the New England Journal of
Education in 1876.

Proof (Garfield). Suppose *a*, *b*, *c* are the sides of a right triangle. Consider the trapezoid in Figure 1.2.

Fig. 1.2 President James
Garfield's proof of the
Pythagorean Theorem

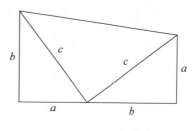

Fig. 1.3 Applying the
Pythagorean Theorem to
analytic geometry

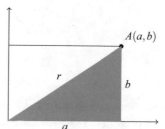

We calculate the area of the trapezoid in two different ways. First recall the standard formula for the area of a trapezoid: If the parallel sides of a trapezoid of height h have lengths x, y, then the area is equal to $h(x + y)/2$. By this formula, the area of our trapezoid is $(a + b)^2/2$. On the other hand, the trapezoid is the union of three right triangles: two with legs equal to a, b, and one with legs equal to c. For this reason the area of the trapezoid is equal to

$$2 \cdot \frac{1}{2}ab + \frac{1}{2}c^2.$$

Setting the two expressions for the area equal to each other gives

$$2 \cdot \frac{1}{2}ab + \frac{1}{2}c^2 = \frac{1}{2}(a + b)^2.$$

Expanding and simplifying the sides of the equality gives the Pythagorean Equation.

□

The Pythagorean Theorem is a fundamental theorem with many applications. For example, the main identity of trigonometry, that for each angle θ

$$\cos^2 \theta + \sin^2 \theta = 1,$$

is nothing but the Pythagorean Theorem in a right triangle with hypotenuse of length 1. The theorem has an interesting interpretation in analytic geometry. Suppose we have a point A with coordinates (a, b) in the xy-plane as in Figure 1.3.

If r is the distance from A to the origin, then applying the Pythagorean Theorem to the gray right triangle gives

$$r^2 = a^2 + b^2.$$

Fig. 1.4 Applying the
Pythagorean Theorem to
three-dimensional analytic
geometry

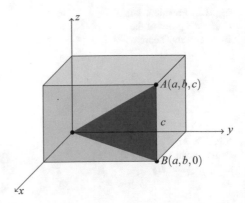

Suppose, on the other hand, we have a fixed number $r > 0$ and we want to identify all points (x, y) which have distance r to the origin. This is of course the circle of radius r centered at the origin with equation

$$x^2 + y^2 = r^2.$$

This picture can be generalized to higher dimensions. Suppose we have a point $A(a, b, c)$ in the three-dimensional space $\mathbb{R}^3$ as in Figure 1.4.

Again let r be the distance from the point $A(a, b, c)$ to the origin $O(0, 0, 0)$. Applying the Pythagorean Theorem to the blue triangle gives

$$r^2 = OB^2 + c^2 = a^2 + b^2 + c^2.$$

As an application, we find that the equation of the sphere of radius r centered at the origin is

$$x^2 + y^2 + z^2 = r^2.$$

Similarly, if we have a point with coordinates $(x_1, \ldots, x_n)$ in $\mathbb{R}^n$, its distance r to the origin satisfies

$$r^2 = x_1^2 + \cdots + x_n^2. \tag{1.2}$$

We can use this result to write down the equation of a sphere in $\mathbb{R}^n$ of radius r centered at the origin.

1.2 Pythagorean triples

In this book we are interested in those solutions of the Pythagorean Equation which are interesting from the number theoretic perspective. This means we will work with solutions x, y, z of Equation (1.1) which are elements of particular subsets of the real numbers, e.g., natural numbers, integers, or rational numbers. In general, a *Diophantine equation* is an equation of the form

$$f(x_1, x_2, \ldots, x_n) = 0$$

where we search for solutions $(x_1, \ldots, x_n) \in \mathbb{Z}^n$, though in some situations we may seek solutions in other sets, e.g., $\mathbb{N}$, $\mathbb{Q}$, $\mathbb{Z}[i]$.

A *Pythagorean triple* is a triple of natural numbers x, y, z satisfying Equation (1.1). A *primitive* Pythagorean triple is one where the three numbers do not share any non-trivial common factors. Such triples are called primitive because if (a, b, c) is some Pythagorean triple, there is a primitive Pythagorean triple (a', b', c') and an integer d such that

$$(a, b, c) = (da', db', dc').$$

The most famous Pythagorean triple is $(3, 4, 5)$, and one can easily check that $5^2 = 25 = 9 + 16 = 3^2 + 4^2$. The next few Pythagorean triples are $(5, 12, 13)$, $(7, 24, 25)$, $(8, 15, 17)$. We will determine all primitive Pythagorean triples in §3.1. A right triangle whose side lengths form a Pythagorean triple is called an *integral right triangle*. We call an integral right triangle *primitive* if its side lengths form a primitive Pythagorean triple.

We can also study the solutions of the Pythagorean Equation in integers x, y, z. Again, we call an integral solution *primitive* if x, y, z do not share any common factors other than $+1$ or -1. If (x, y, z) satisfies the Pythagorean Equation, then we have $x^2 + y^2 = z^2$. If $z \neq 0$, then we divide by z^2 to obtain

$$\left(\frac{x}{z}\right)^2 + \left(\frac{y}{z}\right)^2 = 1,$$

i.e., the point $(x/z, y/z)$ is a point with rational coordinates on the circle of radius 1 centered at the origin. For example, $(3/5, 4/5)$ is a point on the unit circle centered at the origin obtained from the Pythagorean triple $(3, 4, 5)$. In fact the triple $(3, 4, 5)$ gives rise to eight different points on the circle:

$$(\pm 3/5, \pm 4/5), \quad (\pm 4/5, \pm 3/5), \quad (\pm 3/5, \mp 4/5), \quad (\pm 4/5, \mp 3/5).$$

Though we have not yet developed the tools to prove this statement rigorously, the reader should convince herself that there is a correspondence between primitive integral solutions (x, y, z) of the Pythagorean Equation with $z > 0$ and points with rational coordinates on the unit circle center at the origin. We can make similar definitions for higher dimensional Pythagorean Equations

$$x_1^2 + \cdots + x_n^2 = z^2, \tag{1.3}$$

and relate integral solutions to points with rational coordinates on the higher dimensional unit spheres centered at the origin.

1.3   The questions

Understanding the integral solutions of the Pythagorean Equation and exploring the fine properties of integral right triangles have been great sources of inspiration for mathematicians throughout the history of mathematics in general, and number theory in particular. Our purpose in this book is to explore some number theoretic problems that have arisen in relation to right triangles. As we saw a moment ago the study of right triangles and solutions to the Pythagorean Equation is intimately connected with the study of points with rational (or integral) coordinates on circles and spheres. These are some of the questions we address in this book:

1. What are the primitive solutions of the Pythagorean Equation? Does geometry have anything to do with finding the solutions? We study these questions in Chapter 3.
2. What integers are areas of integral right triangles? This is the subject matter of Chapter 4.
3. What numbers are edges of integral right triangles? This question is answered in Chapter 5.
4. How many solutions are there to the Pythagorean Equation modulo various integers? We answer this question in Chapter 8. For what it means to speak of a number modulo an integer, see Chapter 2.
5. How are integral points distributed on big spheres? Some results in this direction are obtained in Chapters 9 and 10.
6. Approximately, how many Pythagorean triples (x, y, z) are there with $z < B$, for a larger number B? The answer to this question occupies Chapter 13.
7. How are points with rational coordinates distributed on the unit circle centered at the origin in $\mathbb{R}^2$? This is discussed in Chapter 14.

The rest of the book is devoted to developing background material for these results, or exploring related topics.

Exercises

1.1 Let a, b, c be the side lengths of a right angle triangle with c the length of the hypotenuse. Use the dissection in Figure 1.5 of a $c \times c$ square into four triangles and a square to give a proof of the Pythagorean Theorem. This proof is due to the famous 12th century Indian mathematician Bhaskara, [9, §3.3].

1.2 Suppose a, b, c are the side lengths of a right triangle. Use Figure 1.6 to give a proof of the Pythagorean Theorem. In the diagram, the three triangles are similar to the original triangle with scaling factors a, b, and c.

1.3 Here is an alternative formulation of the idea exploited in Garfield's proof. Again, suppose a, b, c are the sides of a right triangle. Use Figure 1.7 to give one more proof of the Pythagorean Theorem.

Fig. 1.5 The dissection in
Problem 1.1

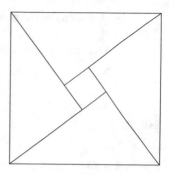

Fig. 1.6 Figure for Problem
1.2

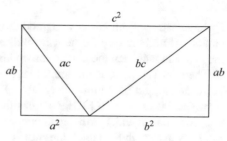

Fig. 1.7 The diagram for
Problem 1.3

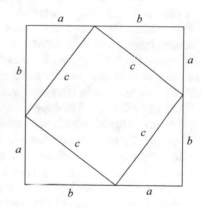

1.4 Let *ABC* be a triangle. Show that

$$\operatorname{sgn}(\angle A + \angle B - \angle C) = \operatorname{sgn}(BC^2 + AC^2 - AB^2).$$

Here sgn is the following function:

$$\operatorname{sgn}(x) = \begin{cases} +1 & x > 0; \\ 0 & x = 0; \\ -1 & x < 0. \end{cases}$$

1.5 (✠) List all Pythagorean triples (a, b, c), with $a \le b < c \le 100$.

1.6 (✠) Let $N(B)$ be the number of Pythagorean triples (a, b, c), with $a, b, c < B$. Compute $N(B)$ for some large values of B like 1000, 15000, 100000. Does $N(B)/B$ approach a limit as B gets large? We will investigate this limit in Chapter 13.

Notes

Pythagoreans

Pythagoreans certainly deserve a good deal of credit for their contributions to mathematics, if nothing else for their formalization of the concept of proof. While they may have in fact been the first people in history to have written down a formal proof of Theorem 1.1, there is no doubt that the theorem itself was known much earlier. For example, the Babylonian clay tablet Plimpton 322 described in [9, §2.6], dated between 1900 and 1600 BCE, contains fifteen pairs of fairly large natural numbers x, z, every one of which is the hypotenuse and a leg of some right triangle with integer sides. Even though the tablet does not contain a diagram showing a right triangle, it is hard to imagine these numbers would have appeared in a context other than the Pythagorean Theorem. Furthermore, given the sizes of the entries, 8161 and 18541, among others, it is only natural to assume that these numbers were not the result of random guesswork, and that the Babylonian mathematicians responsible for the content of the tablet actually had a method to produce integral solutions.

Mathematicians in Egypt too were certainly aware of the Pythagorean Theorem. The Cairo Mathematical Papyrus, described again in [9, §2.6], contains a variety of problems, some of them fairly sophisticated, dealing directly with the Pythagorean Theorem. There is also evidence to suggest that the theorem and something resembling a geometric proof of it were known to Chinese mathematicians some 300 years before Euclid, c.f. [9, §3.3]. Dickson [16, Ch. IV] reports that the Indian mathematicians, Baudhayana and Apastamba, had obtained a number of solutions to the Pythagorean Equation independently of the Greeks around 500 BCE.

At any rate, Pythagoreans were led to irrational numbers from the Pythagorean Theorem. Kline [29, Ch. 3] writes: "The discovery of incommensurable ratios [irrational numbers] is attributed to Hippasus of Metapontum (5th cent. B.C). The Pythagoreans were supposed to have been at sea at the time and to have thrown Hippasus overboard for having produced an element in the universe which denied the Pythagorean doctrine that all phenomena in the universe can be reduced to whole numbers or their ratios."

This most likely refers to the discovery of $\sqrt{2}$. Some historians dispute the story that Hippasus was thrown overboard. The basic argument seems to be that the drowning of the discoverers sounds unlikely—which considering the fact that at the time of this writing fundamentalism in all of its shapes and forms has been eradicated in the world, the skepticism of these historians is justified. There is apparently no historical evidence that Pythagoras himself ever knew of irrational numbers—which, as little as we know of the life of the man, this is not surprising. The earliest reference to irrational numbers is in Plato's Theaetetus [38, Page 200] where it is said of Theodorus: "was writing out for us something about roots, such as the roots of three or five, showing that they are incommensurable by the unit: he selected other examples up to seventeen—there he stopped."

Since Theodorus skips over 2 then presumably this means that the irrationality of root 2 must have already been known. In fact there is mention of this in passing in Aristotle's Prior Analytics [3, §23] and this appears to be the first place this is written down somewhere: "prove the initial thesis from a hypothesis, when something impossible results from the assumption of the contradictory. For example, one proves that the diagonal is incommensurable because odd numbers turn out to be equal to even ones if one assumes that it is commensurable."

To learn more about Pythagoras and his school, we refer the reader to [9], especially Chapter 3. For the philosophical contributions of the Pythagoreans, see Russell's fantastic book [42]. For Greek mathematics in general, see Artman [5]. To see some original writings by the Greek masters, see Thomas [51].

Pythagorean triples throughout history

Proclus, in his commentary on Euclid, states that Pythagoras had obtained the family of Pythagorean triples

$$\begin{cases} x = 2\alpha + 1, \\ y = 2\alpha^2 + 2\alpha, \\ z = 2\alpha^2 + 2\alpha + 1, \end{cases}$$

for α a natural number, c.f. [16, §IV]. As we will see in §3.1 this family does not cover all solutions. Euclid obtained the solutions

$$\begin{cases} x = \alpha\beta\gamma, \\ y = \frac{1}{2}\alpha(\beta^2 - \gamma^2), \\ z = \frac{1}{2}\alpha(\beta^2 + \gamma^2). \end{cases}$$

Diophantus may have been the first person to write the solutions as

$$\begin{cases} x = m^2 - n^2, \\ y = 2mn, \\ z = m^2 + n^2. \end{cases} \tag{1.4}$$

Dickson [16, §IV] mentions an anonymous Arabic text from the tenth century where necessary and sufficient conditions are derived for the integers m, n so that the triple (1.4) is primitive. The same reference contains numerous other works by many mathematicians which provide various formulations of the solutions of the Pythagorean Equation.

Our purpose here is not to review the history of Pythagorean Equation in its entirety—the references [9, 16] do an impressive job at reviewing the history of the subject, though, see *Historical References* in Notes to Chapter 2. Our goal in mentioning the above isolated anecdotes is to highlight the fact that mathematics, as all other branches of human knowledge, progresses very slowly—and sometimes what in hindsight looks completely obvious, takes years, centuries, and sometimes millennia, to develop and mature. We sometimes feel smarter than our predecessors because we have learned their works, but in reality the mathematicians of the antiquity were every bit as brilliant and hardworking as the best of us.

Chapter 2
Basic number theory

In this chapter we cover basic number theory and set up notations that will be used freely throughout the rest of the book. The chapter starts with the basic notions of divisibility and prime numbers with the goal of proving the Fundamental Theorem of Arithmetic, Theorem 2.19. We then prove the Chinese Remainder Theorem (Theorem 2.24), Fermat's Little Theorem (Theorem 2.26), Euler's Theorem (Theorem 2.31), discuss the basic properties of the totient function ϕ, and study polynomials modulo primes, digit expansions, and finally primitive roots. In the Notes at the end of the chapter, we talk about Euclid and his masterpiece the *Elements*; briefly discuss natural numbers and induction; review two standard cryptographic methods based on number theory; and finally, state Artin's conjecture for primitive roots.

2.1 Natural numbers, mathematical induction, and the Well-ordering Principle

The numbers $1, 2, 3, \ldots$ are called *natural numbers*, and we denote the set of all natural numbers by $\mathbb{N}$. A defining property of the set of natural number is the following:

Property 2.1 (Mathematical induction). Let $A \subset \mathbb{N}$ be such that

- $1 \in A$;
- $x \in A$ implies $x + 1 \subset A$.

Then $A = \mathbb{N}$.

The set of natural numbers has the following fundamental property as well:

Property 2.2 (Well-ordering Principle). Every non-empty subset of the set of natural numbers has a smallest element.

For example, if we consider the subset of the set of natural numbers consisting of all even numbers, then the smallest element of this set is the number 2; or, if the

© Springer Nature Switzerland AG 2018
R. Takloo-Bighash, *A Pythagorean Introduction to Number Theory*,
Undergraduate Texts in Mathematics, https://doi.org/10.1007/978-3-030-02604-2_2

subset is the set of all multiples of 75, then the smallest element is 75. Intuitively, the Well-ordering Principle is true because the set of natural numbers does not go *all the way down*, though this is of course not a proof. In fact, the Well-ordering Principle is equivalent to mathematical induction.

Theorem 2.3. *The Well-ordering Principle is logically equivalent to mathematical induction.*

Proof. First we show that mathematical induction implies the Well-ordering Principle. Let P_n be the following statement: Every subset of $\mathbb{N}$ which contains a number x such that $x \leq n$ has a smallest element. Clearly P_1 is true, as in this case the subset will contain 1, and 1 will be the smallest element. So now suppose we know P_k is true, and we wish to show P_{k+1} is true. Suppose $A \subset \mathbb{N}$ is such that A contains some element x with $x \leq k+1$. If A contains some element y with $y \leq k$, then the validity of P_k implies that A must have a smallest element. So assume there are no elements in A which are less than or equal to k. Since we had assumed that A contains some element less than or equal to $k + 1$, but nothing less than or equal to k, we conclude that $k + 1 \in A$, and that $k + 1$ is the smallest element of A.

Next, we show that the Well-ordering Principle implies mathematical induction. Suppose $A \subset \mathbb{N}$ is such that

- $1 \in A$;
- $x \in A$ implies $x + 1 \in A$;
- $A \neq \mathbb{N}$.

Let $B = \mathbb{N} - A$. By assumption B is not empty. By the Well-ordering Principle B has a smallest element b. Since $1 \in A$, $b \neq 1$, and as a result $b - 1 \in \mathbb{N}$. On the other hand, $b - 1 < b$, and as we had assumed that b is the smallest element of B, this means $b - 1 \notin B$. Consequently, $b - 1 \in A$, and this last statement implies that $(b - 1) + 1 \in A$, i.e., $b \in A$, a contradiction. □

2.2 Divisibility and prime factorization

Definition 2.4. For integers a, b with $b \neq 0$, we say b *divides* a if there is a $c \in \mathbb{Z}$ such that $a = bc$. The integer b is then called a *divisor* of a, and a is called a *multiple* of b. In this case, we write $b \mid a$. A natural number p is called *prime* if it has exactly four distinct divisors. For integers a, b, n, with $n \neq 0$, we write $a \equiv b \bmod n$, and say a *is congruent to b modulo n*, if $n \mid a - b$.

For example, $3 \mid (-6)$ as $-6 = 3 \cdot (-2)$. The number 5 is a prime number, since its divisors are $\pm 1, \pm 5$; 6 is not a prime as it is divisible by $\pm 1, \pm 2, \pm 3, \pm 6$, and 1 is not a prime as it only has two divisors ± 1. Finally, $13 \equiv 7 \bmod 3$ as $3 \mid 13 - 7 = 6$. Congruence modulo 0 is equality.

The following lemma is an easy exercise; see Exercise 2.1.

Lemma 2.5. *For an integer n,* congruence modulo *n is an equivalence relation.*

Definition 2.6. The equivalence classes of the congruence relation are called *congruence classes modulo n*. The congruence class of an integer a modulo a non-zero integer n is denoted by $[a]_n$. The set of congruence classes modulo n is denoted by $\mathbb{Z}/n\mathbb{Z}$.

Lemma 2.7. *The set* $\mathbb{Z}/n\mathbb{Z}$ *has a group structure defined by*

$$[a]_n + [b]_n := [a+b]_n.$$

Proof. The identity of the operation is given by $[0]_n$. The inverse of the element $[a]_n$ is $[-a]_n$. Associativity is immediate from the associativity of addition of the group $\mathbb{Z}$. □

Theorem 2.8 (Division Algorithm). *For integers* $a, b,$ *with* $b \neq 0,$ *there are unique integers* q_0, r_0 *with* $0 \leq r_0 < |b|,$ *such that*

$$a = bq_0 + r_0.$$

If we allow negative values of r, we can choose q_0, r_0 *such that*

1. $-\frac{|b|+1}{2} \leq r_0 \leq \frac{|b|-1}{2}$, *if b is odd;*
2. $-\frac{|b|}{2} + 1 \leq r_0 \leq \frac{|b|}{2}$, *if b is even.*

Proof. By replacing q by $-q$ if necessarily, it suffices to prove the theorem for $b > 0$. If a, b, define

$$S = \{a - bq \mid q \in \mathbb{Z}, a - bq \in \mathbb{N}\}.$$

It is clear that $S \subset \mathbb{N}$. We claim that S is non-empty. To see this, we recognize two cases:

- If $a > 0$, then set $q = 0$. In this case $a - 0b = a > 0$, and $a \in S$;
- If $a < 0$ and $b > 0$, let $q = 2a$. We have $a - qb = a - 2ab = -a(2b - 1) > 0$. Again, $S \neq \varnothing$.

Since S is non-empty, Property 2.2 implies that S has a smallest element, call it x. By the definition of S, there is $q \in \mathbb{Z}$ such that $x = a - bq$. We now claim $x \leq b$. If $x = a - bq > b$, then $x - b = a - (b+1)q > 0$. This means $x - b \in S$, and since $x - b < x$, this contradicts the choice of x as the smallest element of S.

Next, if the smallest element $x = b$, then $a - (q+1)x = x - b = 0$, and we set $q_0 = q + 1$ and $r_0 = 0$. If $x < b$, then we set $q_0 = q$ and $r_0 = x$.

Now that we know the first part of the theorem, we can proceed to prove the second part. Suppose b is odd—the proof for the even case is similar. By the first part of the theorem we can write

$$a = bq_0 + r_0$$

with $0 \leq r_0 < |b|$. If $0 \leq r_0 \leq \frac{|b|-1}{2}$ we are done, so assume $\frac{|b|-1}{2} < r_0 < |b|$. We have

$$a = bq_0 + |b| + (r_0 - |b|).$$

Note that $bq_0 + |b|$ is a multiple of b. Next, since $\frac{|b|-1}{2} < r_0 < |b|$ we have

$$\frac{|b| - 1}{2} - |b| < r_0 - |b| < |b| - |b| = 0.$$

To finish the proof we need to verify that

$$\frac{|b| - 1}{2} - |b| \geq -\frac{|b| + 1}{2},$$

but this is clear. □

Note that with the notations of Theorem 2.8, $[a]_b = [r_0]_b$. This observation provides a convenient way to write down representatives for equivalence classes in $\mathbb{Z}/b\mathbb{Z}$. For example, suppose $b = 6$. When we divide an integer a by b, we will have a remainder $0, 1, 2, 3, 4, 5$. Consequently, the set $\{0, 1, 2, 3, 4, 5\}$ will provide a set of representatives for $\mathbb{Z}/6\mathbb{Z}$.

Lemma 2.9. *For every non-zero integer n, $\#(\mathbb{Z}/n\mathbb{Z}) = |n|$.*

Proof. We define a map

$$res_n : \mathbb{Z}/n\mathbb{Z} \to \{0, 1, \cdots, |n| - 1\}.$$

The strategy of the proof is to show that the map res_n is a bijection. We define the function as follows. Let $u \in \mathbb{Z}/n\mathbb{Z}$. Let a be an integer such that $[a]_n = u$. Use Theorem 2.8 to write

$$a = qn + r$$

with $0 \leq r < |n|$. We define $res_n(u) = r$.

Since the definition of res_n involves a choice of the integer a, we need to show that $res_n(u)$ is independent of the choice of a. Suppose the integer b is such that $[b]_n = [a]_n = u$. The assumption on b implies that $a \equiv b$ mod n, i.e., there is an integer k such that $b - a = kn$. If we use the fact that $a = qn + r$, we get $b = a + kn = qn + r + kn = (q + k)n + r$ with $0 \leq r < qn$. As a result, $res_n([b]_n) = r = res_n([a]_n)$.

We now show that res_n is a bijection. That it is a surjective map is obvious. In fact, for every r with $0 \leq r < n$, $res_n([r]_n) = r$. To see that it is injective, we suppose that $res_n(u) = res_n(u') = r$ with $u, u' \in \mathbb{Z}/n\mathbb{Z}$ and some r with the property that $0 \leq r < n$. Write $u = [a]_n$ and $u' = [b]_n$. It follows from the definition of res_n that $a = q_1 n + r$ and $b = q_2 n + r$ for integers q_1, q_2. As a result, $a - b = q_1 n - q_2 n = (q_1 - q_2)n$. Consequently, $n \mid a - b$, or $a \equiv b$ mod n. This means $[a]_n = [b]_n$. □

Definition 2.10. Let n be an integer. By a *complete system of residues modulo n* we mean a collection of n integers $a_1, \ldots, a_n$ such that for each i, j with $1 \leq i, j \leq n$, we have $a_i \equiv a_j$ mod n if and only if $i = j$. Alternatively, a complete system of residues is a complete set of representatives for congruence classes modulo n.

The notion of the *greatest common divisor* described in the following definition is surprisingly important:

Definition 2.11. For integers a, b, the *greatest common divisor of a, b*, denoted $\gcd(a, b)$, is an integer g with the following properties:

- $g \mid a$ and $g \mid b$;
- If d is an integer such that $d \mid a$ and $d \mid b$, then $|d| \leq g$.

Integers a, b are called *coprime* if $\gcd(a, b) = 1$. We also define the *least common multiple of* the non-zero integers a, b, denoted by $\operatorname{lcm}(a, b)$ to be a positive integer l with the following properties:

- $a \mid l$ and $b \mid l$;
- If m is an integer such that $a \mid m$ and $b \mid m$, then $l \leq |m|$.

Basically, the greatest common divisor of integers a and b is precisely what the name suggests: the greatest, common, divisor of a and b, and similarly for the lcm. We similarly define the gcd and lcm of more than two numbers.

Theorem 2.12. *If a, b are integers, then there are integers x, y such that*

$$ax + by = \gcd(a, b).$$

Proof. The theorem is easy if either of a or b is zero. For example, if $a = 0$, then $\gcd(0, b) = b = 1 \times 0 + 1 \times b$. So we may assume that neither a nor b is zero. By changing the signs of x, y, if necessarily, we may assume $a, b > 0$. Define a set S by

$$S = \{ax + by \mid x, y \in \mathbb{Z}, ax + by \in \mathbb{N}\}.$$

Clearly $S \subset \mathbb{N}$ and $S \neq \emptyset$ as, in particular, $a, b \in S$. By Property 2.2, the set S has a smallest element g. By definition, there are integers x_0, y_0 such that $g = ax_0 + by_0$ and $g > 0$.

If d is a common divisor of a, b, then $d \mid ax_0 + by_0 = g$. Consequently, $\gcd(a, b) \mid g$.

Now we claim every element of S is divisible by g. Let $s = ax + by \in S$. Divide s by g, and use Theorem 2.8 to write

$$s - gq \mid r$$

for some $0 \leq r < g$. If $r = 0$, it follows that $g \mid s$ and we are done. Otherwise, we have

$$0 < r = s - gq = (ax + by) - (ax_0 + by_0)q = a(x - x_0 q) + b(y - y_0 q).$$

As a result, $r \in S$. Since $0 < r < g$, this last statement contradicts the assumption that g is the smallest element of S. Consequently, we have established the claim that every element of S is divisible by g. In particular, since $a, b \in S$, we see that $g \mid a$ and $g \mid b$, i.e., g is a common divisor of a, b. As a result, $g \leq \gcd(a, b)$. Since we had already established that $\gcd(a, b) \mid g$, we conclude $g = \gcd(a, b)$. We have proved

$$\gcd(a, b) = ax_0 + by_0.$$ □

A consequence of this theorem is the following interesting result:

Corollary 2.13. *If a, b, d are integers such that $d \mid a$, $d \mid b$, then $d \mid \gcd(a, b)$.*

Proof. Since d is a divisor of both a and b, for all integers x, y we have $d \mid ax + by$. The result now follows from Theorem 2.12. □

Clearly, one way to find the greatest common divisor of a and b is to write the list of all divisors of a and b, look for the common divisors, and find the greatest one. For example, if $a = 12$ and $b = 18$, we have

$$\text{Divisors of } a = \{\pm 1, \pm 2, \pm 3, \pm 4, \pm 6, \pm 12\}$$

and

$$\text{Divisors of } b = \{\pm 1, \pm 2, \pm 3, \pm 6, \pm 9, \pm 18\}.$$

Next,

$$\text{Common divisors of } a \text{ and } b = \{\pm 1, \pm 2, \pm 3, \pm 6\}.$$

Finally,

$$\gcd(a, b) = 6.$$

Note that $6 = (+1) \cdot 18 + (-1) \cdot 12$ in accordance with Theorem 2.12.

This is, of course, inefficient, especially when dealing with large numbers. Euclid presented a clever procedure to compute the greatest common divisor of two integers without listing the divisors of the individual integers. This is known as the *Euclidean Algorithm*. The Euclidean Algorithm is based on the following lemma:

Lemma 2.14. *If $a, b \in \mathbb{N}$ with $a \mid b$, then $\gcd(a, b) = a$. If $a, b \in \mathbb{N}$ with $a > b$, then*

$$\gcd(a, b) = \gcd(a - b, b).$$

Proof. The first statement is easy. In fact, $\gcd(a, b) \leq a$ as the $\gcd(a, b)$ is a divisor of a. On the other hand, a is a common divisor of a and b, hence $a \leq \gcd(a, b)$. Combining these two observations shows that $\gcd(a, b) = a$. Now we prove the second statement by showing that the set of common divisors of a, b is equal to the set of common divisors of $a - b, b$. This statement implies that the greatest elements of the sets are the same, proving the lemma. To see the equality of the two sets, suppose d is a common divisor of a, b. Then $d \mid a$, $d \mid b$, and consequently $d \mid a - b$, i.e., d is a common divisor of b and $a - b$. Hence, the set of common divisors of a, b is a subset of the set of common divisors of b and $a - b$. The reverse inclusion is proved similarly. □

As an example, we compute gcd(18, 12). We have

$$\gcd(18, 12) = \gcd(18 - 12, 12) = \gcd(6, 12) = 6,$$

by applying Lemma 2.14. To see a slightly more interesting example, we examine gcd(57, 12). We have

$$\gcd(57, 12) = \gcd(57 - 12, 12) = \gcd(45, 12) = \gcd(33, 12)$$

$$= \gcd(21, 12) = \gcd(9, 12) = \gcd(12, 9) = \gcd(12 - 9, 9) = \gcd(3, 9) = 3.$$

In the first stage, we needed to subtract 12 from 57 four times. In effect, what we have done is that we have replaced 57 by the remainder of its division by 12. In practice, we do the following: In order to compute $\gcd(a, b)$ with $a > b$, we write $a = bq + r$ with $0 \leq r < b$; if $r = 0$, then $\gcd(a, b) = b$; otherwise, $\gcd(a, b) = \gcd(b, r)$. Since $a > b > r$, we have replaced the pair (a, b) with the "smaller" pair (b, r) with the same gcd. Let us formulate this procedure as a lemma:

Lemma 2.15 (Euclidean Algorithm). *The following procedure computes the* gcd *of a pair of natural numbers (a, b) with $a > b$:*

1. *The pair (a, b) is given with $a > b$;*
2. *Let r be the remainder of the division of a by b;*
3. *If $r = 0$, b is the* gcd *and we are done;*
4. *If $r > 0$, replace (a, b) by (b, r), and go back to (1).*

At the time of this writing, we do not know how to find the prime factors of a large integer n quickly. In contrast, the Euclidean Algorithm is incredibly fast. In fact, by Theorem 12 of [46, Ch. I, §3], originally a theorem of Lamé from 1844, the number of divisions needed is at most five times the number of digits in the decimal expansion of the smaller number b.

The Euclidean Algorithm allows us to make Theorem 2.12 computationally effective. We will illustrate the idea in the following example:

Example 2.16. It is easy to see that gcd(57, 12) = 3. We wish to find integers x, y such that

$$57x + 12y = 3.$$

We write

$$57 = 4 \times 12 + 9;$$

$$12 = 1 \times 9 + 3.$$

Now we write

$$3 = 12 - 9 = 12 - (57 - 4 \times 12) = 12 - 57 + 4 \times 12 = 5 \times 12 - 57,$$

giving $x = -1$ and $y = 5$. We will see more examples of this procedure in the exercises.

A consequence of Theorem 2.12 is the following important theorem:

Theorem 2.17. *If $a \mid bc$ and $\gcd(a, b) = 1$, then $a \mid c$.*

Proof. Since $\gcd(a, b) = 1$, there are integers x, y such that $ax + by = 1$. Multiplying the equality by c gives $c = axc + bcy$. Both terms on the right-hand side of this equation are divisible by a: The term axc is clearly divisible by a, and bcy is divisible by a by assumption. This means c is divisible by a and we are done. $\square$

This theorem implies the following result of Euclid (Elements, Proposition 30, Book VII):

Corollary 2.18 (Euclid's First Theorem). *Let p be a prime number, and $p \mid ab$ for integers a, b. Then either $p \mid a$ or $p \mid b$.*

Proof. Suppose $p \nmid a$. We claim that $\gcd(a, p) = 1$. In fact, if $d = \gcd(a, p)$, then $d \mid p$. This means that either $d = 1$ or $d = p$. We cannot have $d = p$, because then $p = d \mid a$ which is a contradiction. Hence, $d = 1$, and the result follows from Theorem 2.17. $\square$

Euclid's Lemma is the main ingredient in the proof of the uniqueness assertion of the following foundational result:

Theorem 2.19 (Fundamental Theorem of Arithmetic). *Every natural number is a product of prime numbers in an essentially unique fashion.*

In the statement of the theorem, *essentially unique* means up to reordering of the terms. For example, we can write

$$12 = 3 \cdot 2 \cdot 2 = 2 \cdot 3 \cdot 2 = 2 \cdot 2 \cdot 3.$$

Proof. We will prove the existence using induction. Since 1 is the *empty* product of prime numbers, the theorem is true for 1. Now suppose n is a natural number, and suppose we know the existence of a prime factorization for every natural number smaller than n. If n is prime, there is nothing to prove. If n is not prime, then it has a non-trivial divisor y such that $1 < y < n$. Clearly, $1 < n/y < n$. By the induction assumption, $y = p_1 \cdots p_r$ and $n/y = q_1 \cdots q_s$ for primes $p_1, \ldots, p_r$ and $q_1, \ldots, q_s$. Then,

$$n = y \cdot \frac{n}{y} = p_1 \cdots p_r \cdot q_1 \cdots q_s.$$

This gives the existence of a prime factorization.

We now prove the uniqueness. Suppose we have a natural number n which has two different prime factorizations:

$$P_1 \cdots P_k = Q_1 \cdots Q_l.$$

The sets of primes $\{P_1, \ldots, P_k\}$ and $\{Q_1, \ldots, Q_l\}$ may have some common elements. If necessary we simplify the common elements from the sides to obtain an equality of the form

$$P_1 \cdots P_u = Q_1 \cdots Q_v, \tag{2.1}$$

with the sides not having any common factors. Now, we have

$$P_1 \mid Q_1 \cdots Q_v.$$

An easy application of Euclid's First Theorem, Corollary 2.18, says that there is an i such that

$$P_1 \mid Q_i.$$

But since P_1 and Q_i are prime numbers, this divisibility implies that $P_1 = Q_i$, contradicting the assumption that the sides of Equation (2.1) have no common elements.

□

It is convenient to write the prime factorization of a number as a product of prime powers. For example, instead of $12 = 2 \cdot 3 \cdot 2$, we usually write $12 = 2^2 \cdot 3$. We denote the prime factorization of a typical natural number n in the form

$$p_1^{\alpha_1} \cdots p_r^{\alpha_r},$$

or similar expression. In such expressions, even when we do not explicitly mention it, we assume that the prime numbers $p_1, \ldots, p_r$ are distinct. In this case we write $p_i^{\alpha_i} \| n$, meaning $p_i^{\alpha_i} \mid n$ but $p_i^{\alpha_i + 1} \nmid n$, and call α_i the *multiplicity* of p_i in n. It is sometimes convenient to allow the exponents α_i to be equal to zero. For example, if

$$n = p_1^{\alpha_1} \cdots p_r^{\alpha_r},$$

then every divisor of n can be written in the form

$$m = p_1^{\beta_1} \cdots p_r^{\beta_r},$$

where for each $i, 0 \le \beta_i \le \alpha_i$.

The Fundamental Theorem of Arithmetic has many applications. Here we list three consequences. We leave the proofs to the reader; see Exercise 2.4 and Exercise 2.5.

Proposition 2.20. *Let $m = \prod_i p_i^{r_i}$ and $n = \prod_i p_i^{s_i}$. Then*

$$\gcd(m, n) = \prod_i p_i^{\min(r_i, s_i)},$$

and

$$\mathrm{lcm}(m, n) = \prod_i p_i^{\max(r_i, s_i)}.$$

Furthermore,

$$\gcd(m, n) \cdot \mathrm{lcm}(m, n) = mn.$$

The following proposition is used a few times throughout the book:

Proposition 2.21. *Suppose a, b are natural numbers such that $\gcd(a, b) = 1$. If $ab = m^k$ for natural numbers m and k, then $a = m_1^k$ and $b = m_2^k$ for natural numbers m_1, m_2 such that $m_1 m_2 = m$.*

Corollary 2.22. *If $n \in \mathbb{N}$ is not a perfect kth power, there is no rational number γ such that $n = \gamma^k$.*

2.3 The Chinese Remainder Theorem

Theorem 2.12 is a statement about the solvability of the equation

$$ax + by = \gcd(a, b)$$

in integers x, y. More generally, one can ask about the solvability of a general linear Diophantine equation

$$ax + by = c$$

in integers x, y. It is not hard to see that this equation is solvable if and only if $\gcd(a, b) \mid c$. For example if $\gcd(a, b) = 1$, then every equation $ax + by = c$ is solvable. The following is a useful fact:

Theorem 2.23. *Suppose a, b are coprime integers, and let $x_0, y_0 \in \mathbb{Z}$ be such that $ax_0 + by_0 = 1$. Then if $x, y \in \mathbb{Z}$ satisfy $ax + by = 1$, there is $h \in \mathbb{Z}$ such that*

$$x = x_0 + bh, \quad y = y_0 - ah.$$

In general, if the equation $ax + by = c$ is solvable, then since $\gcd(a, b) \mid ax + by$, we see that $\gcd(a, b) \mid c$. Conversely, if $\gcd(a, b) \mid c$, we can write $c = c' \cdot \gcd(a, b)$. By Theorem 2.12 we know that there are integers x_0, y_0 such that $ax_0 + by_0 = \gcd(a, b)$. Multiplying by c' gives $a(x_0 c') + b(y_0 c') = \gcd(a, b)c' = c$, and as a result $x = x_0 c'$ and $y = y_0 c'$ are numbers that satisfy $ax + by = c$.

Formulated in terms of congruence equations, this is equivalent to saying that the equation

$$ax \equiv c \quad \mathrm{mod} \ b \tag{2.2}$$

is solvable if and only if $\gcd(a, b) \mid c$. In particular if $\gcd(a, b) = 1$, the equation is solvable for every c. Back in the general case of Equation (2.2), since $\gcd(a, b) \mid c$, the equation is equivalent to

$$\frac{a}{\gcd(a, b)} x \equiv \frac{c}{\gcd(a, b)} \quad \mathrm{mod} \ \frac{b}{\gcd(a, b)}. \tag{2.3}$$

Now

$$\gcd\left(\frac{a}{\gcd(a, b)}, \frac{b}{\gcd(a, b)}\right) = 1,$$

and as a result Equation (2.2) is solvable with solution

$$x \equiv \left(\frac{a}{\gcd(a, b)}\right)^{-1} \frac{c}{\gcd(a, b)} \quad \mathrm{mod} \ \frac{b}{\gcd(a, b)}.$$

So every equation of the form (2.2), if solvable, has a solution of the form

$$x \equiv k \quad \mathrm{mod}\ m$$

for some $m \mid b$.

For example, the equation $4x \equiv 3 \bmod 6$ is not solvable as $2 = \gcd(4, 6) \nmid 3$. On the other hand, the equation $4x \equiv 2 \bmod 6$ is solvable as $2 = \gcd(4, 6) \mid 2$. To solve the equation $4x \equiv 2 \bmod 6$, we divide by 2 to get $2x \equiv 1 \bmod 3$, which has the solution $x \equiv 2 \bmod 3$.

One can also ask about the solvability of systems of equations

$$\begin{cases} a_1 x \equiv c_1 \quad \mathrm{mod}\ b_1, \\ a_2 x \equiv c_2 \quad \mathrm{mod}\ b_2. \end{cases}$$

Obviously we need each of the equations to be solvable, so our previous considerations apply. In particular the solvability of this system reduces to the solvability of a system of the form

$$\begin{cases} x \equiv k_1 \quad \mathrm{mod}\ m_1, \\ x \equiv k_2 \quad \mathrm{mod}\ m_2. \end{cases} \tag{2.4}$$

It is not hard to see, Exercise 2.22, that this system is solvable if and only if

$$\gcd(m_1, m_2) \mid k_1 - k_2.$$

If x_1, x_2 are solutions of the system (2.4), then $x_1 \equiv x_2 \bmod [m_1, m_2]$.

For a system consisting of more than two equations the exact solvability conditions are fairly painful to state. However, there is a useful special case with many applications:

Theorem 2.24 (The Chinese Remainder Theorem). *Suppose $m_1, \ldots, m_n$ are integers such that for all i, j with $i \neq j$,*

$$\gcd(m_i, m_j) = 1.$$

Then for every string of integers $a_1, \ldots, a_n$ the system of equations

$$\begin{cases} x \equiv a_1 \quad \mathrm{mod}\ m_1, \\ \quad \cdots \\ x \equiv a_n \quad \mathrm{mod}\ m_n, \end{cases}$$

has a solution. If x_1, x_2 are solutions of the system, then

$$x_1 \equiv x_2 \quad \mathrm{mod}\ m_1 \cdots m_n.$$

Example 2.25. Suppose we wish to find all x such that

$$\begin{cases} x \equiv 1 \quad \mathrm{mod}\ 5; \\ x \equiv 2 \quad \mathrm{mod}\ 7; \\ x \equiv 3 \quad \mathrm{mod}\ 9. \end{cases}$$

Every x satisfying the first equation is of the form $1 + 5k$. Insert this expression in the second equation to obtain

$$1 + 5k \equiv 2 \quad \text{mod } 7.$$

This is the same as saying $5k \equiv 1 \bmod 7$, which after multiplying by 3 gives $k \equiv 3 \bmod 7$, i.e., $k = 3 + 7l$ for some l. This means, $x = 1 + 5k = 1 + 5(3 + 7l) = 16 + 35l$. Now we use the third equation to obtain

$$16 + 35l \equiv 3 \quad \text{mod } 9.$$

Since $16 \equiv -2$ and $35 \equiv -1 \bmod 9$, we get $-2 - l \equiv 3 \bmod 9$, from which it follows $l \equiv 4 \bmod 9$. Write $l = 4 + 9r$ for some $r \in \mathbb{Z}$. Going back to x, we have $x = 16 + 35l = 16 + 35(4 + 9r) = 156 + 315r$. Consequently, in order for x to satisfying the system of congruences it is necessary and sufficient that

$$x \equiv 156 \quad \text{mod } 315.$$

2.4 Euler's Theorem

Next, we discuss a beautiful theorem of Fermat:

Theorem 2.26 (Fermat's Little Theorem). *If p is prime, for all integers n, $p \mid n^p - n$.*

First we consider $p = 2$. We know that n is even if and only if n^2 is even. For this reason $n^2 - n$ is always divisible by 2, establishing the theorem for $p = 2$. So we assume that p is an odd prime. In this case it is clear that if the theorem is true for n, it will also be true for $-n$. It suffices to prove the theorem for n a natural number. We proceed by induction. The theorem is trivially true for $n = 0, 1$. Now suppose the theorem is true for n. We wish to prove it is true for $n + 1$. By the Binomial Theorem, Theorem A.4, we have

$$(n + 1)^p - (n + 1) = (n^p - n) + \sum_{k=1}^{p-1} \binom{p}{k} n^k.$$

Since by our induction hypothesis, $p \mid n^p - n$, the theorem follows from the following lemma:

Lemma 2.27. *For each $0 < k < p$,*

$$p \mid \binom{p}{k}.$$

Proof. We have

$$\binom{p}{k} = \frac{p!}{k!(p-k)!} = \frac{p \cdot (p-1)!}{k!(p-k)!}.$$

Since $\binom{p}{k}$ is an integer, this means $k!(p-k)! \mid p.(p-1)!$; but since $\gcd(p, k!(p-k)!) = 1$, Theorem 2.17 implies $k!(p-k)! \mid (p-1)!$. Write $(p-1)! = k!(p-k)! \cdot A$ for an integer A. Then

$$\binom{p}{k} = \frac{p \cdot (p-1)!}{k!(p-k)!} = p \cdot A.$$

The lemma is now obvious. $\square$

We will record one more lemma that will be used in the proof of Theorem 6.8 in Chapter 7.

Lemma 2.28. *Let p be a prime number, and $x_1, \ldots, x_n$ some indeterminates. Then all of the coefficients of the multivariable polynomial*

$$(x_1 + \cdots + x_n)^p - x_1^p - \ldots x_n^p$$

are integers that are multiples of p.

We now describe Euler's generalization of Fermat's Little Theorem. The following proposition is an easy consequence of Theorem 2.12:

Proposition 2.29. *If a and n with $\gcd(a, n) = 1$, then there exists an integer b such that $ab \equiv 1 \bmod n$.*

Proof. Since $\gcd(a, n) = 1$, Theorem 2.12 implies that there are integers b and c such that $ab + cn = 1$. This means $n \mid ab - 1$, i.e., $ab \equiv 1 \bmod n$.

For example if $a = 3$ and $n = 7$, then we may take $b = 5$, as in that case $3 \times 5 \equiv 1 \bmod 7$. The congruence class of the b in the proposition is usually denoted by a^{-1} when there is no confusion about the modulus n. This means that the set of coprime to n congruence classes forms a group under multiplication modulo n. We denote this group by $(\mathbb{Z}/n\mathbb{Z})^\times$.

Definition 2.30. Let $n \in \mathbb{N}$. By a *reduced system of residues modulo n* we mean a set of representatives for $(\mathbb{Z}/n\mathbb{Z})^\times$. For a natural number n, we define the *Euler totient function*, or *Euler's ϕ-function*, by $\phi(n) = \#(\mathbb{Z}/n\mathbb{Z})^\times$.

For every complete system of residues $a_1, \ldots, a_n$ modulo n, the set

$$\{a_i \mid \gcd(a_i, n) - 1\} \tag{2.5}$$

is a reduced system of residues. It is clear that every reduced system of residues modulo n has the same number of elements, $\phi(n)$. Furthermore, if $a_1, \ldots, a_{\phi(n)}$ is a set of distinct residue classes modulo n such that for each i we have $\gcd(a_i, n) = 1$, then the set $a_1, \ldots, a_{\phi(n)}$ is a reduced system of residues modulo n. Note that

$$\phi(n) = \#\{1 \le a \le n \mid \gcd(a, n) = 1\}. \tag{2.6}$$

If, for example, $n = 12$, then the numbers a with $1 \le a \le 12$ which are coprime to 12 are 1, 5, 7, 11, and consequently, $\phi(12) = 4$.

Theorem 2.31 (**Euler**). *Let n be a natural number. For all a with $\gcd(a, n) = 1$ the equation*

$$a^{\phi(n)} \equiv 1 \quad \mod n$$

holds.

In particular when $n = p$ is a prime number, we have $\phi(p) = p - 1$, and we recover Fermat's Little Theorem, Theorem 2.26.

Proof. Suppose $a_1, \ldots, a_{\phi(n)}$ is a reduced system of residues modulo n. Since $\gcd(a, n) = 1$, the set of numbers

$$aa_1, \ldots, aa_{\phi(n)}$$

is another reduced system of residues modulo n. In fact, for each $1 \le i \le \phi(n)$, $\gcd(aa_i, n) = 1$. Furthermore, as $\gcd(a, n) = 1$, $aa_i \equiv aa_j \mod n$ for $1 \le i, j \le \phi(n)$ implies $a_i \equiv a_j \mod n$, which means $i = j$. Next, since

$$a_1, \ldots, a_{\phi(n)}$$

and

$$aa_1, \ldots, aa_{\phi(n)}$$

are both reduced systems of residues, we must have

$$\prod_{i=1}^{\phi(n)} a_i \equiv \prod_{i=1}^{\phi(n)} aa_i \quad \mod n.$$

Rearranging terms gives

$$\prod_{i=1}^{\phi(n)} a_i \equiv a^{\phi(n)} \prod_{i=1}^{\phi(n)} a_i \quad \mod n.$$

Since the a_i's are coprime to n, their product is coprime to n as well. Simplifying $\prod_i a_i$ gives the result. $\square$

The function $\phi(n)$ is explicitly computable. It is easy to see that for each prime p and $\alpha \ge 1$ we have

$$\phi(p^\alpha) = p^\alpha - p^{\alpha-1} = p^\alpha \left(1 - \frac{1}{p}\right).$$

In fact,

$$\phi(p^\alpha) = p^\alpha - \#\{1 \le a \le p^\alpha \mid \gcd(a, p^\alpha) \ne 1\}$$

$$= p^\alpha - \#\{1 \le a \le p^\alpha \mid p|a\}$$

$$= p^\alpha - p^{\alpha-1}.$$

The totient function is famously *multiplicative*:

Theorem 2.32. *For all natural numbers m, n with $\gcd(m, n) = 1$, the identity*

$$\phi(mn) = \phi(m)\phi(n)$$

holds.

Proof. We prove this theorem by constructing a reduced system of residues modulo mn. For $r, s \in \mathbb{Z}$, set

$$f(r, s) = rn + sm.$$

Our first claim is that the set

$$R = \{f(r, s) \mid 1 \le r \le m, 1 \le s \le n\}$$

is a complete system of residues modulo mn. Clearly, we have mn pairs (r, s) as above. We just need to show that for distinct pairs (r, s), the elements $f(r, s)$ are distinct modulo mn. Suppose, for $1 \le r_1, r_2 \le m$ and $1 \le s_1, s_2 \le n$, we have

$$r_1 n + s_1 m \equiv r_2 n + s_2 m \quad \text{mod } mn.$$

Considering this congruence modulo n gives

$$s_1 m \equiv s_2 m \quad \text{mod } n,$$

which, since $\gcd(m, n) = 1$, implies

$$s_1 \equiv s_2 \quad \text{mod } n.$$

Since $1 \le s_1, s_2 \le n$, this gives $s_1 = s_2$. Similarly, we conclude $r_1 = r_2$, and our claim is proved.

Next, we claim that in order for $\gcd(f(r, s), mn) = 1$, it is necessary and sufficient that $\gcd(r, m) = 1$ and $\gcd(s, n) = 1$. In fact, since $\gcd(m, n) = 1$, we have

$$\gcd(f(r, s), mn) = \gcd(rn + sm, m)\gcd(rn + sm, n).$$

Next,

$$\gcd(rn + sm, m) = \gcd(rn, m) = \gcd(r, m).$$

Similarly,

$$\gcd(rn + sm, n) = \gcd(s, n).$$

Consequently,

$$\gcd(f(r, s), mn) = \gcd(r, m) \cdot \gcd(s, n),$$

from which the second claim is immediate. The number of all pairs (r, s) such that $\gcd(r, m) = 1$ and $\gcd(s, n) = 1$ is clearly $\phi(m)\phi(n)$, and the theorem is proved. $\square$

It then follows that for each natural number n with prime factorization $n = p_1^{\alpha_1} \cdots p_k^{\alpha_k}$ we have

$$\phi(n) = \prod_{i=1}^{k} \phi(p_i^{\alpha_i}) = \prod_{i=1}^{k} p_i^{\alpha_i}\left(1 - \frac{1}{p_i}\right) = n \prod_{i=1}^{k}\left(1 - \frac{1}{p_i}\right).$$

We will record this computation as a theorem:

Theorem 2.33. *For every natural number n,*

$$\phi(n) = n \prod_{p \mid n} (1 - \frac{1}{p}).$$

This theorem means that in order to compute the value of $\phi(n)$ we just need to know the prime factors of n, and not the prime factorization. For example, since the prime factors of 12 are 2 and 3, we have

$$\phi(12) = 12 \cdot \left(1 - \frac{1}{2}\right) \cdot \left(1 - \frac{1}{3}\right) = 12 \times \frac{1}{2} \times \frac{2}{3} = 4.$$

Theorem 2.33 has an interesting statistical interpretation. Suppose we have a number n with prime factors $p_1, p_2, \ldots, p_k$. The quotient $\phi(n)/n$ is the probability of choosing a random number a in the set $\{1, \ldots, n\}$ subject to $\gcd(a, n) = 1$. Now, a number a satisfies $\gcd(a, n) = 1$ if and only if for each i, $p_i \nmid a$. The probability of a randomly chosen number to be divisible by p_i is $1/p_i$, and the probability that a randomly chosen number is coprime to p_i is $1 - 1/p_i$. If we pretend that coprimality to distinct primes are independent events, we see that the probability that a number is coprime to $p_1, \ldots, p_k$ is

$$\prod_{i=1}^{k} \left(1 - \frac{1}{p_i}\right),$$

which by Theorem 2.33 is precisely $\phi(n)/n$.

The function ϕ viewed as a function $\mathbb{N} \to \mathbb{R}$ has many surprising properties. Here is an example:

Theorem 2.34. *For all natural numbers n,*

$$\sum_{d \mid n} \phi(d) = n.$$

Proof. By Theorem A.2 there are precisely n distinct complex numbers z such that $z^n = 1$, and they can be expressed as

$$e^{\frac{2\pi i k}{n}}, \quad k = 0, \ldots, n - 1.$$

For a complex number z with $z^n = 1$, we define $o(z)$ to be the smallest positive integer k such that $z^k = 1$. We claim $o(z) \mid n$. If not, by Theorem 2.8 there is an integer q and $0 < r < o(z)$ such that $n = qo(z) + r$. Then

$$1 = z^n = z^{qo(z)+r} = (z^{o(z)})^q z^r = z^r,$$

contradicting the definition of $o(z)$. Next,

$$n = \#\{z \in \mathbb{C} \mid z^n = 1\} = \sum_{d \mid n} \#\{z \in \mathbb{C} \mid z^n = 1, o(z) = d\}. \qquad (2.7)$$

Our next step is to determine $\#\{z \in \mathbb{C} \mid z^n = 1, o(z) = d\}$. In order to do this, we pick $0 \le k \le n - 1$ and determine $o(e^{\frac{2\pi i k}{n}})$. Suppose for $l > 0$ we have

$$\left(e^{\frac{2\pi i k}{n}}\right)^l = 1.$$

This is equivalent to saying

$$e^{\frac{2\pi i k l}{n}} = 1.$$

Consequently, $n \mid kl$. Dividing by $\gcd(n, k)$ gives

$$\frac{n}{\gcd(n, k)} \mid \frac{k}{\gcd(n, k)} \cdot l.$$

Since

$$\gcd\left(\frac{n}{\gcd(n, k)}, \frac{k}{\gcd(n, k)}\right) = 1,$$

Theorem 2.17 implies that

$$\frac{n}{\gcd(n, k)} \mid l.$$

This statement combined with $l > 0$ implies

$$l \ge \frac{n}{\gcd(n, k)}.$$

In particular,

$$o(e^{\frac{2\pi i k}{n}}) \ge \frac{n}{\gcd(n, k)}.$$

We claim that equality holds. To see this, we note

$$\left(e^{\frac{2\pi i k}{n}}\right)^{\frac{n}{\gcd(n,k)}} = e^{\frac{2\pi i k}{n} \cdot \frac{n}{\gcd(n,k)}} = e^{2\pi i \cdot \frac{k}{\gcd(n,k)}} = 1,$$

as $\frac{k}{\gcd(n,k)}$ is an integer. Hence, we have

$$o(e^{\frac{2\pi i k}{n}}) = \frac{n}{\gcd(n, k)}.$$

Now we can go back to determining $\#\{z \in \mathbb{C} \mid z^n = 1, o(z) = d\}$. If $o(e^{\frac{2\pi i k}{n}}) = d$, then we have $\frac{n}{\gcd(n,k)} = d$. It follows, $\gcd(n, k) = \frac{n}{d}$. In particular, $\frac{n}{d} \mid k$. Write $k = \frac{n}{d} \cdot k'$. Note $1 \le k' \le d$. We have

$$\frac{n}{d} = \gcd(n, k) = \gcd\left(\frac{n}{d} \cdot d, \frac{n}{d} \cdot k'\right) = \frac{n}{d} \cdot \gcd(d, k').$$

Hence $\gcd(d, k') = 1$. This means,

$$\#\{z \in \mathbb{C} \mid z^n = 1, o(z) = d\} = \#\{1 \le k' \le d \mid \gcd(d, k') = 1\} = \phi(d).$$

Combining this identity with (2.7) gives the theorem. $\quad\square$

2.5 Polynomials modulo a prime

We often speak of *polynomials modulo p*, with p a prime number. By this we mean
a polynomial $f(x) \in \mathbb{Z}[x]$ where the coefficients and values are considered modulo
p. This is of course nothing but a polynomial in the variable x with coefficients in the
finite field $\mathbb{Z}/p\mathbb{Z}$, but for the purposes of this monograph we can prove the results
we need completely elementarily using the methods presented in this chapter.

Throughout this discussion we fix a prime number p. Let $f(x) = \sum_{j=0}^{n} a_j x^j$,
with $a_j \in \mathbb{Z}$, be a polynomial. We say f is a *non-zero polynomial* modulo p, if there
is a j with $a_j \not\equiv 0 \bmod p$; we say f is of *degree n* if $a_n \not\equiv 0 \bmod p$. We call an
integer k *a root of $f(x)$ modulo p* if $f(k) \equiv 0 \bmod p$. We call the roots k, l distinct
if $k \not\equiv l \bmod p$. For example, if $p = 3$, the polynomial $f(x) = x^5 + 2$ is of degree
5 and has a root $k = 1$ modulo 3. One easily checks that $l = 4$ is another root of
$f(x)$ modulo 3, but 1 and 4 are not distinct modulo 3, as $4 \equiv 1 \bmod 3$.

Remark 2.35. Note that these notions depend on the choice of the prime p. For
example, if $f(x) = 3x^4 + 2x + 5$, then $f(x)$ is of degree 4 if $p \neq 3$, but of degree 1
for $p = 3$. Also, $f(2) = 57 = 3 \times 19$, so $k = 2$ is a root of $f(x)$ modulo 3 and 19,
but not otherwise.

Our goal in this section is to prove the following useful statement:

Theorem 2.36. *Let $f(x)$ be a polynomial of degree n modulo a prime p. Then $f(x)$
has at most n distinct roots modulo p.*

Our proof of this theorem relies on the following lemma the statement of which
the reader should compare with Theorem 2.8:

Lemma 2.37. *Suppose $f(x)$ and $g(x)$ are polynomials with integer coefficients, and
suppose $g(x)$ is a monic polynomial. Then there are unique polynomials $q(x)$ and
$r(x)$ with integral coefficients such that*

$$f(x) = q(x)g(x) + r(x),$$

and either $r(x) = 0$ or $0 \leq \deg r(x) < \deg g(x)$.

Proof. We will prove the lemma by induction on $\deg f$. If $\deg f < \deg g$, then
there is nothing to prove, as we can simply set $q(x) = 0$ and $r(x) = f(x)$. Now
suppose $\deg f \geq \deg g$, and write $f(x) = \sum_{j=0}^{n} a_j x^j$ and $g(x) = x^m + \sum_{l=0}^{m-1} b_l x^l$
with $a_n \neq 0$. Then $\deg(f(x) - a_n x^{n-m} g(x)) < \deg f$. By induction, there are
polynomials $q'(x), r'(x)$ with integer coefficients such that either $r'(x) = 0$ or
$\deg r'(x) < \deg g(x)$, and with the property that

$$f(x) - a_n x^{n-m} g(x) = q'(x)g(x) + r'(x).$$

This equation implies $f(x) = (q'(x) + a_n x^{n-m})g(x) + r'(x)$. Setting $q(x) =
q'(x) + a_n x^{n-m} \in \mathbb{Z}[x]$ and $r(x) = r'(x)$ gives the result. For uniqueness see
Exercise 2.31. $\square$

For example, if we divide the polynomial $f(x) = 3x^3 + 2$ by the polynomial $g(x) = x^2 + 2$, we get $q(x) = 3x$ and $r(x) = -6x + 2$, and $\deg r(x) < \deg g(x)$.

Proof of Theorem 2.36. We prove this theorem by induction on the degree of the polynomial f. If $f(x)$ has no roots modulo p, then there is nothing to prove. So suppose we have a root k. Use Lemma 2.37 to write

$$f(x) = (x - k)q(x) + r(x).$$

The lemma says that either $r(x) = 0$ or $\deg r(x) < \deg(x - k) = 1$. This means that either $r(x) = 0$ or $\deg r(x) = 0$, i.e., $r(x)$ is a constant c which may be zero. In any case, we write

$$f(x) = (x - k)q(x) + c.$$

Insert $x = k$ in this expression to obtain $f(k) = (k - k)q(k) + c = c$. So we obtain

$$f(x) = (x - k)q(x) + f(k).$$

Since $f(k) \equiv 0 \bmod p$, we have

$$f(x) \equiv (x - k)q(x) \quad \bmod p.$$

Consequently, the roots of $f(x)$ modulo p consist of k plus whatever root modulo p that $q(x)$ may have. Since $\deg q(x) = \deg f(x) - 1$, by induction, $q(x)$ has at most $\deg f(x) - 1$ roots. The result is now immediate. $\square$

Remark 2.38. In general, if F is a field, and $f(x) \in F[x]$ a non-zero polynomial, then $f(x) = 0$ has at most $\deg f(x)$ roots in F.

Theorem 2.36 has numerous applications. The following example is a particularly well-known application of this theorem.

Example 2.39. Fix a prime p. By Theorem 2.26, for all $n \in \mathbb{Z}$ we have

$$n^p - n \equiv 0 \quad \bmod p.$$

This means that if we set

$$f(x) = x^p - x,$$

then every integer n is a root of $f(x)$ modulo p. As a result, the elements of a complete system of residues modulo p, e.g., $S = \{0, 1, \ldots, p - 1\}$, are going to be the distinct roots of $f(x)$. On the other hand, we have a polynomial

$$g(x) = x(x - 1)(\cdots)(x - p + 1) = x^p + \text{ terms of lower degree.}$$

with the elements of S as its roots. Now consider the polynomial

$$h(x) = f(x) - g(x).$$

It is clear that $\deg h(x) < p$ as the x^p terms from the polynomial $f(x)$ and $g(x)$ cancel each other out. Now, every element of S is a root of $h(x)$ modulo p. But this would mean that the polynomial h which is of degree less than p has p roots which contradicts Theorem 2.36, unless $h(x) \equiv 0 \bmod p$. Consequently,

$$x^p - x \equiv x(x-1)(\cdots)(x-p+1) \quad \bmod p.$$

We may cancel out an x from the congruence to obtain the identity

$$x^{p-1} - 1 \equiv (x-1)(\cdots)(x-p+1) \quad \bmod p.$$

If we put in $x = p$ in this identity we obtain the following statement known as Wilson's Theorem:

$$(p-1)! \equiv -1 \quad \bmod p. \tag{2.8}$$

2.6 Digit expansions

It is common practice to express real numbers in terms of powers of 10. We call such expressions *decimal expansions*. For example, when we write $x = 347$, what we mean is that x is equal to the following expression

$$3 \times 10^2 + 4 \times 10^1 + 7 \times 10^0.$$

In this expression the numbers $3, 4, 7$ are called the *digits* of x. The digits are always integers larger than or equal to 0 and less than or equal to 9. If we have a non-integral real number, then we use a decimal point to separate the integer part from the fractional part. For instance, when we write $x = 23.6923$ we mean

$$x = 2 \times 10^2 + 3 \times 10^0 + 6 \times 10^{-1} + 9 \times 10^{-2} + 2 \times 10^{-3} + 3 \times 10^{-4}.$$

We wish to generalize this notion. Suppose $g > 1$ is a natural number. In this section we discuss *base g expansions* of real numbers. We will show that every positive real number x is representable in the form

$$\sum_{k \in \mathbb{Z}, k < N} a_k \cdot g^k$$

with $N \in \mathbb{Z}$ and a_k's integers satisfying $0 \le a_k < g$. Once we establish this, we write

$$x = (a_{N-1} \cdots a_1 a_0 . a_{-1} a_{-2} a_{-3} \cdots)_g,$$

if $N > 0$, and

$$x = (0. a_{-1} a_{-2} a_{-3} \cdots)_g$$

if $N = 0$, and

$$x = (0.0 \cdots 0 a_{N-1} a_{N-2} \cdots)_g$$

if $N < 0$, where the number of zeros between the decimal point and a_{N-1} is $-N$. We will also determine the extent to which this representation is unique. Throughout the remainder of this section we will use Exercise 2.47 without explicit mention numerous times.

Suppose $x \in \mathbb{R}$ and $x > 0$. We write $x = n + \xi$ with $n \in \mathbb{N} \cup \{0\}$ and $0 \leq \xi < 1$. Here, $n = [x]$ and $\xi = \{x\}$. Our first step is to construct the base g expansion of n.

If $n = 0$, then we define the base g expansion of n to be 0. So we assume $n \geq 1$. By the Well-ordering Principle, Property 2.2, there is a smallest natural number N such that $n < g^N$. This means that

$$g^{N-1} \leq n < g^N,$$

as otherwise $n < g^{N-1}$ which would contradict the choice of N. By Theorem 2.8 there are integers q and n' such that

$$n = q \cdot g^{N-1} + n'$$

with $0 \leq n' < g^{N-1}$. We claim $0 \leq q < g$. In fact, if $q > g$, then

$$n = q \cdot g^{N-1} + n' > g \cdot g^{N-1} = g^N,$$

contradicting the choice of N; if $q \leq -1$, then

$$n = q \cdot g^{N-1} + n' < -g^{N-1} + n' < 0.$$

Now that we know $0 \leq q < g$, we denote it by a_{N-1}. We have

$$n = a_{N-1} \cdot g^{N-1} + n'$$

with $0 \leq n' < g^{N-1}$. By repeating this process we obtain the representation

$$n = a_{N-1} \cdot g^{N-1} + \cdots + a_1 \cdot g^1 + a_0 \tag{2.9}$$

with each a_i satisfying $0 \leq a_i < g$. The expression on the right-hand side of (2.9) is the *base g expansion of n*, and the a_i's are called the *digits* of n.

Now we show that the base g expansion of a natural number is unique. Suppose a natural number n has two different base g expansions:

$$n = a_{N-1} \cdot g^{N-1} + \cdots + a_1 \cdot g^1 + a_0 = b_{M-1} \cdot g^{M-1} + \cdots + b_1 \cdot g^1 + b_0, \tag{2.10}$$

with $M, N > 0$ and $0 \leq a_i, b_j < g$, and let's assume $a_{N-1} \neq 0$, $b_{M-1} \neq 0$. First we show $M = N$. Suppose $M > N$. We observe

$$b_{M-1} \cdot g^{M-1} + \cdots + b_1 \cdot g^1 + b_0 \geq g^{M-1}.$$

Next,

$$a_{N-1} \cdot g^{N-1} + \cdots + a_1 \cdot g^1 + a_0 \leq (g-1)g^{N-1} + (g-1)g^{N-2} + \cdots + (g-1)g + (g-1)$$

$$= (g-1)(g^{N-1} + \cdots + g + 1) = (g-1)\frac{g^N - 1}{g - 1} = g^N - 1 < g^{M-1}.$$

So on the one hand $n \geq g^{M-1}$ and on the other $n < g^{M-1}$. This is a contradiction, showing that M cannot be larger than N. Similarly, it follows that $N > M$ is impossible as well. As a result $M = N$.

With the equality $M = N$ at hand, Equation (2.10) can be rewritten as

$$a_{N-1} \cdot g^{N-1} + \cdots + a_1 \cdot g^1 + a_0 = b_{N-1} \cdot g^{N-1} + \cdots + b_1 \cdot g^1 + b_0.$$

We will show that for each i, $a_i = b_i$. We will first show that $a_{N-1} = b_{N-1}$. After this has been established, the rest of the argument is an easy induction. Suppose $a_{N-1} \neq b_{N-1}$. Then we have

$$0 = \left| (a_{N-1} \cdot g^{N-1} + \cdots + a_1 \cdot g^1 + a_0) - (b_{N-1} \cdot g^{N-1} + \cdots + b_1 \cdot g^1 + b_0) \right|$$

$$= \left| (a_{N-1} - b_{N-1})g^{N-1} + (a_{N-2} - b_{N-2})g^{N-2} + \cdots + (a_1 - b_1)g + (a_0 - b_0) \right|$$

$$\geq \left| (a_{N-1} - b_{N-1})g^{N-1} \right| - \left| (a_{N-2} - b_{N-2})g^{N-2} + \cdots + (a_1 - b_1)g + (a_0 - b_0) \right|,$$

upon using the following version of *the triangle inequality*: For all real numbers x, y, $|x + y| \geq |x| - |y|$. Since $a_{N-1} \neq b_{N-1}$,

$$\left| (a_{N-1} - b_{N-1})g^{N-1} \right| \geq g^{N-1}.$$

Also, for each i, $|a_i - b_i| \leq g - 1$. This inequality implies

$$\left| (a_{N-2} - b_{N-2})g^{N-2} + \cdots + (a_1 - b_1)g + (a_0 - b_0) \right|$$

$$\leq \left| (a_{N-2} - b_{N-2})g^{N-2} \right| + \cdots + |(a_1 - b_1)g| + |(a_0 - b_0)|$$

$$\leq (g - 1)g^{N-2} + \cdots + (g - 1)g + (g - 1) = g^{N-1} - 1,$$

after using the triangle inequality in the following form: For all $x_1, x_2, \ldots, x_k \in \mathbb{R}$, we have $|x_1 + x_2 + \cdots + x_k| \leq |x_1| + |x_2| + \cdots + |x_k|$. Putting everything together, we have

$$0 \geq \left| (a_{N-1} - b_{N-1})g^{N-1} \right| - \left| (a_{N-2} - b_{N-2})g^{N-2} + \cdots + (a_1 - b_1)g + (a_0 - b_0) \right|$$

$$\geq g^{N-1} - (g^{N-1} - 1) = 1.$$

This is a contradiction, showing that $a_{N-1} = b_{N-1}$. We have proved the following lemma:

Lemma 2.40. *Let $g \in \mathbb{N}$ and $g > 1$. Then every natural number n can be written in a unique way as a sum*

$$n = a_{N-1} \cdot g^{N-1} + \cdots + a_1 \cdot g^1 + a_0$$

with $N \in \mathbb{N}$ and $a_i \in \mathbb{N} \cup \{0\}$ satisfying $0 \leq a_i < g$.

The integers a_j are called the *digits* of n, and we write

$$n = (a_{N-1} \ldots a_1)_g.$$

Now we construct the base g expansion of ξ, the fractional part of the real number x. Set

$$a_{-1} = [g\xi].$$

Next for each $k > 1$, set

$$a_{-k} = \left[g^k \left(\xi - \sum_{j=1}^{k-1} \frac{a_{-j}}{g^j} \right) \right] = [g^k \xi] - g^k \left(\sum_{j=1}^{k-1} \frac{a_{-j}}{g^j} \right).$$

For example,

$$a_{-2} = \left[g^2 \left(\xi - \frac{a_{-1}}{g} \right) \right] = [g^2 \xi] - g \cdot a_{-1},$$

and

$$a_{-3} = \left[g^3 \left(\xi - \frac{a_{-1}}{g} - \frac{a_{-2}}{g^2} \right) \right] = [g^3 \xi] - g^2 \cdot a_{-1} - g \cdot a_{-2}.$$

Now we claim that for each $k > 0$,

$$0 \le \xi - \sum_{j=1}^{k} \frac{a_{-j}}{g^j} < \frac{1}{g^k}. \tag{2.11}$$

If $k = 1$, then by the definition of the integer part we have

$$0 \le g\xi - [g\xi] < 1.$$

Since $a_{-1} = [g\xi]$, this gives $0 < g\xi - a_{-1} < 1$, from which upon dividing by g our inequality follows. For $k > 1$, we have

$$0 \le g^k \left(\xi - \sum_{j=1}^{k-1} \frac{a_{-j}}{g^j} \right) - \left[g^k \left(\xi - \sum_{j=1}^{k-1} \frac{a_{-j}}{g^j} \right) \right] < 1.$$

By definition this means

$$0 \le g^k \left(\xi - \sum_{j=1}^{k-1} \frac{a_{-j}}{g^j} \right) - a_{-k} < 1.$$

Dividing by g^k gives

$$0 \le \xi - \sum_{j=1}^{k-1} \frac{a_{-j}}{g^j} - \frac{a_{-k}}{g^k} < \frac{1}{g^k},$$

and this is the inequality (2.11).

Since $0 < \xi < 1$, $0 < g\xi < g$, and as a result $0 \le [g\xi] < g$. This means $0 \le a_{-1} < g$. If $k > 1$, (2.11) implies

$$0 \leq g^k \left(\xi - \sum_{j=1}^{k-1} \frac{a_{-j}}{g^j} \right) < g,$$

which gives

$$0 \leq a_{-k} < g.$$

Lemma 2.41. *With a_j's defined as above,*

$$\xi = \sum_{j=1}^{\infty} \frac{a_{-j}}{g^j}. \tag{2.12}$$

Proof. Once we note that $g^{-k} \to 0$ as k gets large, this is a consequence of Equation (2.11). □

As before we call the integers a_{-j}'s the *digits* of ξ, and we write

$$\xi = (0.a_{-1}a_{-2}a_{-3}\ldots)_g$$

and call it the *base g expansion* of ξ.

On the other hand we can consider expressions of the form

$$\sum_{j=1}^{\infty} \frac{a_{-j}}{g^j} \tag{2.13}$$

with a_j's integers satisfying $0 \leq a_{-j} < g$ and ask whether they correspond to base g expansions of real numbers. First a lemma:

Lemma 2.42. *Every expression of the form (2.13) is convergent.*

Proof. In order to see this, set

$$s_N = \sum_{j=1}^{N} \frac{a_{-j}}{g^j}.$$

By the definition of convergence, for $\varepsilon > 0$, we need to show there is N_0 such that if $M, N > N_0$, then

$$|s_N - s_M| < \varepsilon.$$

Without loss of generality suppose $N > M$. Then,

$$|s_N - s_M| = \sum_{j=M+1}^{N} \frac{a_{-j}}{g^j} \leq \sum_{j=M+1}^{N} \frac{g-1}{g^j} = \frac{g-1}{g^{M+1}} \sum_{k=0}^{N-M-1} \frac{1}{g^k}$$

$$= \frac{g-1}{g^{M+1}} \cdot \frac{1 - \frac{1}{g^{M+N}}}{1 - \frac{1}{g}} = \frac{1}{g^M} \cdot \frac{g^{M+N} - 1}{g^{M+N}} < \frac{1}{g^M}.$$

So given $\varepsilon > 0$, we pick N_0 such that

$$\frac{1}{g^{N_0}} < \varepsilon.$$

Once this is done, the above computation shows that as soon as $N > M > N_0$, then

$$|s_N - s_M| < \frac{1}{g^M} < \frac{1}{g^{N_0}} < \varepsilon,$$

establishing the convergence. $\square$

Now we ask whether distinct series of the sort considered in Equation (2.13) can give the same real number. Suppose we have an identity

$$\sum_{j=1}^{\infty} \frac{a_{-j}}{g^j} = \sum_{j=1}^{\infty} \frac{b_{-j}}{g^j},$$

where each side is a series of the type considered above: For each j, a_{-j}, b_{-j} are integers satisfying $0 \leq a_{-j}, b_{-j} < g$. Let N be the smallest natural number such that $a_{-N} \neq b_{-N}$. Then we have

$$0 = \sum_{j=1}^{\infty} \frac{a_{-j}}{g^j} - \sum_{j=1}^{\infty} \frac{b_{-j}}{g^j} = \left| \sum_{j=1}^{\infty} \frac{a_{-j}}{g^j} - \sum_{j=1}^{\infty} \frac{b_{-j}}{g^j} \right|$$

$$= \left| \sum_{j=N}^{\infty} \frac{a_{-j} - b_{-j}}{g^j} \right| \geq \left| \frac{a_{-N} - b_{-N}}{g^N} \right| - \sum_{j=N+1}^{\infty} \left| \frac{a_{-j} - b_{-j}}{g^j} \right|$$

$$\geq \frac{1}{g^N} - \sum_{j=N+1}^{\infty} \frac{g-1}{g^j} = 0,$$

using an easy computation involving geometric series. As a result all the inequalities appearing here should be equalities. This means that either $a_{-N} - b_{-N} = 1$ and for each $j > N$, $a_{-j} = 0, b_{-j} = g - 1$, or $a_{-N} - b_{-N} = -1$, and for each $j > N$, $a_{-j} = g - 1, b_{-j} = 0$. What this means, for example, is that if we have a sequence of integers b_{-j}, with $0 \leq b_{-j} < g$ such that for some N, and for all $j > N, b_{-j} = g - 1$, then we can define a real number ξ by setting

$$\xi = \sum_{j=1}^{\infty} \frac{b_{-j}}{g^j}.$$

Now if we write the base g digit expansion of ξ according to Lemma 2.41 we obtain

$$\xi = \frac{b_{-1}}{g} + \cdots + \frac{b_{-N+1}}{g^{N-1}} + \frac{1 + b_{-N}}{g^N}. \tag{2.14}$$

We call such a base g expansion a *finite expansion*. We say an expansion of the form

$$\sum_{j=1}^{\infty} \frac{b_{-j}}{g^j}$$

to be *unacceptable* if there is M such that for all $j \geq M$, $b_{-j} = g - 1$. We say an expansion is *acceptable* if it is not unacceptable.

It is clear that the number ξ with expansion as in Equation (2.14) can be written as

$$\xi = \frac{r}{g^N}$$

for some natural number r. By canceling out every common factor between r and g^N we arrive at a fraction of the form A/B where all the prime factors of B are prime factors of g. Conversely, suppose we have a fraction of the form $\xi = A/B$ with

$$B = \prod_{p|g, p \text{ prime}} p^{e_p}$$

with integers $e_p \geq 0$. Let $M = \max_p e_p$, the largest number among the e_p's. Let

$$C = \prod_{p|g, p \text{ prime}} p^{M-e_p}.$$

Then $BC = g^M$. We have

$$\xi = \frac{A}{B} = \frac{AC}{BC} = \frac{r}{g^M}$$

with $r = AC$. Now we write the base g expansion of r using Lemma 2.40 in the form

$$r = \sum_{k=0}^{N-1} a_k \cdot g^k$$

for some $N \in \mathbb{N}$, $0 \leq a_k < g$. We then have

$$\xi = \frac{\sum_{k=0}^{N-1} a_k \cdot g^k}{g^M} = \sum_{k=0}^{N-1} a_k \cdot g^{k-M}.$$

We summarize this discussion as the following proposition:

Proposition 2.43. *An expression of the form*

$$\sum_{j=1}^{\infty} \frac{b_{-j}}{g^j}$$

is the base g expansion of some real number $0 \leq \xi < 1$ if and only if it is acceptable. The base g expansion of a real number ξ is finite if and only if it is a rational number expressible in the form A/B with B a divisor of g^M for some M.

Putting everything together, we have the following theorem:

Theorem 2.44. *Let $g > 1$ be a natural number. Every positive real number can be written as*

$$\sum_{k\in\mathbb{Z},k<N} a_k \cdot g^k$$

with $N \in \mathbb{N}$, $a_k \in \mathbb{N} \cup \{0\}$, $0 \le a_k < g$, subject to the additional requirement that

$$\sum_{k\in\mathbb{Z},k<0} a_k \cdot g^k$$

be acceptable.

2.7 Digit expansions of rational numbers

In Proposition 2.43 we determined the base g expansions of rational numbers A/B with B certain special numbers. In this section we determine what base g expansions of arbitrary rational numbers look like.

The reader will probably remember from elementary school that decimal expansions of rational numbers are eventually periodic, in the sense that there will be blocks of digits that will repeat exactly. For example:

$$\frac{7}{15} = 0.46666666\ldots;$$

$$\frac{2}{7} = 0.285714285714285714\ldots;$$

$$\frac{7}{12} = 0.583333333\ldots;$$

$$\frac{1}{19} = 0.052631578947368421052631578947368421052631578947368421\ldots.$$

In the first example, the repeating block is the single digit 6; in the second one, it is 285714; in the third one, 3; and in the last one, 052631578947368421. The common practice is to draw a line above the repeating block so as to save space and avoid confusion, e.g.,

$$\frac{7}{17} = 0.4\overline{6}; \quad \frac{2}{7} = \overline{285714}; \quad \frac{7}{12} = 0.58\overline{3}; \quad \frac{1}{19} = 0.\overline{052631578947368421}.$$

We will see that similar results hold for base g expansions of rational numbers for arbitrary natural numbers $g > 1$. We say a base g expansion is *repeating* if from some point onward, the sequence of digits is the back to back repetitions of some fixed finite sequence of numbers. The examples we gave above are all repeating expansions for the base 10. In general, a repeating base g expansion will look like this:

$$(a_{N-1}\ldots a_1.a_{-1}a_{-2}\ldots a_{-k}\overline{b_1 b_2 \ldots b_t})_g, \tag{2.15}$$

where as before the line on top of $b_1 b_2 \ldots b_t$ means that this is the repeating sequence of digits. We call the sequence $b_1 \ldots b_k$ the *repeating block*, and the number k, the *period*. A base g expansion of the form $(a_{N-1} \ldots a_1.\overline{b_1 b_2 \ldots b_k})_g$ is called *purely periodic*.

Our goal is to prove the following theorem:

Theorem 2.45. *Let $g > 1$ be a natural number. A positive real number is rational if and only if its base g expansion is repeating.*

Proof. Note that a finite base g expansion is repeating: The repeating sequence of numbers is simply 0. We already saw in Proposition 2.43 that finite base g expansions give rational numbers.

Let $g > 1$ be a natural number. Our first step is to show that repeating base g expansions give rational numbers. Suppose we have a repeating base g expansion as in Equation (2.15):

$$x = (a_{N-1} \ldots a_1.a_{-1}a_{-2} \ldots a_{-k}\overline{b_1 b_2 \ldots b_t})_g$$

$$= (a_{N-1} \ldots a_1.a_{-1}a_{-2} \ldots a_{-k})_g + \frac{(0.\overline{b_1 b_2 \ldots b_t})_g}{g^{k+1}}.$$

By Proposition 2.43, or just by direct inspection, the number $(a_{N-1} \ldots a_1.a_{-1}a_{-2} \ldots a_{-k})_g$ is rational. So in order to show that x is rational, we just need to show that

$$\gamma := (0.\overline{b_1 b_2 \ldots b_t})_g$$

is a rational number. In order to see this we observe

$$\gamma = \sum_{j=0}^{\infty} \left(\frac{b_1}{g^{1+jk}} + \frac{b_2}{g^{2+jk}} + \cdots + \frac{b_k}{g^{k+jk}} \right)$$

$$= \left(\frac{b_1}{g^1} + \frac{b_2}{g^2} + \cdots + \frac{b_k}{g^k} \right) \sum_{j=0}^{\infty} \frac{1}{g^{jk}}$$

$$= \frac{b_1 \cdot g^{k-1} + b_2 \cdot g^{k-2} + \cdots + b_k}{g^k (1 - g^{-k})},$$

after using Exercise 2.47. We conclude that

$$\gamma = \frac{b_1 \cdot g^{k-1} + b_2 \cdot g^{k-2} + \cdots + b_k}{g^k - 1}, \tag{2.16}$$

clearly showing that γ is a rational number. We have shown that every repeating base g expansion gives a rational number.

Next we show that the base g expansion of every positive rational number is repeating. Suppose we have a rational number

$$x = \frac{A}{B}$$

with $A, B \in \mathbb{N}$, and $\gcd(A, B) = 1$. The starting point of the argument is to write

$$B = B_1 \cdot B_2$$

with B_1 the largest divisor of B which is coprime to g. This means that every prime factor of B_2 is prime factor of g. By an argument similar to the one used in the paragraph preceding Proposition 2.43 there is an integer C and a natural number M such that $B_2 C = g^M$. We then have

$$x = \frac{A}{B} = \frac{AC}{BC} = \frac{AC}{B_1 B_2 C} = \frac{AC}{B_1 g^M}.$$

Note that if we show the base g expansion of AC/B_1 is repeating, then we will be done, as dividing by g^M only introduces a shift in the base g expansion. So without loss of generality, we may assume that

$$x = A/B$$

with $A, B \in \mathbb{N}$, $\gcd(A, B) = 1$, $\gcd(B, g) = 1$. By Theorem 2.8 we can write

$$A = qB + r$$

with $0 \leq r < B$. This means

$$x = q + \frac{r}{B}.$$

If $r = 0$ there is nothing to prove, so suppose $r \neq 0$. It suffices to show that the base g expansion of r/B is repeating. The key to the argument is the expression we found in Equation (2.16). Suppose there is an integer D such that $BD = g^k - 1$ for some $k \in \mathbb{N}$. Then we have

$$\frac{r}{B} = \frac{rD}{BD} = \frac{rD}{g^k - 1}.$$

Now since $rD < g^k - 1$, reversing the steps of the first part of the proof shows that the base g expansion of r/B is repeating. So in order to finish the proof we just need to prove the following assertion: If B with $\gcd(B, g) = 1$, then there is a $k \in \mathbb{N}$ such that $B \mid g^k - 1$, i.e., $g^k \equiv 1 \mod B$. By Theorem 2.31 $k = \phi(B)$ works and we are done. $\square$

2.8 Primitive roots

In the proof of Theorem 2.45 we observed that if g, B with $\gcd(g, B) = 1$, then for each $0 < r < B$, the fraction r/B has a purely periodic base g expansion with period $\phi(B)$. In general the fraction r/B may have a smaller period. For example, let's consider the fraction $1/7$. The base 2 expansion of the fraction $1/7$ can be computed as follows:

$$\frac{1}{7} = \frac{1}{2^3 - 1} = \frac{1}{2^3} \cdot \frac{1}{1 - 2^{-3}} = \frac{1}{2^3} \cdot \sum_{k=0}^{\infty} \frac{1}{2^{3k}} = \sum_{k=0}^{\infty} \frac{1}{2^{3k+3}}.$$

From this computation it follows that

$$\frac{1}{7} = (0.001001001001\ldots)_2 = (0.\overline{001})_2.$$

The period is 3, which is half of $\phi(7) = 6$. Now we compute the base 3 expansion of $1/7$:

$$\frac{1}{7} = \frac{1}{3^6 - 1} = \frac{1}{3^6} \cdot \frac{1}{1 - 3^{-6}} = \frac{1}{3^6} \cdot \sum_{k=0}^{\infty} \frac{1}{3^{6k}} = \sum_{k=0}^{\infty} \frac{1}{3^{6k+6}}.$$

Consequently,

$$\frac{1}{7} = (0.000001000001000001\ldots)_3 = (0.\overline{000001})_3,$$

and in this case the period is 6. So depending on the base g, sometimes the period of the base g expansion of $1/7$ is $\phi(7) = 6$, and sometimes it is not. In fact, it follows from the proof of Theorem 2.45 that the minimal period of the base g expansion of $1/n$, if $\gcd(g, n) = 1$, is the smallest positive integer k such that $g^k \equiv 1 \mod n$. We make the following definition.

Definition 2.46. For a natural number n, and an integer a, with $\gcd(a, n) = 1$, the *order of a modulo n*, denoted by $o_n(a)$, is the smallest positive integer k such that

$$a^k \equiv 1 \mod n.$$

Note that by Theorem 2.31, $o_n(a) \leq \phi(n)$. Also, the congruence classes of the elements

$$a^j, \quad 1 \leq j \leq o_n(a)$$

are distinct modulo n.

Lemma 2.47. *If for some integer* k, $a^k \equiv 1$ *mod* n, *then* $o_n(a) \mid k$. *In particular,* $o_n(a) \mid \phi(n)$.

Proof. Write $k = q o_n(a) + r$ with $0 \leq r < o_n(a)$. We have,

$$1 \equiv a^k \equiv a^{q o_n(a)+r} \equiv \left(a^{o_n(a)}\right)^q a^r \equiv (1)^q a^r \equiv a^r \mod n.$$

Consequently, $a^r \equiv 1 \mod n$. Since $0 \leq r < o_n(a)$, this last equation implies $r = 0$. The last assertion follows from Theorem 2.31. $\square$

Definition 2.48. A number g is called a *primitive root modulo n* if $o_n(g) = \phi(n)$.

In terms of fractions, this means that the base g expansion of $1/n$ is purely periodic of period $\phi(n)$, the largest possible value. The existence of a primitive root is equivalent to the cyclicity of the Abelian group $(\mathbb{Z}/n\mathbb{Z})^\times$. Note that primitive roots may not exist. For example, if $n = 8$, then $\phi(n) = 4$. However, for all odd numbers a, $a^2 \equiv 1 \bmod 8$. In fact, $1^2 \equiv 1$, $3^2 = 9 \equiv 1$, $5^2 = 25 \equiv 1$, and $7^2 = 49 \equiv 1 \bmod 8$. In contrast, if $n = 7$, then $3^1 \equiv 3 \bmod 7$, $3^2 \equiv 2 \bmod 7$, $3^3 \equiv 6 \bmod 7$, $3^4 \equiv 4 \bmod 7$, $3^5 \equiv 5 \bmod 7$, and $3^6 \equiv 1 \bmod 7$, implying that 3 is a primitive root modulo 7.

The following theorem provides an extremely important class of situations where we know primitive roots exist.

Theorem 2.49. *If p is a prime, there is a primitive root modulo p.*

Proof. If $p = 2$, then the result is trivial. So let's assume p is odd. Let $a_1, \ldots, a_{p-1}$ be a reduced system of residues modulo p. Since by Lemma 2.47 for each j, $o_p(a_j) \mid p - 1$, we have

$$p - 1 = \sum_{d \mid p-1} \#\{1 \le j \le p - 1 \mid o_p(a_j) = d\}. \tag{2.17}$$

In our case, Theorem 2.34 says

$$p - 1 = \sum_{d \mid p-1} \phi(d). \tag{2.18}$$

Our strategy is to show that for each $d \mid p - 1$,

$$\#\{1 \le j \le p - 1 \mid o_p(a_j) = d\} = \phi(d). \tag{2.19}$$

Once this is established, letting $d = p - 1$ gives

$$\#\{1 \le j \le p - 1 \mid o_p(a_j) = p - 1\} = \phi(p - 1) \ne 0.$$

This means there are primitive roots modulo p, and in fact $\phi(p - 1)$ of them.

We now proceed to prove (2.19). Our first step is to show that for $d \mid p - 1$, if $\#\{1 \le j \le p - 1 \mid o_p(a_j) = d\} \ne 0$, then it is equal to $\phi(d)$. So, let us assume that this quantity is non-zero and pick a congruence class a modulo p such that $o_p(a) = d$. Since the congruence classes of the d elements

$$a^j, \quad 1 \le j \le d$$

are distinct and all satisfy the equation $x^d \equiv 1 \bmod p$, Theorem 2.36 implies that these are all the solutions of the equation.

Now we determine which of these elements a^j have the property $o_p(a^j) = d$. In order to do this, for an integer k, with $1 \le k \le d$, let us determine $o_p(a^k)$. If for a positive integer l, $(a^k)^l \equiv 1 \bmod p$, we get $a^{kl} \equiv 1 \bmod p$. Lemma 2.47 implies that $o_p(a) \mid kl$, or $d \mid kl$. This implies

$$\frac{d}{\gcd(d, k)} \mid \frac{k}{\gcd(d, k)} \cdot l.$$

Since

$$\gcd\left(\frac{d}{\gcd(d, k)}, \frac{k}{\gcd(d, k)}\right) = 1,$$

Theorem 2.17 implies $\frac{d}{\gcd(d,k)} \mid l$. We conclude

$$l \geq \frac{d}{\gcd(d,k)}.$$

In particular,

$$o_p(a^k) \geq \frac{d}{\gcd(d,k)}.$$

As in the proof of Theorem 2.34, we claim that equality holds. It suffices to check that

$$(a^k)^{\frac{d}{\gcd(d,k)}} \equiv 1 \mod p.$$

But this is immediate, as

$$(a^k)^{\frac{d}{\gcd(d,k)}} \equiv (a^d)^{\frac{k}{\gcd(d,k)}} \equiv 1 \mod p.$$

We have used the fact that $a^d \equiv 1 \mod p$ and $k/\gcd(d,k) \in \mathbb{N}$. Now that we have established

$$o_p(a^k) = \frac{d}{\gcd(d,k)},$$

we determine under what conditions on k, $o_p(a^k) = d$. In order for this to happen we need to have

$$\frac{d}{\gcd(d,k)} = d,$$

or, what is the same, $\gcd(d,k) = 1$. Consequently, if $1 \leq k \leq d$ with $\gcd(d,k) = 1$, $o_p(a^k) = d$. As a result, if $o_p(a) = d$, then

$$\{a^k \mid 1 \leq k \leq d, \gcd(d,k) = 1\}$$

is the set of elements whose congruence classes have order d modulo p. Since the latter set has $\phi(d)$ elements, we conclude that if

$$\#\{1 \leq j \leq p - 1 \mid o_p(a^j) = d\} \neq 0,$$

then

$$\#\{1 \leq j \leq p - 1 \mid o_p(a^j) = d\} = \phi(d).$$

As a result, for each $d \mid p - 1$,

$$\phi(d) - \#\{1 \leq j \leq p - 1 \mid o_p(a^j) = d\} \geq 0.$$

Summing up over all $d \mid p - 1$ gives

$$\sum_{d \mid p-1} \left(\phi(d) - \#\{1 \leq j \leq p - 1 \mid o_p(a^j) = d\} \right) = 0,$$

after using (2.17) and (2.18). Since each term of the sum is nonnegative this means every term has to be zero, establishing (2.19). The proof of the theorem is complete. □

Remark 2.50. If p is a small prime number, then it is easy to check whether a number g is a primitive root. For example, one can check easily, by direct computation, that $g = 2$ is a primitive root modulo 11. For large primes it is in general difficult to decide if a natural number g is a primitive root modulo p. Later in Lemma 2.57 we present a criterion to decide whether a number g is a primitive root modulo a prime p. This criterion, unfortunately, requires the knowledge of the prime factors of $p - 1$.

Next we use the above theorem to determine all numbers n for which there is a primitive root modulo n.

Theorem 2.51. *There is a primitive root modulo n if and only if $n = 1, 2, 4, p^\alpha, 2p^\alpha$, for an odd prime p.*

We present the proof of this theorem as a series of lemmas.

Lemma 2.52. *Suppose n can be written as mk, with $\gcd(m, k) = 1$ and $m, k > 2$. Then there are no primitive roots modulo n. In particular, if there is a primitive root modulo n, then $n = 2^\alpha, p^\alpha, 2p^\alpha$ for some odd prime p.*

Proof. Let a be an integer such that $\gcd(a, n) = 1$. Then $\gcd(a, m) = \gcd(a, k) = 1$. By Theorem 2.31, we have $a^{\phi(m)} \equiv 1 \bmod m$ and $a^{\phi(k)} \equiv 1 \bmod k$. Since $\phi(m) \mid \mathrm{lcm}(\phi(m), \phi(k))$,

$$a^{\mathrm{lcm}(\phi(m),\phi(k))} \equiv 1 \quad \bmod m.$$

Similarly,

$$a^{\mathrm{lcm}(\phi(m),\phi(k))} \equiv 1 \quad \bmod k.$$

By the uniqueness assertion of Theorem 2.24 we have

$$a^{\mathrm{lcm}(\phi(m),\phi(k))} \equiv 1 \quad \bmod mk. \tag{2.20}$$

Next, we observe that $\mathrm{lcm}(\phi(m), \phi(k)) < \phi(mk)$. Indeed,

$$\mathrm{lcm}(\phi(m), \phi(k)) = \frac{\phi(m)\phi(k)}{\gcd(\phi(m), \phi(k))} = \frac{\phi(mk)}{\gcd(\phi(m), \phi(k))},$$

after using Proposition 2.20 and Theorem 2.32. Since $m, k > 2$, Exercise 2.39 shows that $\phi(m)$ and $\phi(k)$ are both even, and consequently, $\gcd(\phi(m), \phi(k))$ is a non-zero even number, hence $\mathrm{lcm}(\phi(m), \phi(k)) < \phi(mk)$. Equation (2.20) now shows that there is an integer $0 < u < \phi(mk)$ such that for all a with $\gcd(a, mk) = 1$ we have $a^u \equiv 1 \bmod mk$. This proves the lemma. $\square$

It is clear that if $n = 1, 2$ then 1 is a primitive root modulo n. Also, if $n = 4$ then there is a primitive root, namely $g = 3$. Now we show that there are no primitive roots for higher powers of 2.

Lemma 2.53. *If $n = 2^\alpha$ with $\alpha > 2$, then there are no primitive roots modulo n.*

Proof. We have already seen that for all odd numbers a, $a^2 \equiv 1 \mod 8$. Since $\phi(8) = 4$, this means that for all a with $\gcd(a, 8) = 1$ we have

$$a^{\frac{\phi(2^3)}{2}} \equiv 1 \mod 2^3.$$

Our goal is to show that for all $\alpha > 2$, and for all a with $\gcd(a, 2^\alpha) = 1$, we have

$$a^{\frac{\phi(2^\alpha)}{2}} \equiv 1 \mod 2^\alpha. \tag{2.21}$$

Note that this identity proves the lemma. Since $\phi(2^\alpha) = 2^\alpha(1 - 1/2) = 2^{\alpha-1}$, (2.21) is equivalent to saying

$$a^{2^{\alpha-2}} \equiv 1 \mod 2^\alpha. \tag{2.22}$$

We will prove this assertion via mathematical induction. We already checked the validity of the claim for $\alpha = 3$. Now suppose we know (2.21) for α. This means there is an integer k such that

$$a^{2^{\alpha-2}} = 1 + k2^\alpha.$$

Next,

$$a^{2^{\alpha-1}} = \left(a^{2^{\alpha-2}}\right)^2 = (1+k2^\alpha)^2 = 1+2{\cdot}k2^\alpha+2^{2\alpha} = 1+k2^{\alpha+1}+2^{2\alpha} \equiv 1 \mod 2^{\alpha+1},$$

proving (2.22). The lemma has been proved. □

Remark 2.54. Compare the above proof with the proof of Lemma 8.5.

With these lemmas in place, we just need to prove the existence of primitive roots for $n = p^\alpha, 2p^\alpha$ for p an odd prime number. The key input is Theorem 2.49. First we prove the existence of primitive roots for the powers of an odd prime.

Lemma 2.55. *If p is an odd prime and $\alpha \in \mathbb{N}$, there is a primitive root modulo p^α.*

Proof. By Theorem 2.49 we know the result for $\alpha = 1$. Let g be a primitive root modulo p. We will show that either g or $g + p$ is a primitive root modulo p^2. We know that $o_{p^2}(g) \mid \phi(p^2) = p(p - 1)$. On the other hand, since $g^{o_{p^2}(g)} \equiv 1 \mod p^2$, we have $g^{o_{p^2}(g)} \equiv 1 \mod p$. Consequently, $p - 1 = o_p(g) \mid o_{p^2}(g)$. This means that $o_{p^2}(g) \mid p(p - 1)$ and $p - 1 \mid o_{p^2}(g)$. Consequently, there are two possibilities for $o_{p^2}(g)$: Either $o_{p^2}(g) = p(p - 1) = \phi(p^2)$ in which case we have already found a primitive root modulo p^2, or $o_{p^2}(g) = p - 1$. Suppose we are in this latter situation. Since $g+p \equiv g \mod p$, $g+p$, too, is a primitive root modulo p, and again $o_{p^2}(g+p)$ is either $p - 1$ or $p(p - 1)$. We will show that $o_{p^2}(g + p) \neq p - 1$. In order to see this we compute $(g + p)^{p-1}$ by using the Binomial Theorem (Theorem A.4) and we will show that it is not congruent to 1 modulo p^2. By Theorem A.4 we have

$$(g + p)^{p-1} = \sum_{k=0}^{p-1} \binom{p - 1}{k} g^{p-1-k} p^k.$$

Now we examine this identity modulo p^2, noting that if $k \geq 2$, $p^k \equiv 0 \bmod p^2$. We have

$$(g + p)^{p-1} \equiv g^{p-1} + (p-1)g^{p-2}p \equiv 1 + (p-1)pg^{p-2} \quad \bmod p^2.$$

This expression is not congruent to 1 modulo p^2, since otherwise,

$$1 + (p-1)pg^{p-2} \equiv 1 \quad \bmod p^2$$

implies $p^2 \mid (p-1)pg^{p-2}$, or $p \mid (p-1)g^{p-2}$ which is impossible. Consequently, $o_{p^2}(g + p) = p(p-1) = \phi(p^2)$.

Now suppose that for $\alpha \geq 2$ we have a primitive root g modulo p^α. We will show that g is also a primitive root modulo $p^{\alpha+1}$. As before, $o_{p^{\alpha+1}}(g) \mid \phi(p^{\alpha+1}) = p^\alpha(p-1)$ and $p^{\alpha-1}(p-1) = \phi(p^\alpha) \mid o_{p^{\alpha+1}}(g)$. Again, there are two possibilities for $o_{p^{\alpha+1}}(g)$: Either it is equal to $\phi(p^{\alpha+1})$ in which case we are done, or it is equal to $\phi(p^\alpha)$. To reach a contradiction, let us assume

$$o_{p^{\alpha+1}}(g) = \phi(p^\alpha) = p^{\alpha-1}(p-1).$$

In particular,

$$g^{p^{\alpha-1}(p-1)} \equiv 1 \quad \bmod p^{\alpha+1}. \tag{2.23}$$

Let m be the largest nonnegative integer such that $p^m \mid g^{p^{\alpha-2}(p-1)} - 1$, so that

$$g^{p^{\alpha-2}(p-1)} = 1 + up^m \tag{2.24}$$

for some integer u with $\gcd(u, p) = 1$. Note that this is indeed a sensible definition as $\alpha \geq 2$. Furthermore, since by Theorem 2.26, $g^{p-1} \equiv 1 \bmod p$, $m \geq 1$. Next,

$$g^{p^{\alpha-1}(p-1)} = (g^{p^{\alpha-2}(p-1)})^p = (1 + up^m)^p.$$

Applying Theorem A.4 gives

$$g^{p^{\alpha-1}(p-1)} = 1 + p \cdot up^m + \sum_{k=2}^{p} \binom{p}{k}(up^m)^k = 1 + u'p^{m+1}$$

with u' an integer satisfying $(u', p) = 1$. Going back to (2.23) we obtain

$$1 + u'p^{m+1} \equiv 1 \quad \bmod p^{\alpha+1}.$$

It then follows that $p^{m+1} \equiv 0 \bmod p^{\alpha+1}$, or what is the same, $m \geq \alpha$. Equation (2.24) now shows

$$g^{p^{\alpha-2}(p-1)} \equiv 1 \quad \bmod p^\alpha.$$

This is a contradiction as g was assumed to be a primitive root modulo p^α, and $p^{\alpha-2}(p-1) < p^{\alpha-1}(p-1) = \phi(p^\alpha)$. This contradiction shows that $o_{p^{\alpha+1}}(g) = \phi(p^{\alpha+1})$ and we are done. $\square$

Lemma 2.56. *If p is an odd prime and $\alpha \in \mathbb{N}$, then there is a primitive root modulo $2p^\alpha$.*

Proof. Note that $\phi(2p^\alpha) = \phi(2)\phi(p^\alpha) = \phi(p^\alpha)$. By the above lemma, there is a primitive root g modulo p^α. Theorem 2.24 shows the existence of a number h such that

$$h \equiv 1 \mod 2, \quad h \equiv g \mod p^\alpha.$$

Clearly, h is coprime to $2p^\alpha$. Furthermore, since the congruence classes of the numbers

$$g^j, \quad 1 \le j \le \phi(p^\alpha)$$

are distinct modulo p^α, the congruence classes of the numbers

$$h^j, \quad 1 \le j \le \phi(p^\alpha) = \phi(2p^\alpha)$$

are distinct modulo $2p^\alpha$. This observation proves the lemma. $\square$

Combining these lemmas gives Theorem 2.51. $\square$

Next we discuss the problem of finding primitive roots when they exist. It is a consequence of Lemma 2.55 and Lemma 2.56 that once we know primitive roots modulo odd prime numbers, we can find primitive roots for odd prime powers and twice prime powers. The following lemma is easy to prove:

Lemma 2.57. *Let p be an odd prime number. Then a number g is a primitive root modulo p if and only if for all prime factors q of $p - 1$, we have*

$$g^{\frac{p-1}{q}} \not\equiv 1 \mod p.$$

Proof. Suppose g is a primitive root. Then since for each prime factor q of $p - 1$, $(p - 1)/q < p - 1$, we have $g^{(p-1)/q} \not\equiv 1 \mod p$. For the other direction, if g is not a primitive root, then there is a divisor d of $p - 1$ such that $1 < d < p - 1$ and $g^d \equiv 1 \mod p$. Let q be a prime divisor of $(p - 1)/d$. Then $d \mid (p - 1)/q$, which clearly implies $g^{(p-1)/q} \equiv 1 \mod p$. $\square$

As we will see momentarily this lemma gives a nice method to determine whether a given integer g is a primitive root modulo a prime number p, provided that $p - 1$ has easily detectable prime factors. This can be a real challenge for a randomly chosen large prime number p. See the Notes at the end of this chapter for some comments on how this idea has been applied to cryptography.

Example 2.58. In this example we will use the lemmas proved above to determine primitive roots for the moduli $n = 17^\alpha, 2 \cdot 17^\alpha$. The proofs of Lemma 2.55 and Lemma 2.56 show that the key step is to find a primitive root modulo 17. In order to apply Lemma 2.57, we note $17 - 1 = 2^4$. Since the only prime factor of $17 - 1$ is 2, and $(17 - 1)/2 = 8$, Lemma 2.57 says that an integer g is a primitive root modulo 17 if and only if $17 \nmid g$ and $g^8 \not\equiv 1 \mod 17$. The easiest way to search for candidates is by testing natural numbers in order starting with 2, jumping over squares. In our case, it is easy to check that

$$2^8 = 256 \equiv 1 \mod 17,$$

so 2 is not a primitive root modulo 17. Next, we check $g = 3$. We have

$$3^8 \equiv 16 \mod 17.$$

Hence $g = 3$ is indeed a primitive root modulo 17. Next, we check to see if $g = 3$ is a primitive root modulo 17^2. By the proof of Lemma 2.55, since $3^{16} \equiv 171 \not\equiv 1 \mod 17^2$, $g = 3$ is a primitive root modulo 17^2, and consequently a primitive modulo 17^α for all $\alpha \in \mathbb{N}$. Also, since $3 \equiv 1 \mod 2$, the proof of Lemma 2.56 implies that $g = 3$ is also a primitive root for $2 \cdot 17^\alpha$ for every $\alpha \in \mathbb{N}$.

Example 2.59. Using the method of the above example one can show that $g = 2$ is a primitive root modulo 19^α for every $\alpha \in \mathbb{N}$. Note that in this case since $19 - 1 = 2 \cdot 3^2$, a number g is a primitive root modulo 19 if and only if $g^9 \not\equiv 1 \mod 19$ and $g^6 \not\equiv 1 \mod 19$. Since 2 is even, it cannot be a primitive root modulo $2 \cdot 19^\alpha$ for any α. In this case $g = 2 + 19^\alpha$ is a primitive root modulo $2 \cdot 19^\alpha$ for all α.

Exercises

2.1 Prove Lemma 2.5.

2.2 Show that the alternative definitions in Definition 2.10 are equivalent.

2.3 Use the Euclidean Algorithm to give another proof for Theorem 2.12.

2.4 Prove Proposition 2.20.

2.5 Prove Proposition 2.21.

2.6 For the following pairs of integers (a, b), find integers x, y such that $\gcd(a, b) = ax + by$:

   a. $(13, 15)$;

   b. $(398, 270)$;

   c. $(162, 65)$.

2.7 (✠) Find the gcd of 6437 and 12675. Find integers x, y such that $6437x + 12675y = \gcd(6437, 12675)$.

2.8 (✠) Find the gcd of 2594876242943772804330 and 11446995929696298.

2.9 Write the following number as a fraction $\frac{a}{b}$ with $a, b \in \mathbb{N}$ and $\gcd(a, b) = 1$:

$$10^{50} \left(\frac{1025}{1024}\right)^5 \left(\frac{1048576}{1048575}\right)^8 \left(\frac{6560}{6561}\right)^3 \left(\frac{15624}{15626}\right)^8 \left(\frac{9801}{9800}\right)^4.$$

Determine the prime factorizations of a, b without the use of a computer. Mossaheb [34] attributes this problem to Gauss.

2.10 Determine all natural numbers n such that $\prod_{d \mid n} d = n^2$.

2.11 Suppose for integers a, m, n, k we have $a^m \equiv 1 \mod k$ and $a^n \equiv 1 \mod k$. Show that $a^{\gcd(m,n)} \equiv a^{\mathrm{lcm}(m,n)} \equiv 1 \mod k$.

2.12 Show that if a rational number $\frac{a}{b}$, with $a, b \in \mathbb{Z}$ and $\gcd(a, b) = 1$, satisfies the equation

$$a_n x^n + a_{n-1} x^{n-1} + \dot{+} a_1 x + a_0 = 0,$$

with $a_0, a_1, \ldots, a_n \in \mathbb{Z}$, then $a \mid a_0$ and $b \mid a_n$. Use this result to find the rational roots of the following equations:

a. $5x^3 + 8x^2 + 6x - 4 = 0$;
b. $x^5 - 7x^3 - 12x^2 + 6x + 36 =$;
c. $6x^6 - x^5 - 23x^4 - x^3 - 2x^2 + 20x - 8 = 0$.

2.13 Use the previous exercise to show $\sqrt{2} + \sqrt[3]{3}$ is irrational.

2.14 Show that for all integers a, b, c, d satisfying $ad - bc = 1$ we have $\gcd(a + b, c + d) = 1$.

2.15 Show that for all integers $n > 1$, $1 + 1/2 + 1/3 + \cdots + 1/n$ is not an integer.

2.16 If f is a non-constant polynomial with integer coefficients, $f(n)$ is composite for infinitely many values of n.

2.17 Show that if p, q are prime numbers larger than 3, then the remainder of division of $p^2 + q^2$ by each of the numbers 3, 4, 6, 12, and 24 is equal to 2.

2.18 (✠) Show that a number n is prime if and only if it is not divisible by any natural numbers m with $1 \le m \le n^{1/2}$. This result is known as the *Sieve of Eratosthenes*. Use this idea to list all prime numbers between 1 and 1000.

2.19 (✠) Find five natural numbers k such that $22 + 37k$ is a prime number.

2.20 Show that for all $m, n \in \mathbb{N}$,

$$\frac{\gcd(m, n)}{n} \binom{n}{m}$$

is an integer.

2.21 Suppose $F_n = 2^{2^n} + 1$. Show that for all $m > n$, $F_n \mid F_m - 2$.

2.22 Find necessary and sufficient conditions for the solvability of the system (2.4). Find the general solution of the system.

2.23 Solve the system of congruence equations

$$\begin{cases} 3x \equiv 1 & \bmod 4, \\ 3x \equiv 1 & \bmod 13, \\ 5x \equiv 11 & \bmod 21. \end{cases}$$

2.24 Find the general integral solution of the Diophantine equation

$$239x - 111y = 1.$$

2.25 Find all pairs of integers (x, y) satisfying the equation $6x + 9y = 12$.

2.26 (✠) Find all x such that $85x \equiv 970 \bmod 64322$.

2.27 (✠) Find all solutions of $37x \equiv 217 \bmod 8600$.

2.28 (✠) Find all x that satisfy the following system of congruence equations:

$$\begin{cases} x \equiv 12 & \text{mod } 64; \\ x \equiv 1 & \text{mod } 173; \\ x \equiv 5 & \text{mod } 715. \end{cases}$$

2.29 Show that $5!25! \equiv 1$ mod 31.

2.30 Show that if $p \equiv 3$ mod 4, then

$$\left((\frac{p-1}{2})! \right)^2 \equiv 1 \quad \text{mod } p.$$

2.31 Prove the uniqueness assertion of Lemma 2.37.

2.32 Give two different proofs for the statement that for all integers n, $n^5/5+n^3/3+7n/15 \in \mathbb{Z}$. Generalize.

2.33 Find all the solutions to the congruence $x^2 \equiv 1$ mod 264.

2.34 By examining the solutions of the equation $x^2 \equiv 1$ mod p, show that for all primes p, $(p-1)! \equiv -1$ mod p. Show that if $n > 4$ is not prime, $(n-1)! \equiv 0$ mod n.

2.35 Let $n \in \mathbb{N}$. Compute the product

$$\prod_{\substack{1 \le d \le n \\ d^2 \equiv 1 \text{ mod } n}} d.$$

Use your formula to determine

$$\prod_{\substack{1 \le d \le n \\ \gcd(d,n)=1}} d.$$

2.36 Find the roots of the polynomials $x^2 - x + 1$ and $x^2 - x + 2$ modulo 7.

2.37 (✠) Find an integer x such that $x^2 \equiv 1879121$ mod 3698963.

2.38 (✠) Find the last four digits of 2^{4000}.

2.39 Show that $\phi(n)$ is even if and only if $n > 2$.

2.40 Determine all n such that $\phi(n) = 6$.

2.41 Determine all n such that $\phi(n) = 40n/77$.

2.42 Show that for every odd integer $n > 1$ we have $\phi(n) > \sqrt{n}$.

2.43 Determine all n with $\phi(n) \mid n$.

2.44 Give a different proof for Theorem 2.32 using the *Inclusion–Exclusion Principle*.

2.45 Prove the following generalization of Theorem 2.32: If $\gcd(m, n) = d$, then

$$\phi(mn) = \phi(m)\phi(n)\frac{d}{\phi(d)}.$$

2.46 Use Theorem 2.33 to give another proof for Theorem 2.34.

2.47 Show that for each complex number $\alpha \neq 1$ and for each natural number n, we have

$$\sum_{k=0}^{n-1} \alpha^k = \frac{1 - \alpha^n}{1 - \alpha};$$

Show that if $|\alpha| < 1$, then

$$\sum_{k=0}^{\infty} \alpha^k = \frac{1}{1 - \alpha}.$$

2.48 Show that for each $g > 4$, the number $(4.41)_g$ is the square of a rational number. Find its square root. Repeat the same problem for $(148.84)_g$ for $g > 8$.

2.49 For what values of g, are the numbers $(0.16)_g$, $(0.20)_g$, $(0.28)_g$ the consecutive terms of a geometric sequence?

2.50 If $25/128 = (0.0302)_g$, find g.

2.51 Find the base 5 expansion of $2877/3125$.

2.52 Find the base 9 expansion of $(200.211)_3$.

2.53 Determine the rational number with base 7 expansion $(0.\overline{130})_7$. Solve the same problem for $(0.1\overline{296})_{12}$.

2.54 Find all primitive roots modulo 38.

2.55 (✠) Find all primes $p < 1000$ for which 3 is a primitive root.

2.56 (✠) Find five primitive roots modulo 100003.

2.57 (✠) Find five primitive roots modulo 987654103^2.

2.58 If $p = 4q + 1$ and $q = 3r + 1$ are prime, show that 3 is a primitive root modulo p.

2.59 Let p, q be distinct primes. Find the number of solutions of $x^p \equiv 1 \bmod q$ in terms of p, q.

2.60 Without using a computer, prove that $2^{17} - 1$ is prime. Hint: Show that if it is not prime, it must be divisible by one of the numbers $103, 137, 239$, or 307.

2.61 Let p be an odd prime such that $(p - 1)/2$ is an odd prime. Prove that if a is a positive integer with $1 < a < p - 1$, then $p - a^2$ is a primitive root modulo p.

2.62 Show that n is a square if and only if $d(n)$ is odd.

2.63 Show that for all $n, s, t \in \mathbb{N}$, with $s \neq t$, $s - t \mid d(n^s) - d(n^t)$.

2.64 Show that for all $m, n, s \in \mathbb{N}$, we have $s \mid d(m^s) - d(n^s)$.

2.65 Find necessary and sufficient conditions on integers a, b, c, d so that there are integers x, y, z satisfying the system of Diophantine equations

$$\begin{cases} 4x + y + az = b, \\ x + y + cz = d. \end{cases}$$

2.66 Let p be an odd prime. Show that if we write $1 + 1/2 + \cdots + 1/(p - 1)$ as a fraction a/b with $a, b \in \mathbb{N}$, then $p \mid a$. A far more interesting problem is to show that if $p > 5$, $p^2 \mid a$.

Notes

Historical references

The standard reference for the history of classical number theory is Dickson's *History of the theory of numbers* in three volumes. Most of the material in this chapter has been reviewed in the first volume [15], especially Ch. III, V, and VIII. A more current reference for the history of mathematics is [9]. As impressive as these books are, like many other books on the history of science, they are unfortunately very Eurocentric. The history of mathematics as told through these and other similar texts runs like this: The Greeks invented mathematics; then as Europe was falling into the Dark Ages, Muslims ran to the rescue; Muslims carefully guarded mathematics for a few centuries; with the arrival of the Renaissance, the Muslims handed mathematics back to the Europeans who gracefully accepted the gift, and who have ever since been championing the progress of mathematics. This Eurocentricity does not stop at the history, and in fact it permeates every aspect of the practice of mathematics. In reality the history of mathematics is far more complicated and far more multicultural than a simple straight line connecting Athens of the antiquity to the North America and Europe of 21st century.

In this book I have made a conscious effort to highlight contributions by non-Europeans to number theory. However—and this is far from an acceptable excuse—because of my lack of expertise as well my own Eurocentric education I am not able to do justice to the subject. Getting the history right is not just a matter of intellectual curiosity. Those of us who work as educators in North America are acutely aware of the fact that a good portion of our students are not of European descent. To many of our students mathematics is a European invention, and will continue to be practiced by Europeans and people of European descent. Nothing could be further from the truth. Mathematics has been practiced on every continent, by all sorts of people, for thousands of years, and there are distinguished mathematicians of every imaginable background today doing fantastic mathematics—and this should be emphasized in our teaching. There is, unfortunately, a shortage of modern, easily accessible texts putting in the correct historical perspective the progress of mathematics through the millennia. Even in cases where a serious mathematician such as van der Waerden [55] has attempted to write a history of mathematics as inspired by the progress made by non-Europeans, the works of these non-Europeans are described in relation to and within the framework of modern European mathematics, or the Greek mathematics of the antiquity, in the sense that, what of the works of non-Europeans that has not been superseded and swallowed by some mathematical work developed by a European mathematician is often not considered worthy of review. The same problem exists in most works written by European or North American historians of mathematics, with a notable exception being Plofker [39]. Writings by historians like Roshdi Rashed, especially the second volume of *Encyclopedia of the history of Arabic sciences* [40] which covers mathematics, and Joseph [28] are good alternatives to standard Eurocentric narratives that saturate the literature.

On a personal note, growing up in Iran, I never felt that mathematics was a European invention or practice—I knew of Iranian mathematicians like Omar Khayyam, Mohammad Al-Khwarizmi, and Mohammad Karaji, and these were people I identified with. I credit Iranian education pioneers like G. H. Mossahab, M. Hashtroodi, M. Hessabi in the 1940s and 1950s, and more recently S. Shahshahani, P. Shahriari, O. A. Karamzadeh, Y. Tabesh, and others starting in the 1970s, for initiating the effort to instill the notion in the minds of the Iranian youth that mathematics, along with other sciences, was as Iranian as apple pie is American. It is because of their efforts that Iran has enjoyed a revitalization of mathematics in the last 25 years. Culture building takes time, and, as in the case of those Iranian pioneers, one may not live long enough to see the fruits of one's labor, but with patience and perseverance great things are possible.

Euclid and his Elements

Euclid (325–265 BCE) was the person who transformed mathematics from a number of uncoordinated and loosely proven theorems into an articulated and surely grounded science. Some of the theorems in Euclid's Elements were previously known by other mathematicians: Thales (624–546 BCE) who was according to Aristotle the first Greek philosopher, Eudoxus (410–355 BCE), Pythagoras and other Pythagoreans, etc. A predecessor to Euclid was Hippocrates (470-410 BCE) who wrote the first *Elements* around 430 BCE. Euclid was extremely rigorous in his treatment of mathematics. (Though as noted by David Hilbert [26], Euclid should have augmented his postulates by adding a few more.) E. T. Bell argues that if the world had followed Archimedes as opposed to Euclid, Calculus would have been discovered before the birth of Christ. This is a harsh criticism of the Euclidean rigor, and of the course of history, but it is nonetheless most likely true that the sort of rigor that Euclid brought into mathematics slowed down progress in some sense. Archimedes was a master problem solver who was interested in the applications of mathematics in the real world. Euclid, on the other hand, was interested in gaining a deep understanding of concepts via systematic study. For what it is worth, almost 2500 years later, we still practice mathematics the way Euclid did mathematics in his magnum opus. An interesting feature of *the Elements* is that the writing is extremely homogeneous. Euclid makes no distinction between trivial facts and deep theorems, and everything is proved with the same degree of care. Was Euclid really not aware that some of his results are more important than others? We will never know.

The theory of numbers is treated by Euclid in books 7-10 of the *Elements*. At the beginning of Book 7 Euclid lists definitions: unit, numbers, multiple, even and odd number, prime and composite numbers, square, proportional, perfect number, etc. These are very much in the Pythagorean style, but with some modifications. We refer the reader to the excellent commentary in Sir Thomas L. Heath's "The Thirteen Books of Euclid's Elements" [20] published in 1926. In this book Sir Heath compares Euclid's definitions to those given by his predecessors. In the case of prime numbers, Euclid's definition varies slightly from the one written by the Pythagorean Philolaus

(480–390 BCE) who seems to have been the first person to give a definition of prime numbers.

For all their aura of naturalness, prime numbers almost never appear in nature for reasons of primality. The only example of such a process is the life cycles of a certain genus of cicadas. These insects spend most of their lives underground and emerge to daylight every 13 or 17 years. The fact that 13 and 17 are prime numbers gives these insects a computable but small evolutionary edge over their predators. Over millions of years the evolutionary edge of these insects has helped them not go extinct. Beyond this, we are not aware of any cosmic or earthly processes that produce prime numbers for reasons of primality. Even within mathematics, as practiced by human beings, it appears that prime numbers were an invention of the Greeks, and that no one else in the ancient world had a notion of prime numbers. Mathematicians in Babylon, India, China, and the Americas investigated very sophisticated mathematical theories, including those applicable to astronomy and other sciences, but as far as we can tell none of these mathematicians had a theory of prime numbers.

For more on Euclid's work on prime numbers, see Notes, Chapter 6.

Natural Numbers and mathematical induction

In this book we will treat natural numbers in a common sense, intuitive fashion. We assume the set of natural numbers $\mathbb{N}$ consists of positive integers $1, 2, 3, \ldots$, equipped with the standard addition and multiplication operations, enjoying the familiar properties of commutativity and associativity for addition and multiplication, and distribution laws for multiplication over addition. We also *know* that we can prove statements in the set of natural numbers using *mathematical induction*, accepted as an axiom. In reality, however, all of these statements are non-trivial and require close examination. The axiomatic study of the set of natural numbers has a long, rich history. We refer the reader to [18, Ch. 1] for an accessible introduction to this beautiful subject.

Number-theory-based cryptography

Many modern cryptographic methods are based on the material presented in this chapter. Here we will explain two standard techniques. For an elementary treatment of these methods and other number theoretic cryptosystems we refer the reader to [53].

The RSA Cryptosystem, named after Ron Rivest, Adi Shamir, and Leonard Adleman, is based on the notion that while multiplying numbers is easy, finding the prime factors of a large number is difficult. More specifically, if we know the prime factors of a natural number n, then Theorem 2.33 tells us how to compute the value of $\phi(n)$. However, without knowing the prime factors of n, we do not have a fast algorithm to compute $\phi(n)$. Presumably, one can take (2.6) as the definition of $\phi(n)$. This requires going through the list of numbers 1 to n and examining the gcd of each one with n, which, if the number n is of the order of 10^{500}, would be impossible.

RSA is an example of a *public key cryptosystem*. In such a cryptographic scheme an individual A sets up a public key K, which is available to everyone, and keeps a private piece of information S, which is kept secret. The idea is that anyone who wants to communicate with A will encrypt the message using the publicly available key K but decrypting the encrypted message requires the secret information S. In the case of RSA, the public key is a large natural number n which is the product of prime numbers p, q. The prime numbers p and q are kept secret..

This is how RSA works. Suppose Azadeh wants to set up a public key. She picks large prime numbers p, q. She computes $n = pq$, $\phi(n) = (p-1)(q-1)$, and she picks a natural number e such that that $\gcd(e, \phi(n)) = 1$. She also finds an integer d such that $ed \equiv 1 \bmod \phi(n)$, i.e, $ed = 1 + u\phi(n)$ for some integer u. She will keep p, q, d, and $\phi(n)$ secret, but publishes the pair (n, e). Now suppose Azadeh's friend, Behnam, wants to communicate with Azadeh. Suppose the message that Behnam wants to send has numerical value m, obtained using ASCII or some other method (technically speaking, Behnam will have to make sure that $\gcd(m, n) = 1$). Behnam downloads the pair (n, e) from Azadeh's public profile, and computes $y := m^e \bmod n$, i.e., the remainder of the division of m^e by n which will be a number between 0 and n. Behnam keeps the message m secret, but sends the message y to Azadeh over some public channel, e.g., Facebook or SMS. Azadeh receives the message y, and deciphers it by computing

$$y^d \equiv (m^e)^d \equiv m^{1+u\phi(n)} \equiv (m^{\phi(n)})^u \cdot m \equiv m \mod n,$$

after using Theorem 2.31. On the other hand, Esmat, an evil person, is listening to the conversation happening between Azadeh and Behnam. Esmat downloads the message y. She also knows (n, e) as these are publicly available. However, at present there is no reasonably fast way to get from the data y, (n, e) to m without knowing d, and knowing d requires $\phi(n)$. As noted above computing $\phi(n)$, at the time of this writing, requires knowing the prime factors of n, which Azadeh is keeping secret.

For example, suppose Azadeh picks the prime numbers $p = 101$ and $q = 113$ (this is just a prototype; in practice the prime numbers are a few hundred digits long). Hence $n = 101 \times 113 = 11413$. We have $\phi(n) = (101 - 1)(113 - 1) = 11200$. She also picks $e = 3$. Note that $\gcd(3, 11200) = 1$. Azadeh's public key is the pair $(11413, 3)$. What Azadeh is not sharing with the public are the prime numbers 101 and 113. She also keeps secret the number d such that $3d \equiv 1 \mod 11200$. Azadeh can easily compute, for example using SageMath, Appendix C, that $d = 7467$ works. Now suppose Behnam wants to transmit a message m with numerical value 77 to Azadeh. Behnam computes $m^e \bmod n$. In this case since $m = 77$, $e = 3$, and $n = 11413$, he computes

$$77^3 \equiv 13 \mod 11413.$$

So Behnam's message, which he can communicate over a public channel, is $y = 13$. Anyone can read this message x, and everyone knows Azadeh's public key

(11413, 3). So the problem that Esmat, the evil person, needs to solve is this: Find m such that $m^3 \equiv 13 \bmod 11413$. For Azadeh, this is easy. All she needs to do is compute

$$13^{7467} \equiv 77 \quad \bmod 11413,$$

which she can easily do using SageMath.

The ElGamal Cryptosystem, named after the Egyptian computer scientist Taher ElGamal, is based on the difficulty of the Discrete Log problem. As mentioned earlier RSA cryptography is based on the idea that it is difficult to go from $(m^e \bmod n, e, n)$ to m. The flip side of this idea is the *Discrete Log* problem. Let n be a natural number for which we have a primitive root g. Let $1 < x < n$ be a natural number that is coprime to n. The *Discrete Log* problem asks for the determination of an integer $0 < l < \phi(n)$ such that $x \equiv g^l \bmod n$.

In the ElGamal Cryptosystem, Azadeh picks a large prime p, a primitive root g modulo p, a random number l, with $1 < l < p - 1$, and computes $e = g^l \bmod n$. Azadeh's public key is (p, g, e) which she publishes. She keeps l secret. Behnam wants to send a message m to Azadeh. Benham picks a random integer $u, 1 < u < p - 1$, and computes $x := g^u \bmod p$, and $y := m \cdot e^u \bmod p$. Behnam sends the pair (x, y) over a public channel to Azadeh. Azadeh recovers m by computing

$$y \cdot x^{-l} \equiv m \cdot (g^l)^u \cdot (g^u)^{-l} \equiv m \quad \bmod p.$$

We refer the reader to [53], especially Ch. 6 for RSA and Ch. 7 for ElGamal.

Primitive roots and Artin's conjecture

The notion of the order of a modulo n made an appearance in Gauss's book [21, articles 315-317], when he considered the decimal expansion of $1/p$ for a prime number p, $p \neq 2, 5$. In this case, the fraction $1/p$ is purely periodic and its period is equal to $o_p(10)$. In general, we saw in this chapter that if m, n are natural numbers with $\gcd(m, n) = 1$, then the base n expansion of $1/m$ is purely periodic with minimal period equal to $o_m(n)$. In particular the minimal period is at most equal to $\phi(m)$. In the case where $m = p$ is a prime number, $\phi(p) = p - 1$. The following is a natural question: For a natural number n, are there infinitely many prime numbers p such that the base n-expansion of $1/p$ has period $p - 1$? Note that in this case n will have to be a primitive root modulo p. While the answer to the question is expected to be yes, it is not known for any n, not even $n = 10$.

Conjecture 2.60 (Artin 1927). Fix an integer $g \neq -1, 0, 1$ which is not a perfect square. Then there are infinitely many primes p such that g is a primitive root modulo p.

In fact, Artin conjectured an asymptotic formula for $\#\{p \text{ prime} \mid p \leq X, o_p(g) = p-1\}$ of the form $\delta(g)X/\log X$ as $X \to \infty$, for some constant $\delta(g) > 0$. Artin gave a heuristic argument to derive a formula for $\delta(g)$; however, in 1957 Derrick and Emma

Lehmer observed that Artin's predicted formula did not match numerical data. Artin was then able to pinpoint the error in the original heuristic reasoning and corrected the prediction. In 1967 Hooley [81] gave a proof of the predicted asymptotic formula which relied on some version of Riemann's Hypothesis, not yet proved; see Notes to Chapter 13. See Murty's expository article [89] for an accessible account of the progress made toward the conjecture up until the time of its publication. For a more up-to-date report on the conjecture and the methods and techniques used in its study, see Moree's survey [88].

Chapter 3
Integral solutions to the Pythagorean Equation

In this chapter we present two different methods to find the solutions of the Pythagorean Equation, one algebraic and one geometric. We then apply the geometric method to find solutions of some other equations. The first class of non-Pythagorean Equations that we will apply this method to is Pell's equation, and the second class, equations of degree three. As an application of our solution to the Pythagorean Equation we will prove a special case of *Fermat's Last Theorem*. In the Notes, we briefly review some classical works related to Pell's Equation over integers; explain why some cubic equations are called *elliptic*; give some references related to Fermat's Last Theorem; and discuss the *abc* Conjecture.

3.1 Solutions

Suppose (a, b, c) is a triple of integer solutions to the Pythagorean Equation. Then by definition

$$a^2 + b^2 = c^2. \tag{3.1}$$

If a, b, c have a common factor λ, then

$$\left(\frac{a}{\lambda}\right)^2 + \left(\frac{b}{\lambda}\right)^2 = \left(\frac{c}{\lambda}\right)^2.$$

So without loss of generality we may assume that a, b, c have no common factors. These are the triples we called *primitive* in Chapter 1. The Pythagorean triples we consider in this chapter are all primitive. A quick computer search produces the following list of the first few Pythagorean triples:

```
(3, 4, 5)
(5, 12, 13)
(8, 15, 17)
(7, 24, 25)
```

© Springer Nature Switzerland AG 2018
R. Takloo-Bighash, *A Pythagorean Introduction to Number Theory*,
Undergraduate Texts in Mathematics, https://doi.org/10.1007/978-3-030-02604-2_3

```
(20,  21,  29)
(12,  35,  37)
(9,  40,  41)
(28,  45,  53)
. . .
```

Our goal in this section is to find all primitive solutions of Equation (3.1). Since $\gcd(a, b, c) = 1$, it is clear that not all a, b, c are even. We recognize several possibilities.

- a, b, c odd. This is impossible, as one side will be odd and the other side even.
- a, b odd, c even. If a is odd, then $a^2 \equiv 1$ mod 8, and $b^2 \equiv 1$ mod 8; hence $a^2 + b^2 \equiv 2$ mod 8. But since c is even, $4 \mid c^2$, so $c^2 \equiv 0, 4$ mod 8. So this case is impossible as well.
- a even, b odd, c odd.
- a odd, b even, c odd.

We will see momentarily that these last two cases are in fact possible. By symmetry we may assume that a is even, and b odd. Write

$$b^2 = c^2 - a^2 = (c - a)(c + a).$$

We claim that $\gcd(c - a, c + a) = 1$. To see this, we have

$$\gcd(c - a, c + a) = \gcd(c + a, c + a - (c - a)) = \gcd(c + a, 2a).$$

But since $c + a$ is odd, $\gcd(c + a, 2a) = \gcd(c + a, a) = \gcd(c, a)$. If there is a prime number $p \mid \gcd(c, a)$, then $p \mid a^2$ and $p \mid c^2$ so $p \mid b^2 = c^2 - a^2$, and consequently, $p \mid b$. This statement contradicts the assumption that $\gcd(a, b, c) = 1$. Since the product of the coprime numbers $c + a$ and $c - a$ is a square b^2, by Proposition 2.21 each of them individually is a square, i.e., there are odd coprime integers x, y such that

$$c + a = x^2, c - a = y^2.$$

Solving for c, and a, gives

$$\begin{cases} a = & \frac{x^2 - y^2}{2}, \\ b = & xy, \\ c = & \frac{x^2 + y^2}{2}. \end{cases}$$

It is of course true that

$$\left(\frac{x^2 - y^2}{2}\right)^2 + (xy)^2 = \left(\frac{x^2 - y^2}{2}\right)^2$$

as one can easily check. For example, if $(x, y) = (3, 1)$ we recover the well-known triple $(4, 3, 5)$, and if $(x, y) = (5, 1)$, then we get $(12, 5, 13)$. In general, instead of writing a triple as ordered vector, we write the triple as a set. So instead of $(12, 5, 13)$ we write $\{12, 5, 13\}$, and our general solution will be written as

$$\left\{ \frac{x^2 - y^2}{2}, xy, \frac{x^2 + y^2}{2} \right\}.$$

We summarize this discussion in the following theorem:

Theorem 3.1. *Let a, b, c be the three sides of a primitive integral right triangle. There are odd coprime integers x, y such that*

$$\{a, b, c\} = \left\{ \frac{x^2 - y^2}{2}, xy, \frac{x^2 + y^2}{2} \right\}.$$

3.2 Geometric method to find solutions

In this section we present a geometric method to find the solutions of the Pythagorean Equation. First a piece of notation: For a point $(x, y, z) \in \mathbb{R}^3$ with $z \neq 0$ we define $R(x, y, z)$ be the point $(x/y, x/z) \in \mathbb{R}^2$. If it is clear that if $(x, y, z) \in \mathbb{Z}^3$ with $z \neq 0$, then $R(x, y, z)$ be a *rational point*, i.e., a point with coordinates that are rational numbers in $\mathbb{R}^2$. Suppose (a, b, c) is a primitive solution of the Pythagorean Equation. We have

$$\left(\frac{a}{c} \right)^2 + \left(\frac{b}{c} \right)^2 = 1.$$

This means that $R(a, b, c)$ is a point with rational coordinates on the unit circle $x^2 + y^2 = 1$. Now suppose we have a rational point $(a/b, c/d)$, $a, b, c, d \in \mathbb{Z}$, on the unit circle centered at the origin, and suppose that the rational numbers a/b and c/d are in reduced form, meaning $\gcd(a, b) = \gcd(c, d) = 1$. We wish to show that there is a primitive solution (x, y, z) of the Pythagorean Equation such that $R(x, y, z) = (a/b, c/d)$. This claim is obvious if one of the coordinates $a/b, c/d$ is zero. So we assume that $ac \neq 0$. After changing the signs if necessary we assume $a, b, c, d > 0$. Since $(a/b)^2 + (c/d)^2 = 1$, $a^2 d^2 + c^2 b^2 = b^2 d^2$. Since $b^2 | c^2 b^2$ and $b^2 | b^2 d^2$ we conclude $b^2 | a^2 d^2$, but since we have assume $\gcd(a, b) = 1$, by Theorem 2.17, we have $b^2 | d^2$. This means $b | d$. Similarly, $d | b$. Consequently, $b = d$. As a result every rational point in the first quadrant on the unit circle will be of the form $(a/b, c/b)$ with a, b, c natural numbers and $\gcd(a, b) = 1$ and $\gcd(c, b) = 1$. Also, we have $a^2 + c^2 = b^2$, i.e., (a, c, b) is a solution of the Pythagorean Equation. It is also easy to see that $\gcd(a, c) = 1$. In fact, if u is a common factor of a and c, then $u^2 | a^2 + c^2 = b^2$, giving $u^2 | b^2$, from which it follows $u | b$. This implies $u | \gcd(a, b) = 1$. Hence $u = 1$. Summarizing, for a rational point (x, y) on the unit circle with $x, y > 0$, there are pairwise coprime natural numbers a, b, c such that $x = a/b$, $y = c/b$ and $a^2 + c^2 = b^2$. This means $R(a, c, b) = (x, y)$. Note that $R(-a, -c, -b) = (x, y)$ as well, and (a, c, b) and $(-a, -b, -c)$ are the only primitive Pythagorean triples whose R is (x, y). Finally if either of x or y is negative, we can adjust the sign of a or c to get the correct sign. The map R is always 2-to-1 from primitive Pythagorean triples to the set of rational points on the unit circle.

Fig. 3.1 Finding rational
points on the unit circle. Here
we have connected the point
$(-1, 0)$ to the point (x_0, y_0)

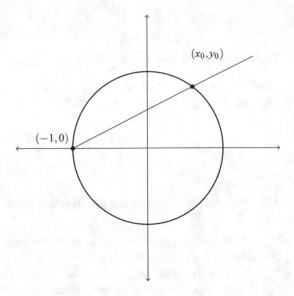

A consequence of this discussion is that in order to find Pythagorean triples it is
sufficient to determine rational points on the unit circle.

We proceed to determine the set of rational points on the unit circle. The circle
$x^2 + y^2 = 1$ in Figure 3.1 has some obvious solutions, e.g., $(\pm 1, 0)$ or $(0, \pm 1)$. Let's
pick one of these points, say $(-1, 0)$. The main observation is that if (x_0, y_0) is a
point with rational coordinates, then the slope of the line connecting this point to the
base point $(-1, 0)$ is

$$m = \frac{y_0}{x_0 + 1}$$

is a rational number.

Our idea is to do the opposite of this, i.e., pass a line with rational slope through
$(-1, 0)$, look at the point of intersection of the line with the circle $x^2 + y^2 = 1$, and
hope that the resulting point is a rational point. The equation of the line with slope
m through $(-1, 0)$ is

$$y = m(x + 1).$$

To find the point of intersection of this line with the circle we need to solve the system
of equations

$$\begin{cases} y = m(x + 1), \\ x^2 + y^2 = 1. \end{cases}$$

Inserting the value of y from the first equation in the second equation gives

$$x^2 + m^2(x + 1)^2 = 1.$$

Simplifying gives

$$(m^2 + 1)x^2 + 2m^2 x + (m^2 - 1) = 0.$$

Since the product of the roots of the equation is $(m^2 - 1)/(m^2 + 1)$ and one of the roots is -1, we see that the second root is

$$x = \frac{1 - m^2}{1 + m^2}.$$

By using the equation $y = m(x + 1)$, we see that $y = 2m/(1 + m^2)$. This means that the point of intersection is

$$P_m := \left(\frac{1 - m^2}{1 + m^2}, \frac{2m}{1 + m^2} \right). \tag{3.2}$$

Now we would like to derive a triple of integers (a, b, c) from this pair of rational numbers. Let $m = r/s$ with r, s coprime integers. Then we get

$$P_m = \left(\frac{s^2 - r^2}{s^2 + r^2}, \frac{2rs}{s^2 + r^2} \right).$$

Now we find a primitive Pythagorean triple (u, v, w) such that

$$R(u, v, w) = \left(\frac{s^2 - r^2}{s^2 + r^2}, \frac{2rs}{s^2 + r^2} \right).$$

We need to calculate

$$\gcd(s^2 - r^2, s^2 + r^2), \quad \gcd(2rs, s^2 + r^2).$$

Lemma 3.2. *For coprime integers r, s, define a function*

$$\delta(r, s) = \gcd(2, s^2 + r^2).$$

Then

$$\gcd(s^2 - r^2, s^2 + r^2) = \gcd(2rs, s^2 + r^2) = \delta(r, s).$$

Proof. Since $\gcd(r, s) = 1$, we have

$$\gcd(rs, s^2 + r^2) = 1.$$

Indeed, if p is a prime number and $p \mid \gcd(rs, s^2 + r^2)$, then either $p \mid r$ or $p \mid s$. If $p \mid r$, then since $p \mid s^2 + r^2$, we have $p \mid s^2$, and as a result $p \mid s$. So $p \mid r, p \mid s$, contradicting the coprimality assumption. As a result

$$\gcd(2rs, s^2 + r^2) = \gcd(2, s^2 + r^2).$$

Next,

$$\gcd(s^2 - r^2, s^2 + r^2) = \gcd(s^2 + r^2, s^2 + r^2 + (s^2 - r^2))$$

$$= \gcd(s^2 + r^2, 2s^2) = \gcd(2, s^2 + r^2) = \delta(r, s),$$

again as $\gcd(s^2, s^2 + r^2) = 1$.

Note that

$$\delta(r, s) = \begin{cases} 2 & \text{if } r \equiv s \text{ mod } 2; \\ 1 & \text{otherwise.} \end{cases}$$

It follows from the lemma that

$$P_m = \left(\frac{\frac{s^2 - r^2}{\delta(r,s)}}{\frac{s^2 + r^2}{\delta(r,s)}}, \frac{\frac{2rs}{\delta(r,s)}}{\frac{s^2 + r^2}{\delta(r,s)}} \right),$$

and in this representations the coordinates of P_m are in reduced form. Consequently, if we set

$$(u, v, w) = \left(\frac{s^2 - r^2}{\delta(r, s)}, \frac{2sr}{\delta(r, s)}, \frac{s^2 + r^2}{\delta(r, s)} \right),$$

then $R(u, v, w) = P_m$. If we do not care about the order, we may write $\{u, v, w\} = \tau(r, s)$, where

$$\tau(s, r) = \left\{ \frac{s^2 - r^2}{\delta(r, s)}, \frac{2sr}{\delta(r, s)}, \frac{s^2 + r^2}{\delta(r, s)} \right\}.$$

The trouble with this parametrization of Pythagorean triples is that it is not a bijection with the set of coprime integers r, s. For example, if the pairs $(r, s) = (1, 2)$ and $(1, 3)$ both give the famous Pythagorean triple 3, 4, 5. In fact, in general, if r, s are both odd, we obtain

$$\{u, v, w\} = \left\{ \frac{s^2 - r^2}{2}, sr, \frac{s^2 + r^2}{2} \right\}.$$

So the question that we now need to answer is: What happens to the cases where either r or s is even. This has an amusing explanation.

Lemma 3.3. *Let r, s be coprime integers of different parity. Then $r + s$ and $r - s$ are coprime odd numbers and*

$$\tau(s, r) = \tau(s + r, s - r).$$

Proof. An easy check shows that

$$\frac{(s + r)^2 + (s - r)^2}{2} = s^2 + r^2;$$

$$\frac{(s + r)^2 - (s - r)^2}{2} = 2sr;$$

and

$$(s + r)(s - r) = r^2 - s^2.$$

We have proved the following theorem:

Theorem 3.4. *Let u, v, w be the three sides of a primitive integral right triangle. There are coprime integers x, y of different parity such that*

$$\{u, v, w\} = \{x^2 - y^2, 2xy, x^2 + y^2\}.$$

For example, if $x = 2$, $y = 1$, we obtain 3, 4, 5; if $x = 3$, $y = 2$, we have 5, 12, 13; if $x = 4$, $y = 3$, we find the triple 7, 24, 25.

Remark 3.5. It is important to compare the statement of Theorem 3.4 with Theorem 3.1.

Remark 3.6. The interesting thing about Equation (3.2) is that we do not have to assume that $m \in \mathbb{Q}$. In fact the same computation works over any field, e.g., $\mathbb{R}$, $\mathbb{C}$, or even finite fields. In general care is needed to ensure the denominator $1 + m^2$ is not zero. For fields like $\mathbb{Q}$ and $\mathbb{R}$ this is not an issue, but as soon as we work over a field like $\mathbb{C}$, then $1 + m^2$ can in fact be zero. We will return to this point in Chapters 8 and 14.

3.3 Geometric method to find solutions: Non-Pythagorean examples

It might seem superfluous to use the geometric method of §3.2 to find the solutions of the Pythagorean Equation in light of the much easier methods of §3.1. However, the geometric methods of §3.2 have applications to situations where the elementary methods of §3.1 give little or no information. To demonstrate this method we discuss two examples in this section.

The first example we discuss is Pell's Equation:

$$x^2 - Dy^2 = 1, \tag{3.3}$$

where we assume D is a square-free positive integer. Typically this equation is considered as a Diophantine equation with integral solutions where the solutions are determined using the continued fraction expansion of the quadratic surd $\sqrt{D}$, cf. [33, Ch. 7]. Here we would like to consider this equation over the rational numbers. There are some obvious solutions, namely $(+1, 0)$ and $(-1, 0)$. We will use one of these, say $(-1, 0)$, to find the other rational solutions.

The equation of the straight line passing through $(-1, 0)$ with slope m is

$$y = m(x + 1).$$

We find the points of intersection of this line with the curve with equation $x^2 - Dy^2 = 1$ by inserting the value of y from the equation of the straight line in the equation of the curve. We obtain

$$x^2 - Dm^2(x + 1)^2 = 1.$$

Expanding $(x + 1)^2$ and collecting terms gives

$$(1 - Dm^2)x^2 - 2Dm^2x - (Dm^2 + 1) = 0.$$

Since we know that $x = -1$ is a solution of this equation, we find the other solution to be

Fig. 3.2 The graph of
$x^2 - Dy^2 = 1$. Here we
have drawn a straight line
with slope m through the
point $(-1, 0)$

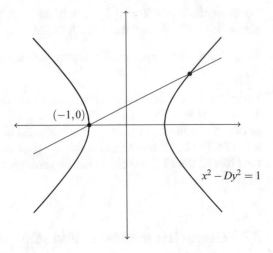

$$x = \frac{1 + Dm^2}{1 - Dm^2}.$$

With this at hand we find the corresponding y as

$$y = m\left(\frac{1 + Dm^2}{1 - Dm^2} + 1\right) = \frac{2m}{1 - Dm^2}.$$

Consequently, we have proved the following theorem:

Theorem 3.7. *Every solution of Pell's Equation over the rational numbers other than the pair $(x, y) = (-1, 0)$ is expressible as*

$$\begin{cases} x = \frac{1+Dm^2}{1-Dm^2}, \\ y = \frac{2m}{1-Dm^2} \end{cases}$$

for some $m \in \mathbb{Q}$.

One can easily find *integral* solutions to the equation

$$X^2 - DY^2 = Z^2 \tag{3.4}$$

by using the above rational parametrization; see Exercise 3.1. As an example, let's consider the case where $D = 3$. Then Theorem 3.7 says that the rational solutions of the equation $x^2 - 3y^2 = 1$ are of the form

$$x = \frac{1 + 3m^2}{1 - 3m^2}, \quad y = \frac{2m}{1 - 3m^2}$$

for $m \in \mathbb{Q}$. If we put $m = 2$, then we get the pair $(x, y) = (-13/11, -4/11)$, from which the solution $(X, Y, Z) = (-13, -4, 11)$ for the equation $X^2 - 3Y^2 = Z^2$ is obtained. If on the other hand we put $m = 1/2$, we obtain $(x, y) = (7, 4)$. This pair gives the solution $(X, Y, Z) = (7, 4, 1)$ of $X^2 - 3Y^2 = Z^2$.

The above method works for any quadratic polynomial. Indeed, suppose $f(x, y)$ is a quadratic polynomial of degree two in the variables x, y with rational coefficients. In the examples we have discussed so far, $f(x, y) = x^2 + y^2 - 1$ in the Pythagorean case, or $f(x, y) = x^2 - Dy^2 - 1$ in the Pell case. Then the graph of $f(x, y) = 0$ either contains infinitely many points with rational coordinates, or none at all. The proof of this fact is identical to our arguments for the examples we have discussed so far.

In our next example, we consider the important case where the degree of the polynomial f is equal to 3. The most general polynomial $f(x, y)$ of degree 3 with rational coefficients can be written as

$$a_1 x^3 + a_2 y^3 + a_3 x^2 y + a_4 x y^2 + a_5 x^2 + a_6 y^2 + a_7 xy + a_8 x + a_9 y + a_{10}.$$

Here we assume that the a_i's are rational numbers and at least one of a_1, a_2, a_3, and a_4 is non-zero. Let C be the graph of f. If we try and imitate what we did for quadratic polynomials, we run into trouble. Indeed, suppose (a, b) is a point on the curve C. Then the equation of the line passing through (a, b) with slope m is

$$y = m(x - a) + b.$$

If we insert this expression for y in the equation $f(x, y) = 0$ we obtain a degree 3 equation in x which has three roots. By construction, one of the roots of this equation is $x = a..$ In general there is no reason that the resulting equation should have two more rational solutions. We can see this in an example as follows.

Suppose, for example, that

$$f(x, y) = y^2 + x^3 + 1.$$

Then there is an obvious solution of $(-1, 0)$. The line through this point with slope m has equation

$$y = m(x + 1).$$

We then obtain the equation

$$m^2(x + 1)^2 + x^3 + 1 = 0.$$

Expanding and simplifying give

$$x^3 + m^2 x^2 + 2m^2 x + (1 + m^2) = 0.$$

Since this equation a priori has a root $x = -1$, the polynomial on the left should be divisible by $(x + 1)$. One easily sees that the polynomial factors as $(x + 1)$ multiplied by

$$x^2 + (m^2 - 1)x + (m^2 + 1). \tag{3.5}$$

This quadratic equation will have rational roots if its discriminant is a rational square t^2, for some $t \in \mathbb{Q}$. We calculate the discriminant as

$$\Delta = (m^2 - 1)^2 - 4(m^2 + 1) = m^4 - 2m^2 + 1 - 4m^2 - 4 = m^4 - 6m^2 - 3.$$

So the equation we need to find rational solutions for is

$$t^2 = m^4 - 6m^2 - 3$$

which is of higher degree than the original equation $y^2 + x^3 + 1 = 0$.

The above discussion suggests the following strategy: Instead of using one rational point on the curve and a rational slope, use two rational points on the curve. Once we have two points, connect the points using a straight line; look at the intersection of the resulting line with the curve. This last point is then a new point with rational coordinates on the curve. We demonstrate this idea with a couple of examples.

Example 3.8. Consider the curve $y^2 = x^3 + 17$. An inspection reveals the points $(-1, 4)$, $(2, 5)$ with rational coordinates on the curve. The equation of the line connecting the points is

$$y = \frac{1}{3}x + \frac{13}{3}.$$

The intersection point of the line with the curve is the point determined by solving the system of equations

$$\begin{cases} y^2 = x^3 + 17, \\ y = \frac{1}{3}x + \frac{13}{3}. \end{cases}$$

To solve, we insert the value of y from the second equation in the first equation to obtain

$$\left(\frac{1}{3}x + \frac{13}{3} \right)^2 = x^3 + 17.$$

Simplifying gives

$$x^3 - \frac{1}{9}x^2 - \frac{26}{9}x - \frac{16}{9} = 0.$$

We already know two of the roots of this equation, namely -1 and 2. Since the product of the three roots of the equation is $16/9$, we find that the third root is

$$x = -\frac{8}{9}.$$

Now we use the equation of the straight line to find y:

$$y = \frac{109}{27}.$$

It is easy to check that the point $(-8/9, 109/27)$ is indeed on the curve. There are in fact infinitely many pairs of rational numbers satisfying $y^2 = x^3 + 17$, but the proof of this fact is beyond the scope of this book.

Fig. 3.3 The cubic curve $y^2 = x^3 + 1$ with the colinear points $(-1, 0)$, $(0, 1)$, and $(2, 3)$

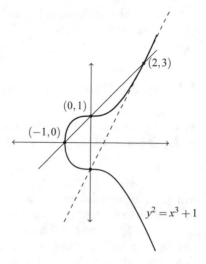

What we did in the above example was choosing points A, B on the curve, connecting them, and looking at the point of the intersection of the resulting line with the curve. Now suppose we choose the points A, B very close to each other. As the points get close to each other, the line connecting them approaches the tangent line to the curve at the point obtained from identifying A and B. So one way to obtain rational points on a cubic curve is by starting from a rational point and drawing the tangent line to the curve at that point. The other point of intersection of the tangent line with the curve must then be a rational point. In the next example we show how this idea is used in practice.

Example 3.9. The equation $y^2 = x^3 + 1$ in Figure 3.3 has the obvious solutions $(0, 1)$ and $(-1, 0)$.

The straight line connecting the points is

$$y = x + 1.$$

The intersection of this line with the curve is the point $(2, 3)$. Now that we have a new point, we can draw the tangent line at the point $(2, 3)$ to find more points. By implicit differentiation we have

$$2yy' = 3x^2.$$

Hence the slope of the tangent line at the point $(2, 3)$ is

$$m = 2.$$

The equation of the tangent line is $y = 2x - 1$. This is the dashed line in the figure. The intersection of this line with the graph of $y^2 = x^3 + 1$ is the point $(0, -1)$ which is a new point, though not very interesting. In fact, by using more advanced techniques than what is discussed in this book, one can show that the equation $y^2 = x^3 + 1$ has only finitely many solutions in pairs of rational numbers.

See the Notes at the end of this chapter for more on these cubic curves and the connections to the theory of elliptic curves.

3.4   Application: $X^4 + Y^4 = Z^4$

At some point around 1637, Pierre de Fermat famously declared in the margin of a book that if $n \in \mathbb{N}$ is larger than 2, the Diophantine equation

$$X^n + Y^n = Z^n$$

will not have any solutions in integers X, Y, Z, except for those satisfying $XYZ = 0$. He went on to say that he had an amazing proof of the fact, but that the margin was too small to fit the proof. This claim is now known as Fermat's Last Theorem even though its proof was finally completed by Sir Andrew Wiles, then a professor at Princeton University, in a joint work with Richard Taylor in 1994. Wiles' work was a crowning achievement of modern mathematics which built on works by many, many mathematicians spanning, literally, hundreds of years. Nowadays very few mathematicians believe that Fermat actually had a proof for the general case, neither does anyone hope that one might ever be able to give a reasonably short, elementary proof of the theorem accessible to Fermat. It is, however, possible to prove many special cases of the theorem using elementary methods. Here we present a proof of the special case for $n = 4$ discovered by Fermat. The proof we give uses our knowledge of the solutions of the Pythagorean Equation.

Theorem 3.10 (Fermat). *If the integers X, Y, Z satisfy $X^4 + Y^4 = Z^2$, then $XY = 0$.*

Proof. Suppose our claim is wrong, i.e., there are solutions (X, Y, Z) with $X > 0$, $Y > 0$, and $Z > 0$. Property 2.2 allows us to choose among these solutions the triple (x, y, z) with the smallest possible z. Clearly then $\gcd(x, y, z) = 1$. By Theorem 3.4 there are coprime integers m, n such that

$$\begin{cases} x^2 = 2mn, \\ y^2 = m^2 - n^2, \\ z = m^2 + n^2. \end{cases}$$

The second equation in the list can be rewritten as $n^2 + y^2 = m^2$. Again, Theorem 3.4 tells us that there are coprime integers u, v of different parity such that

$$\begin{cases} n = 2uv, \\ y = u^2 - v^2, \\ m = u^2 + v^2. \end{cases}$$

Next, we have

$$x^2 = 2mn = 4uv(u^2 + v^2).$$

Note that since u, v are coprime and of different parity, the integers u, v, and $u^2 + v^2$ are pairwise coprime; since their product is a square, each of them individually is a square, i.e., there are integers r, s, and t such that

$$\begin{cases} u = r^2, \\ v = s^2, \\ u^2 + v^2 = t^2. \end{cases}$$

Combining these three equations gives

$$r^4 + s^4 = t^2.$$

By construction $r > 0, s > 0, t > 0$. We now observe

$$0 < t \le t^2 = u^2 + v^2 = m < z.$$

Hence we have found a solution (r, s, t) of the equation $X^4 + Y^4 = Z^2$ with $0 < t < z$. This contradicts our assumption that (x, y, z) was the solution with the smallest possible z. □

The theorem has the following immediate corollary:

Corollary 3.11. *If the integers X, Y, Z satisfy*

$$X^4 + Y^4 = Z^4, \tag{3.6}$$

then $XY = 0$.

This is how the proof of Theorem 3.10 works: Suppose we have some integral solution (X, Y, Z) of the equation $x^4 + y^4 = z^2$ with $XYZ \ne 0$. Since z appears in the equation with even exponent, we conclude that $(X, Y, |Z|)$ will satisfy the equation as well. This means that the equation will then have solutions (X, Y, Z) with $Z \in \mathbb{N}$. Now let S be the set of all such Z's. Since S is assumed to be non-empty, Property 2.2 shows that S must have a smallest element Z_0. The main piece of the proof of the theorem consists of showing that there is another number $Z_1 \in S$ such that $Z_1 < Z_0$, and this is a contradiction as we had assumed that Z_0 was the smallest element of S.

The method used in the proof of Theorem 3.10 is called *infinite descent*. The method of infinite descent relies on the Well-ordering Principle, Property 2.2. As we saw in Theorem 2.3, the Well-ordering Principle is nothing but mathematical induction. Infinite descent was of the most powerful methods in Fermat's arsenal of tools and tricks. We will see some more applications of this method in the exercises. We will also use this method in the proof of Theorem 4.4.

Exercises

3.1 For an integer D, find the integral solutions of Pell's Equation (3.4).

3.2 Find the rational solutions to $x^2 - y^2 = 1$ by writing $x - y = m/n$ and $x + y = n/m$.

3.3 Find every integral solution of the equation

$$a^2 + b^2 + c^2 = d^2.$$

3.4 Prove that the only integral solution to the equation $x^2 + y^2 + z^2 = 2xyz$ is $x = y = z = 0$.

3.5 Find all the rational solutions of $x^2 + y^2 = z^2 + t^2$.

3.6 Show that for all natural numbers n, the equation $x^2 - y^2 = n^3$ is solvable in integers x, y. Determine the number of solutions if n is odd.

3.7 Show that the equation

$$x^2 + (x+1)^2 + (x+2)^2 + (x+3)^2 + (x+4)^2 = y^2$$

has no solutions in integers x, $y \in \mathbb{Z}$.

3.8 Find all the solutions of the equation

$$3(x^2 + y^2) + 2xy = 664$$

in integers x, y.

3.9 Show that for every $t \in \mathbb{Z}$ the triple

$$(x, y, z) = (9t^4, 1 - 9t^3, 3t - 9t^4)$$

satisfies

$$x^3 + y^3 + z^3 = 1.$$

Also verify that for each $t \in \mathbb{Z}$

$$(x, y, z) = (1 + 6t^3, 1 - 6t^3, -6t^3)$$

is a solution of the equation $x^3 + y^3 + z^3 = 2$. Show that the equation $x^3 + y^3 + z^3 = 4$ has no solutions in $\mathbb{Z}$. It is in general not known how to solve equations of the form $x^3 + y^3 + z^3 = n$ with $x, y, z \in \mathbb{Z}$.

3.10 Find all integral right triangles whose hypotenuse is a square.

3.11 Find all right triangles one of whose legs is a square.

3.12 Find all primitive right triangles with square perimeter.

3.13 Show that for every $n \in \mathbb{N}$, there are at least n distinct primitive right triangles which share a leg.

3.14 Show that for every $n \in \mathbb{N}$, there are at least n distinct primitive right triangles which share their hypotenuse.

3.15 Find all integral right triangles whose side lengths form an arithmetic progression.

3.16 Show that for every n there are n points in the plane, not all of which are on a straight line, such that the distance between every two of them is an integer. How about infinitely many points?

3.17 Show that for every Pythagorean triple (u, v, w) we have

$$(uv)^4 + (vw)^4 + (wu)^4 = (w^4 - u^2v^2)^2.$$

Conclude that the equation

$$x^4 + y^4 + z^4 = t^2$$

has infinitely many solutions in integers x, y, z, t such that $\gcd(x, y, z) = 1$.

3.18 Solve the system of Diophantine equations

$$\begin{cases} x^2 + t = u^2, \\ x^2 - t = v^2. \end{cases}$$

3.19 Verify that the points $(1, 0)$ and $(0, 2)$ satisfy the equation

$$y^2 = x^3 - 5x + 4.$$

Use the geometric method of this chapter to find more solutions.

3.20 Verify that the point $(-3, 9)$ satisfies the equation $y^2 = x^3 - 36x$. Use this point to produce more solutions.

3.21 Use *infinite descent* to show that there is no rational number y such that $y^2 = 2$.

3.22 Show that there are no non-zero integral solutions to the following equations:

a. $2x^4 - 2y^4 = z^2$;
b. $x^4 + 2y^4 = z^2$;
c. $x^4 - y^4 = 2z^2$;
d. $8x^4 - y^4 = z^2$.

3.23 Show that the only solutions to $x^4 + y^4 = 2z^2$ in integers are $z = \pm x^2$ and $|y| = |x|$.

3.24 ($\maltese$) Find the number of solutions (x, y, z) in integers of the equation $x^2 - 5y^2 = z^2$ with $|x|, |y|, |z| < 1000$.

3.25 ($\maltese$) Find 25 pairs of integers (x, y) such that $x^2 - 2y^2 = 1$. You might want to use Equation (3.7) of the Notes.

3.26 ($\maltese$) Find ten pairs of rational numbers (x, y) such that $y^2 = x^3 + 3$.

Notes

Pell's Equation

Traditionally, Pell's Equation is Equation (3.3) with the extra assumption that x, y are integers. The equation

$$x^2 - Dy^2 = -1,$$

too, is called Pell's Equation. Calling any of these equations Pell's Equation is a famous mischaracterization by Euler. Historically these equations were of interest to mathematicians for hundreds of years before Euler and his contemporaries; see, for example, [27, Ch. 2]. This last reference states that in 628 the great Indian mathematician Brahmagupta (598–670 CE) discovered the identity

$$(a^2 - Db^2)(p^2 - Dq^2) = (ap + Dbq)^2 - D(aq + bp)^2.$$

An immediate consequence of this fact is the remarkable statement that if Pell's Equation $x^2 - Dy^2 = \pm 1$ has a non-trivial integral solution, i.e., one where $y \neq 0$, it will have infinitely many integral solutions. In fact, let (x_1, y_1) be the solution of the equation

$$x^2 - Dy^2 = \pm 1,$$

with $x_1, y_1 > 0$, and x_1 the smallest possible. We call (x_1, y_1) the *fundamental solution*. Then, there are two possibilities:

1. If $x_1^2 - Dy_1^2 = +1$, then the equation $x^2 - Dy^2 = -1$ has no solutions. Furthermore, every solution of the equation $x^2 - Dy^2 = +1$ is of the form $(\pm x_N, \pm y_N)$ with

$$x_N + \sqrt{D}y_N = (x_1 + \sqrt{D}y_1)^N \tag{3.7}$$

for some $N \in \mathbb{Z}$.

2. If $x_1^2 - Dy_1^2 = -1$, then the equation $x^2 - Dy^2 = -1$ has solutions $(\pm x_N, \pm y_N)$ determined by Equation (3.7) with $N \in \mathbb{Z}$ odd. The solutions of $x^2 - Dy^2 = +1$ are the pairs $(\pm x_N, \pm y_N)$ with $N \in \mathbb{Z}$ even.

For example when $D = 2$, the fundamental solution to $x^2 - 2y^2 = \pm 1$ is $(1, 1)$ which satisfies $1^2 - 2 \cdot 1^2 = -1$. If $N = 2$, we compute

$$(1 + \sqrt{2})^2 = 3 + 2\sqrt{2},$$

and it is clear that $(3, 2)$ satisfies $3^2 - 2.2^2 = +1$. If $N = 3$,

$$(1 + \sqrt{2})^3 = 7 + 5\sqrt{2},$$

and $7^2 - 2 \cdot 5^2 = -1$.

Because of these observations, finding the solutions of Pell's Equation reduces to the search for the fundamental solution. Note that even though the fundamental solution (x_1, y_1) is the smallest solution of the equation, it does not have to be *small* in any reasonable sense. For example, the smallest solution of $x^2 - 61y^2 = 1$ is $(x, y) = (1766319049, 226153980)$. The most effective way to write down the fundamental solution is via *continued fractions*. This method was originally discovered by the Indian mathematicians Jayadeva (c. 950–~ 1000 CE) and Bhaskara (c. 1114 –1185 CE) who completed Brahmagupta's method, though they gave no formal proof of this. The formal proof was provided by Lagrange in the 18th century. For a complete history of this subject we refer the reader to Weil's book [57]. For details of this method, see [27, Ch. 3] or [33, Ch. 7], especially §7.6.3.

Fig. 3.4 Ellipse with
equation $\frac{x^2}{a^2} + \frac{y^2}{b^2} = 1$

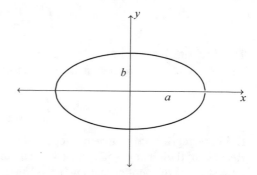

Elliptic curves

The cubic curves considered in §3.3 are called *elliptic curves*. These are some of the most important objects in all of mathematics, and they have been the subject of intense research for a few hundred years. The genesis of the adjective in the name of these curves goes back to 17th and 18th centuries. Let us briefly explain the connection; see [92] for details and references.

Consider the ellipse with the equation

$$\frac{x^2}{a^2} + \frac{y^2}{b^2} = 1,$$

with $a > b$. It is an easy integration exercise to show that the area of the ellipse is equal to πab. Now suppose we want to compute the perimeter of the ellipse.

A parametrization for the ellipse is given by

$$\begin{cases} x = a \sin t \\ y = b \cos t \end{cases} \quad 0 \le t \le 2\pi.$$

By the arc length formula, itself an application of the Pythagorean Theorem, the perimeter ℓ of the ellipse is equal to

$$\ell = \int_0^{2\pi} \sqrt{\left(\frac{dx}{dt}\right)^2 + \left(\frac{dy}{dt}\right)^2} \, dt$$

$$= 4 \int_0^{\pi/2} \sqrt{a^2 \cos^2 t + b^2 \sin^2 t} \, dt$$

$$= 4a \int_0^{\pi/2} \sqrt{1 - k^2 \sin^2 t} \, dt,$$

with $k^2 = 1 - b^2/a^2$. A change of variables with $u = \sin t$ gives

$$\ell = 4a \int_0^1 \frac{\sqrt{1 - k^2 u^2}}{\sqrt{1 - u^2}} \, du. \tag{3.8}$$

This is a special value of an *elliptic integral of second kind*. In general, *elliptic integrals of the second kind* are defined as follows: For $0 \le w \le 1$ we define

$$E(w) = \int_0^w \frac{\sqrt{1 - k^2 u^2}}{\sqrt{1 - u^2}} \, du.$$

Elliptic integrals are in general not expressible in terms of elementary functions. Because of their many applications in mathematical physics these types of integrals attracted a lot of attention starting in the 18th century. It was Abel in the 19th century who realized that the correct object of study is the inverse of the function E. The motivation for this point of view is the $\sin^{-1}$ integral: We know

$$\sin^{-1} w = \int_0^w \frac{du}{\sqrt{1 - u^2}},$$

but the more natural function to work with is the inverse function of $\sin^{-1}$, the ubiquitous sine. Going back to Equation (3.8), we make one more change of variable $z = 1 - k^2 u^2$ to obtain

$$\ell = 2a \int_\lambda^1 \frac{z}{\sqrt{z(1 - z)(z - \lambda)}} \, dz,$$

with $\lambda = 1 - k^2$. Upon setting $z = q + \frac{1+\lambda}{3}$ the integral transforms to

$$\ell = 2a \int_{\frac{2\lambda-1}{3}}^{\frac{2-\lambda}{3}} \frac{q + \frac{1+\lambda}{3}}{\sqrt{-q^3 + \frac{1}{3}(\lambda^2 + \lambda - 1)q + \frac{1}{27}(2\lambda^3 + 3\lambda^2 - 3\lambda - 2)}} \, dq.$$

Finally (!), set $q = -\sqrt[3]{4}v$ to get

$$\ell = 2\sqrt[3]{4}a \int_{-\frac{2-\lambda}{3\sqrt[3]{4}}}^{-\frac{2\lambda-1}{3\sqrt[3]{4}}} \frac{-\sqrt[3]{4}v + \frac{1+\lambda}{3}}{\sqrt{4v^3 - \frac{\sqrt[3]{4}}{3}(\lambda^2 + \lambda - 1)v + \frac{1}{27}(2\lambda^3 + 3\lambda^2 - 3\lambda - 2)}} \, dv.$$

Let

$$g_2 = \frac{\sqrt[3]{4}}{3}(\lambda^2 + \lambda - 1),$$

and

$$g_3 = -\frac{1}{27}(2\lambda^3 + 3\lambda^2 - 3\lambda - 2).$$

Karl Weierstrass defined a function $\wp(u)$ with the property that

$$u = \int_{\wp(u)}^\infty \frac{dv}{\sqrt{4z^3 - g_2 z - g_3}}.$$

So clearly ℓ is related to the function $\wp$. A remarkable property of the $\wp$-function is that it satisfies the functional equation

$$\wp'(z)^2 = 4\wp(z)^3 - g_2\wp(z) - g_3,$$

i.e., the point $(\wp(z), \wp'(z))$ lies on the curve

$$y^2 = 4x^3 - g_2x - g_3. \tag{3.9}$$

In fact, the points $(\wp(z), \wp'(z))$ give a full parametrization for the points with complex coordinates on the curve. Furthermore,

$$\wp(u + v) = -\wp(u) - \wp(v) + \frac{1}{4}\left(\frac{\wp'(u) - \wp'(v)}{\wp(u) - \wp(v)}\right)^2,$$

and

$$\wp(-u) = \wp(u), \quad \wp'(-u) = -\wp'(u).$$

These formulae have an interesting interpretation for the points on the curve. We define a group law $\oplus$ on the set of points of the curve as follows: For a point A on the curve, define $-A$ to be the reflection of A with respect to the x-axis; for three points A, B, C, we say $A \oplus B = C$ if A, B, and $-C$ are colinear; and O, the identity point, is the point at infinity in the direction of the y axis, i.e. $A \oplus (-A)$ for any point A.

The work we did in §3.3 shows that if $g_2, g_3 \in \mathbb{Q}$, then the $\oplus$ of any two points with coordinates in $\mathbb{Q}$ will be again a point with coordinates in $\mathbb{Q}$. Clearly, also, for a point A with rational coordinates, $-A$ will have rational coordinates. This means that the collection of points on the curve with rational coordinates forms a group. It is a truly surprising fact that, by a theorem of Mordell, this group is finitely generated. We refer the reader to [48, Ch. 3] or [47, Ch. VIII] for details.

Fermat's Last Theorem

Fermat's Last Theorem is an esoteric statement with no applications as such, but despite its obscurity it has given rise to an enormous amount of mathematics. Edwards [19] presents algebraic number theory as it was originally motivated by false proofs of Fermat's Last Theorem. "The Proof", a NOVA documentary [114] on Wiles' work, is an excellent account of the last steps toward the proof. Charles Mozzochi's endearing photo essay "The Fermat Diary" [36] is a photo album of all those whose works contributed to the proof of the theorem in the last fifty years. Finally, even though it is written for experts, Sir Andrew Wiles' introduction to his masterful paper in the Annals of Mathematics [110] is a delight to read.

The abc Conjecture

The *abc* Conjecture is an easy to state conjecture with many surprising consequences in number theory. The conjecture was formulated by D. W. Masser and J. Oesterlé in the 80's. This is the statement:

Conjecture 3.12 *(The abc Conjecture).* If $\varepsilon > 0$, then the number of triples (a, b, c) of coprime natural numbers such that $c = a + b$ and

$$c > \left(\prod_{p \mid abc} p \right)^{1+\varepsilon}$$

is finite.

The conjecture could also be formulated as follows: For every $\varepsilon > 0$, there is a constant $\kappa_\varepsilon > 0$ such that for every triple (a, b, c) of coprime natural numbers satisfying $c = a + b$ we have

$$c \leq \kappa_\varepsilon \left(\prod_{p \mid abc} p \right)^{1+\varepsilon}.$$

To see a quick application, let us apply the *abc* Conjecture to Fermat's Last Theorem. Suppose we have three coprime natural numbers x, y, z such that $x^n + y^n = z^n$. If $\varepsilon > 0$ is given, then applying the *abc* Conjecture with $a = x^n$, $b = y^n$, and $c = z^n$ shows that with the exception of finitely many choices of x, y, z we have

$$z^n \leq \left(\prod_{p \mid x^n y^n z^n} p \right)^{1+\varepsilon}$$

Next, $p \mid x^n y^n z^n$ if and only if $p \mid xyz$. So we have

$$\prod_{p \mid x^n y^n z^n} p = \prod_{p \mid xyz} p.$$

Now we observe that if n is a natural number, $\prod_{p \mid n} p \leq n$. Using this observation we have

$$\prod_{p \mid xyz} p \leq xyz < z^3.$$

In the last step we have used the fact that $x < z$ and $y < z$. Putting everything together, we conclude that except for finitely many choices of x, y, z we have

$$z^n < (z^3)^{1+\varepsilon} = z^{3(1+\varepsilon)}.$$

This implies that $n < 3(1 + \varepsilon)$. Since the choice of ε is arbitrary, this means $n \leq 3$. What we have proved is the following:

Corollary 3.13 (Assuming *abc* Conjecture). *For each $n > 3$, Fermat's equation $x^n + y^n = z^n$ has at most finitely many solutions in coprime natural numbers x, y, z.*

The statement of the *abc* Conjecture is ineffective. This means that for a fixed $\varepsilon > 0$ the conjecture does not provide any estimate for the number or the size of triples (a, b, c) satisfying the conditions of the conjecture. There are several explicit versions of the *abc* Conjecture in literature. Here we state one of these explicit conjectures which is due to Alan Baker [63].

To state Baker's *abc* Conjecture we need some notation. For a natural number n, we set rad (n) to be the product of the prime divisors of n, i.e.,

$$\operatorname{rad}(n) = \prod_{p|n} p.$$

For example, rad $(1) = 1$, rad $(12) = 2 \times 3 = 6$ and rad $(25) = 5$. We also let $\omega(n) = \sum_{p|n} 1$, i.e., the number of prime divisors of n. With this definition we have $\omega(1) = 0$, $\omega(12) = 2$, $\omega(25) = 1$. Using this notation, the original *abc* Conjecture asserts that for $\varepsilon > 0$, there is $\kappa_\varepsilon > 0$ such that for a triple (a, b, c) of coprime natural numbers, we have

$$c < \kappa_\varepsilon (\operatorname{rad}(abc))^{1+\varepsilon}.$$

Conjecture 3.14 (Baker's abc Conjecture). Let (a, b, c) be a triple of coprime natural numbers such that $c = a + b$. Let $N = \operatorname{rad}(abc)$ and $r = \omega(N)$. Then

$$c < \frac{6}{5} N \frac{(\log N)^r}{r!}.$$

We leave it to the reader to verify that Baker's *abc* Conjecture in fact implies the *abc* Conjecture. The papers by Granville and Tucker [77] and Waldschmidt [107] outline various applications of the *abc* Conjecture. In April of 2012, Shinichi Mochizuki of Kyoto University announced a proof of the *abc* Conjecture occupying hundreds of pages. At the time of this writing it is still not known if Mochizuki's proof is correct, and for that reason the *abc* Conjecture is still considered open.

Chapter 4
What integers are areas of right triangles?

In this chapter we study the set of integers that are the area of a right triangle with integer sides. We define a *congruent number* to be a natural number which is the area of a right triangle with rational sides. After verifying some easy properties of congruent numbers, we prove a theorem of Fermat (Theorem 4.4) that asserts no square is a congruent number. Later in the chapter, we explain the connection between congruent numbers and cubic equations. In the Notes, we review the history of congruent numbers and state a celebrated theorem of Tunnell.

4.1 Congruent numbers

If a, b, c are the three sides of an integral right triangle, with c the hypotenuse, since at least one of a, b is even then the area $ab/2$, is a natural number.

Question 4.1. Is there a criterion to decide whether a natural number n is the area of some integral right triangle?

It will become apparent very quickly that this is a difficult problem. In fact at the time of this writing, there is still no complete characterization of the set of areas of integral right triangles. It is, however, possible to obtain some information using elementary methods. We start with a definition.

Definition 4.2 (Congruent number). A natural number which is the area of a right triangle with rational sides is called a *congruent number*. We denote the set of all congruent numbers by $\mathscr{S}$.

Let's take a moment and clarify the connection between Question 4.1 and Definition 4.2. It is clear that if we have integral right triangles T and T' with side lengths a, b, c and $\lambda a, \lambda b, \lambda c$, respectively, with $\lambda \in \mathbb{N}$, then the area of T' is λ^2 times the area of T. This means if n is the area of some integral right triangle, then if $\lambda \in \mathbb{N}$, $\lambda^2 n \in \mathscr{S}$. This suggests that one should not be too concerned with the

© Springer Nature Switzerland AG 2018
R. Takloo-Bighash, *A Pythagorean Introduction to Number Theory*,
Undergraduate Texts in Mathematics, https://doi.org/10.1007/978-3-030-02604-2_4

square factors that show up in areas of integral right triangles. There is a bit of trouble here: Suppose we have a natural number n which is the area of some integral right triangle with side lengths a, b, c, and suppose n has a square factor u^2, $u \in \mathbb{N}$, so that $n = u^2 \cdot m$ with $m \in \mathbb{N}$. As tempted as we might be to scale down the triangle by a factor of u to get an integral triangle with side lengths $a/u, b/u, c/u$, sometimes these latter quotients are not integers. For example, the right triangle with side lengths $(8, 15, 17)$ has area $60 = 2^2 \cdot 15$, but the triangle with half the size, with area 15, has side lengths $(4, 15/2, 17/2)$ which are *rational*, and not integral.

We define a function sqf, *the square-free part of n*, by defining its value for a natural number n to be the smallest natural number m such that $n = m \cdot k^2$ for some natural number k. For example, $sqf(6) = 6$, $sqf(12) = 3$, and $sqf(9) = 1$. The following lemma is easy to prove.

Lemma 4.3. *For $n \in \mathbb{N}$, $n \in \mathscr{S}$ if and only if $sqf(n) \in \mathscr{S}$.*

The lemma shows that in order to determine the elements of the set $\mathscr{S}$ we just need to determine its square free elements. An important point to note is that a square-free element of $\mathscr{S}$ is not necessarily the area of a right triangle with integral sides. For example, the right triangle with sides $(8, 15, 17)$ has area 60, so $60 \in \mathscr{S}$. We have $sqf(60) = 15$, so $15 \in \mathscr{S}$. However, as we see in Exercise 4.7, there are no integral right triangles with area 15.

We saw in Theorem 3.4 that if a, b, c are the three sides of a primitive right triangle, with c the hypotenuse, then there are co-prime integers x, y of different parity such that

$$\{a, b, c\} = \{x^2 - y^2, 2xy, x^2 + y^2\}.$$

The area S of this triangle is then equal to

$$S = \frac{1}{2}ab = xy(x^2 - y^2).$$

For this reason, one way to produce congruent numbers is to define a function

$$f(x, y) = sqf(xy(x^2 - y^2))$$

with domain being the set of pairs of integers (x, y), $\gcd(x, y) = 1$, $x > y$, and x, y of different parity. Then a natural number is a congruent number if its square free part is $f(x, y)$ for some (x, y) as above. For example, the values of $f(x, 1)$ for x larger than 1 and even are as follows: $6, 15, 210, 14, 110, \ldots$.

It is very hard to know which numbers appear as values of f. In fact, even when we know a number is a congruent number, it is not clear how one should go about finding the pair (x, y) such that $f(x, y)$ is equal to that number. For example, as noted in [70], 53 is a congruent number, but the first time it appears as $f(x, y)$ is when

$$x = 1873180325, \quad y = 1158313156.$$

In fact,

$$xy(x^2 - y^2) = 53 \times 2978556542849787902^2.$$

4.2 Small numbers

In general it is fairly difficult to determine with bare hands if a natural number is congruent. We will see in a moment that 1 is not congruent. Lemma 4.3 then shows that no perfect square is congruent. We will see in Exercise 4.9 that 2 and 3 are not congruent. The smallest congruent number is 5: 5 is the area of the right triangle with rational sides $20/3$, $3/2$, $41/6$. We already saw that 6 is a congruent number as it is the area of the right triangle with side lengths $3, 4, 5$. As already stated at this point despite all the progress made in the last few hundred years there are still many basic questions about congruent number which we do not know how to answer; see, however, the Notes to this chapter where we state a theorem of Tunnell and explain some recent progress.

The following theorem goes back to Fermat. The proof of this theorem, like the proof of Theorem 3.10, uses *infinite descent*.

Theorem 4.4 (Fermat). $1 \notin S$.

Proof. By Lemma 4.3 we need to show that there are no right triangles with rational sides whose area is the square of a natural number. Suppose we have a right triangle with rational sides a, b, c, with a, b the legs and c the hypotenuse, and suppose that the area $ab/2$ is a perfect square t^2. Let $\lambda \in \mathbb{N}$ be the common denominator between a, b, c. If we scale the triangle by λ, we obtain a new triangle with integral sides $a\lambda, b\lambda, c\lambda$, and area $\lambda^2 t^2$ which is still a perfect square. So we may assume, without loss of generality, that $a, b, c \in \mathbb{N}$. If $\gcd(a, b) = \delta$, then we write $a = a'\delta, b = b'\delta$ with $(a', b') = 1$. Clearly, $\delta \mid c$ and we can write $c = c'\delta$. Then $t^2 = ab/2 = a'b'\delta^2/2$. This implies that $a'b'/2 = t'^2$ for some integer t'. Consequently, we have a right triangle with side lengths a', b', c' such that $\gcd(a', b', c') = 1$ and whose area is a perfect square. So, we may without loss of generality assume that our original numbers a, b, c are coprime. We have

$$\begin{cases} c^2 = a^2 + b^2, \\ ab = 2t^2. \end{cases}$$

Observe that one of the a, b is even, so we may assume that $a = 2k$ is even. So we have

$$kb = t^2.$$

Since $\gcd(a, b) = 1$, we have $\gcd(k, b) = 1$. Since the product of k and b is a perfect square, by Proposition 2.21 each of k, b individually is a perfect square, i.e., $k = m^2$ and $b = n^2$, for some natural numbers m, n. Going back to a, b, we have $a = 2m^2$, $b = n^2$. Now the Pythagorean Equation becomes

$$4m^4 + n^4 = c^2. \tag{4.1}$$

This equation resembles Equation (3.6) which was studied in the proof of Theorem 3.10, and, in fact we use the method of infinite descent that was used in the proof of Theorem 3.10 to show every solution to Equation (4.1) satisfies $mn = 0$.

As in the proof of Theorem 3.10, suppose we have a solution of (4.1) such that c is the smallest possible. We use Theorem 3.4 to write

$$\begin{cases} 2m^2 = 2uv, \\ n^2 = u^2 - v^2, \\ c = u^2 + v^2, \end{cases}$$

for coprime integers u, v with different parities. Since $m^2 = uv$ and $\gcd(u, v) = 1$, Proposition 2.21 implies that $u = r^2, v = s^2$ for natural numbers r, s. If on the other hand we write the middle equation as a Pythagorean Equation $n^2 + v^2 = u^2$, we see that u is odd and v is even, and also that there are coprime integers x, y of different parity such that

$$\begin{cases} n = x^2 - y^2, \\ v = 2xy, \\ u = x^2 + y^2. \end{cases}$$

Suppose x is even, $x = 2\alpha$. Then we write the middle equation as $s^2 = 4\alpha y$. Since s is even, we write $\alpha y = (s/2)^2$. Again, we conclude that $\alpha = \beta^2, y = \gamma^2$ for integers β, γ. With these substitutions, the equation $u = x^2 + y^2$ becomes

$$4\beta^4 + \gamma^4 = r^2,$$

i.e., the numbers β, γ, r are another set of solutions of Equation (4.1). It is clear that $r < c$, and this is a contradiction. $\square$

4.3   Connection to cubic equations

The problem of determining congruent numbers is intimately related to the study of rational solutions to the cubic equations considered in §3.3.

Theorem 4.5. *Let $n \in \mathbb{N}$ be fixed. There is a one-to-one correspondence between the following sets:*

$$V_1 = \{(a, b, c) \in \mathbb{Q}^3 \mid a^2 + b^2 = c^2, ab/2 = n\}$$

and

$$V_2 = \{(x, y) \in \mathbb{Q}^2 \mid y^2 = x^3 - n^2x, xy \neq 0\}.$$

The correspondence is given by

$$f_1 : V_1 \to V_2,$$

$$(a, b, c) \mapsto \left(\frac{nb}{c-a}, \frac{2n^2}{c-a} \right),$$

and

$$f_2 : V_2 \to V_1,$$

$$(x, y) \mapsto \left(\frac{x^2 - n^2}{y}, \frac{2nx}{y}, \frac{x^2 + n^2}{y} \right).$$

The proof of this theorem is straightforward, and not too tedious, Exercise 4.10. Tunnell in the introduction of [105] attributes this construction to Don Zagier. The wonderful paper [70] has a fun appendix where it is explained how one might have found the above correspondence. One important point to note is that the correspondence described in the theorem is valid over every field, not just $\mathbb{Q}$. Furthermore, it gives a bijection between pairs of positive rational numbers (x, y), and positive rational numbers a, b, c described in the theorem, see Exercise 4.11.

The equation $y^2 = x^3 - n^2 x$ has very few solutions with $xy = 0$. In fact, by the easy Exercise 4.12, the only solutions of $y^2 = x^3 - n^2 x$ with $xy = 0$ are $(0, 0)$ and $(\pm n, 0)$. We call these solutions the *trivial solutions*. Hence we have the following corollary:

Corollary 4.6 (Stephens [97]). *A natural number n is a congruent number if and only if the equation $y^2 = x^3 - n^2 x$ has some non-trivial solution.*

We now consider an explicit example. The paper [70] contains many numerical examples of this nature.

Example 4.7. We start with the Pythagorean triple $(5, 12, 13)$. The area of the triangle with these side lengths is $5 \times 12/2 = 30$. In this case, Theorem 4.5 says that the pair

$$(x, y) = \left(\frac{30 \times 12}{13 - 5}, \frac{2 \times 30^2}{13 - 5} \right) = (45, 225)$$

is a solution of the equation $y^2 = x^3 - 30^2 x$. Now we proceed as in Example 3.9. Implicit differentiation gives

$$y' = \frac{3x^2 - 30^2}{2y},$$

and consequently the slope of the tangent line at the point $(45, 225)$ is

$$m = \frac{23}{2}.$$

A computation shows that the equation of the tangent line is

$$y = \frac{23}{2}x - \frac{585}{2}.$$

The points of intersection of this line with the curve $y^2 = x^3 - 30^2 x$ must satisfy the system

$$\begin{cases} y^2 = x^3 - 30^2 x, \\ y = \frac{23}{2}x - \frac{585}{2}. \end{cases}$$

Inserting y from the second equation into the first equation gives

$$\left(\frac{23}{2}x - \frac{585}{2}\right)^2 = x^3 - 30^2 x,$$

and this implies

$$x^3 - \left(\frac{23}{2}\right)^2 x^2 + Ax + B = 0$$

with numbers A, B the exact value of which is of no significant importance. Since we obtained this equation using a tangent line, two of the solutions are $x = 45$. The third solution must then satisfy

$$45 + 45 + x = \left(\frac{23}{2}\right)^2.$$

This gives $x = 169/4$, and we obtain the point

$$(x, y) = \left(\frac{169}{4}, \frac{1547}{8}\right)$$

on the curve $y^2 = x^3 - 30^2 x$. Now we apply the bijection f_2 from Theorem 4.5 to this pair to obtain a right triangle with rational sides whose area is 30. Explicitly, we have

$$(a, b, c) = \left(\frac{(\frac{169}{4})^2 - 30^2}{\frac{1547}{8}}, \frac{2 \times 30 \times \frac{169}{4}}{\frac{1547}{8}}, \frac{(\frac{169}{4})^2 + 30^2}{\frac{1547}{8}}\right)$$

$$= \left(\frac{119}{26}, \frac{1560}{119}, \frac{42961}{3094}\right).$$

A quick computation shows that this triple in fact satisfies the Pythagorean Equation, and that

$$\frac{1}{2} \times \frac{119}{26} \times \frac{1560}{119} = 30.$$

We have obtained a new triangle with area 30.

Fig. 4.1 The diagram for
Problem 4.6

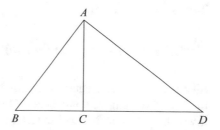

Exercises

4.1 Determine all right triangles with integral sides such that the perimeter and the area are equal.

4.2 Show that two right triangles with equal hypotenuse and area are congruent.

4.3 Show that for every $n \in \mathbb{N}$, there are n distinct integral right triangles with the same area.

4.4 A *Heronian triangle* is a triangle with rational sides whose area is a rational number. Show that triangles with side lengths (13, 14, 15) and (65, 119, 180) are Heronian.

4.5 Show that there are infinitely many isosceles Heronian triangles.

4.6 Let ABC and ACD be right triangles with rational sides which share a side AC as in Figure 4.1. Show that the triangle ABD is Heronian. Conversely, suppose ABD is a Heronian triangle with $\angle BAD$ the largest angle of the triangle. Draw the altitude AC and show that the triangles ABC and ACD are right triangles with rational sides.

4.7 Show $15 \notin \mathscr{S}$.

4.8 Show that a square-free natural number n is a congruent number if and only if there is a rational number x such that $x^2 - n$ and $x^2 + n$ are squares of rational numbers.

4.9 Show that 2 and 3 are not congruent numbers.

4.10 Prove Theorem 4.5 by direct computation.

4.11 Show that in Theorem 4.5, for $n \in \mathbb{N}$, x, y are positive rational numbers, if and only if a, b, c are positive rational numbers.

4.12 Show that the only solutions of $y^2 = x^3 - n^2 x$ with $xy = 0$ are $(0, 0)$ and $(\pm n, 0)$.

4.13 Find three rational right triangles with area 6.

4.14 (✖) Find fifty congruent numbers.

4.15 (✖) Find ten rational right triangles with area 30.

4.16 (✖) Use Tunnell's Theorem 4.8 from the Notes to find all congruent numbers less than 100.

Notes

The history of congruent numbers

Like many other concepts in elementary number theory, the standard reference for the history of congruent numbers is Dickson's classic book [16], especially Chapter XVI. The definition that Dickson uses is different from ours. He defines a congruence number to be a natural number n if there is a rational number x such that $x^2 - n$ and $x^2 + n$ are squares of rational numbers; that this definition is equivalent to our definition is Exercise 4.8. Let us mention here that if S is the area of the right triangle with sides a, b, c, with c the hypotenuse, then

$$c^2 \pm 4S = c^2 \pm 2ab = a^2 + b^2 \pm 2ab = (a \pm b)^2.$$

This means, we have a three term arithmetic progression

$$\left(\frac{c}{2}\right)^2 - S, \quad \left(\frac{c}{2}\right)^2, \quad \left(\frac{c}{2}\right)^2 + S$$

consisting of rational squares. This is perhaps the reason for the name *congruent*. Dickson mentions that in tenth century an Iranian mathematician and this author's fellow townsman Mohammad Ben Hossein Karaji (953–1029) stated that the problem of determining congruent numbers was the "principal object of the theory of rational right triangles." Dickson [16, Ch. XVI] is a wonderful review of work by various mathematicians on the problem of characterizing congruent numbers over the millennium up to its publication. For a modern treatment of this subject we refer the reader to [30, Ch. 1].

Tunnell's theorem

The theory of rational points on cubic curves, the *theory of elliptic curves*, is a rich active area of research with connections to many parts of modern mathematics [47]. In the last three decades many results about congruent numbers have been obtained that use methods and techniques involving elliptic curves. It appears that Stephens's very short paper [97] was the first paper that made the connection to elliptic curves explicit. Tunnell's paper [105] pushed the theory far. Among other results, Tunnell proved the following surprising theorem:

Theorem 4.8 (Tunnell). *Define a formal power series in the variable q by*

$$g = q \prod_{n=1}^{\infty} (1 - q^{8n})(1 - q^{16n}),$$

and for each $t \in \mathbb{N}$ set $\theta_t = \sum_{n \in \mathbb{Z}} q^{tn^2}$. Define integers $a(n)$ and $b(n)$ via the identities

$$g\theta_2 = \sum_{n=1}^{\infty} a(n)q^n,$$

and

$$g\theta_4 = \sum_{n=1}^{\infty} b(n)q^n.$$

Then, we have

- *If $a(n) \neq 0$, then n is not a congruent number;*
- *If $b(n) \neq 0$, then 2n is not a congruent number.*

Conjecturally, both statements in the theorem should be *if and only if*. The coefficients $a(n)$, $b(n)$ are computable in terms of the number of solutions in integers of equations of the form

$$Ax^2 + By^2 + Cz^2 = n$$

for $A, B, C \in \mathbb{N}$. (We advise the reader to do this as an exercise!) Tunnell recovers a number of previously known results from his numerical criterion. For example, he shows a prime p of the form $8k + 3$ is not congruent, as for such primes $a(p) \equiv 2 \bmod 4$, or that if p, q are primes of the form $8k + 5$, then $2pq$ is not congruent. It is an easy exercise to derive Theorem 4.4 from Theorem 4.8.

At least conjecturally one expects the existence of many congruent numbers. For example, we have the following conjecture which is a consequence of the Birch and Swinnerton-Dyer Conjecutre [47, Conjecture 16.5]:

Conjecture 4.9 ([59, 60]). Every positive integer congruent to 5, 6, or 7 modulo 8 is a congruent number.

Recently some impressive results have been obtained in this direction [101, 102]. Smith [95] has proved that at least 55.9% of positive square free integers $n \equiv 5, 6, 7 \bmod 8$ are congruent numbers. In contrast, Smith [96] has proved that congruent numbers are rare among natural numbers $n \equiv 1, 2, 3 \bmod 8$.

Chapter 5
What numbers are the edges of a right triangle?

In this chapter we study numbers that appear as the side lengths of primitive right triangles. We use rings of Gaussian integers to prove our main theorems. We give a quick review of the basic properties of the ring of Gaussian integers. We then prove that the ring of Gaussian integers is a Euclidean domain, leading to the analogue of the Fundamental Theorem of Arithmetic in this context. We also determine the irreducible elements and units of $\mathbb{Z}[i]$. For a more thorough exposition of the theory of Gaussian integers we refer the reader to the classical text by Sierpinski [46] or Conrad [69]. In this chapter we also determine what numbers are a sum of two squares (Theorem 5.2) and determine the numbers which appear as the hypotenuse of a primitive right triangle (Theorem 5.1). In the Notes we state a famous theorem of Dirichlet (Theorem 5.11) and say a couple of words about algebraic number theory.

5.1 The theorem

If n is an odd number, then it is the side length of some right triangle. In fact, we can always write $n = xy$ with x, y coprime and odd. Then the following set

$$\tau(x, y) = \left\{ \frac{x^2 - y^2}{2}, n = xy, \frac{x^2 + y^2}{2} \right\}$$

is the set of side lengths of a primitive right triangle.

If n is an even number, then we can write $n = 2xy$ with x, y coprime. If one of x or y is even, i.e., if n is divisible by 4, then again the set $\{x^2 - y^2, 2xy, x^2 + y^2\}$ is the set of side lengths of a primitive right triangle. If on the other hand $2||n$, then n cannot be the side length of a primitive right triangle.

© Springer Nature Switzerland AG 2018
R. Takloo-Bighash, *A Pythagorean Introduction to Number Theory*,
Undergraduate Texts in Mathematics, https://doi.org/10.1007/978-3-030-02604-2_5

It turns out that the question of whether a natural number can occur as the length of the hypotenuse of a primitive right triangle is more subtle. Recall the list of the first few Pythagorean triples at the beginning of §3.1. The hypotenuse lengths that occur in the list are $5, 13, 17, 25, 29, 29, 37, 41, 53$. The prime numbers occurring in the prime factorization of these numbers are $5, 13, 17, 29, 37, 41, 53$, and a quick check reveals that all of these prime numbers are of the form $4k + 1$. In this chapter we prove the following theorem:

Theorem 5.1. *A number n is the length of the hypotenuse of some primitive right triangle if and only if all of its prime factors are of the form $4k + 1$.*

Recall the formula for the hypotenuse of a primitive right triangle, $x^2 + y^2$ if x, y are coprime of different parity, or $(x^2 + y^2)/2$ if x, y are coprime and both odd. Clearly something is going on with sums of two squares! And in fact the first step to prove the theorem is understanding what numbers can be written as a sum of two squares. We start by examining the sequence of natural numbers. Clearly, $1 = 1^2 + 0^2, 2 = 1^2 + 1^2, 3$ is not a sum of two squares, $4 = 2^2 + 0^2, 5 = 2^2 + 1^2$, 6 is not, 7 is not, $8 = 2^2 + 2^2, 9 = 3^2 + 0, 10 = 3^2 + 1^2$, 11 is not, 12 is not, $13 = 3^2 + 2^2$, 14 is not, 15 is not, $16 = 4^2 + 0^2, 17 = 4^2 + 1^2, 18 = 3^2 + 3^2$, $20 = 4^2 + 2^2$, 21 is not, 22 is not, 23 is not, 24 is not, $25 = 5^2 + 0^2, 26 = 5^2 + 1^2$, 27 is not, 28 is not, $29 = 5^2 + 2^2$, 30 is not, 31 is not, $32 = 4^2 + 4^2$, etc. While it is not immediately clear that one should do this next thing, but we look at the prime factorization of the integers that are *not* sums of two squares: $3, 6 = 2 \cdot 3, 7, 11$, $12 = 2^2 \cdot 3, 14 = 2 \cdot 7, 15 = 3 \cdot 5, 21 = 3 \cdot 7, 22 = 2 \cdot 11, 23, 24 = 2^3 \cdot 3, 27 = 3^3$, $28 = 2^2 \cdot 7, 30 = 2 \cdot 3 \cdot 5, 31$. The common feature of all of these numbers is that they all have at least one prime factor of the form $4k + 3$ which appears with an odd exponent. In fact, we will prove the following theorem:

Theorem 5.2. *A number n is the sum of two squares if and only if every one of its prime factors of the form $4k + 3$ has even exponent in its prime factorization.*

We refer to Theorem 5.2 as the *Two Squares Theorem*. As we noted it is clear that 3 is not the sum of two squares. If on the other hand some number which is a multiple of 3 is a sum of two squares $a^2 + b^2$, then this means $3 \mid a^2 + b^2$. Now, $0^2 \equiv 0, 1^2 \equiv 0, 2^2 \equiv 1 \mod 3$. Consequently, in order for $a^2 + b^2 \equiv 0 \mod 3$, we need to have $a \equiv 0, b \equiv 0 \mod 3$, i.e., both a, b are divisible by 3. This implies that $a^2 + b^2$ is actually divisible by 3^2. Note that 3 is a prime of the form $4k + 3$. The sort of situation we just described does not happen for primes of the form $4k + 1$ such as 5 as for example $5 = 1^2 + 2^2$, and neither 1 nor 2 is divisible by 5.

The proof we present for these theorems is best expressed in terms of the arithmetic of complex integers which we present in the next section. The idea is that if we have a complex number $z = x + iy$, with $x, y \in \mathbb{R}$, then $|z|^2 = x^2 + y^2$, and these are the sorts of expressions that we wish to study. Since in our theorem we need to look at those cases where $x, y \in \mathbb{Z}$, we are led to study complex numbers $z = x + iy$ with $x, y \in \mathbb{Z}$.

5.2 Gaussian integers

A *Gaussian integer* is a complex number $x + iy$ with $x, y \in \mathbb{Z}$. We define the *ring of Gaussian integers* to be

$$\mathbb{Z}[i] = \{a + bi \mid a, b \in \mathbb{Z}\},$$

where here and elsewhere $i^2 = -1$. Equipped with the standard addition and multiplication of complex numbers, $\mathbb{Z}[i]$ is a commutative ring with identity. For $z \in \mathbb{Z}[i]$, we define $\bar{z}$ to be the complex conjugate of z, i.e.,

$$\overline{a + bi} = a - bi,$$

and we define the *norm* of z, $N(z)$, to be

$$N(z) = z \cdot \bar{z}.$$

A computation shows that

$$N(a + ib) = a^2 + b^2.$$

We let $|z| = N(z)^{1/2}$.

Lemma 5.3. *For all $z, w \in \mathbb{Z}[i]$,*

$$N(zw) = N(z)N(w).$$

Proof. It is easy to check that $\overline{z \cdot w} = \bar{z} \cdot \bar{w}$. Then we have

$$N(zw) = zw \cdot \overline{zw} = zw \cdot \bar{z} \cdot \bar{w} = (z\bar{z}) \cdot (w\bar{w}) = N(z)N(w).$$

$\square$

An element $u \in \mathbb{Z}[i]$ is called a *unit*, if there is a $v \in \mathbb{Z}[i]$ such that $uv = 1$. Taking norms gives $N(u) \cdot N(v) = 1$. Since the norm is always nonnegative, this identity implies $N(u) = 1$. It is easy to check that these are indeed units. It is also easy to check if $u \in \mathbb{Z}[i]$ satisfies $N(u) = 1$, then u is a unit, because then $u \cdot \bar{u} = N(u) = 1$. An easy examination shows that u can only be one of the following elements: $+1$, -1, i, and $-i$. Gaussian integers x, y are called *associates* if $x = uy$ for a unit u.

Divisibility and unique factorization

There is a division algorithm in $\mathbb{Z}[i]$:

Theorem 5.4. *If $a, b \in \mathbb{Z}[i]$ with $b \neq 0$, then there are $q, r \in \mathbb{Z}[i]$ such that*

$$a = bq + r$$

with $N(r) < N(b)$. Consequently, $\mathbb{Z}[i]$ is a Euclidean domain.

Proof. We wish to write

$$\frac{a}{b} = q + \frac{r}{b}$$

with $q \in \mathbb{Z}[i]$ and $N(r/b) < 1$. Write a/b as $a\bar{b}/N(b)$, and set $a\bar{b} = u + iv$. By the last statement in Theorem 2.8 we can write

$$u = q_1 N(b) + r_1,$$

$$v = q_2 N(b) + r_2,$$

with $|r_1|, |r_2| \le N(b)/2$. Consequently,

$$\frac{a}{b} = \frac{a\bar{b}}{N(b)} = \frac{u + iv}{N(b)} = q_1 + iq_2 + \frac{r_1 + ir_2}{N(b)}.$$

If we set $q = q_1 + iq_2$, we get

$$a = qb + \frac{r_1 + ir_2}{\bar{b}}.$$

Since a and qb are in $\mathbb{Z}[i]$, we see that

$$r := \frac{r_1 + ir_2}{\bar{b}} \in \mathbb{Z}[i].$$

We just need to show that

$$N(r) < N(b).$$

We have

$$N(r) = N(\frac{r_1 + ir_2}{\bar{b}}) = \frac{r_1^2 + r_2^2}{N(\bar{b})}$$

$$\le \frac{N(b)^2/4 + N(b)^2/4}{N(b)} = \frac{N(b)}{2} < N(b).$$

$\square$

Here is an alternative, geometric way to *see* the above theorem. Let's fix the non-zero Gaussian integer b as in the theorem, and examine the set of all Gaussian integers of the form qb with $q \in \mathbb{Z}[i]$. Write $q = q_1 + q_2 i$ with $q_1, q_2 \in \mathbb{Z}$, to obtain

$$qb = (q_1 + q_2 i)b = q_1 \cdot b + q_2 \cdot ib.$$

The Gaussian integer ib is obtained from b via a counterclockwise 90-degree rotation around the origin as in Figure 5.1.

This means that $0, b, ib$, and $ib + b$ are the four vertices of a square. Furthermore, since every Gaussian integer of the form qb is an integral linear combination of b and ib, the set of all such points qb is going to be a square grid in the plane as in the diagram. Now, every Gaussian integer a falls in one of these squares. The distance between a to the closest vertex of the square in which it lives is at most the side

Fig. 5.1 The geometric
proof of Theorem 5.4

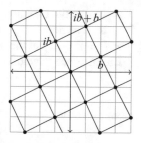

length of the square, $|b|$, i.e., $|a - qb| < |b|$ for some Gaussian integer q. Squaring
this inequality gives $N(a - bq) < N(b)$, as claimed.

Since $\mathbb{Z}[i]$ is a Euclidean domain, it is a *principal ideal domain* (PID), and therefore a *unique factorization domain*; see [25, Ch. 3, §7]. The latter means that every
element of $\mathbb{Z}[i]$ is a product of *irreducible elements* in an essentially unique way.
Recall that we call an element ϖ of $\mathbb{Z}[i]$ an *irreducible element* if any identity of the
form $\varpi = xy$ with $x, y \in \mathbb{Z}[i]$ implies either x or y is a unit. Since $\mathbb{Z}[i]$ is a UFD,
every irreducible element is prime. Recall that an element p of a domain R is called
prime if the principal ideal (p) is prime.

5.3 The proof of Theorem 5.2

The proof of Theorem 5.2 uses three ingredients:

Lemma 5.5 (Ingredient 1). *If m and n are expressible as sums of squares, then so
is mn.*

Proof. The easiest way to see this is by using complex numbers. If $m = a^2 + b^2$, then
$m = N(a + ib)$, the norm of the complex number $a + ib$. Similarly, if $n = c^2 + d^2$,
then $n = N(c + id)$. Next, by Lemma 5.3,

$$mn = N(a + ib)N(c + id) = N((a + ib)(c + id))$$

$$= N((ac - bd) + i(ad + bc)) = (ac - bd)^2 + (ad + bc)^2.$$

$\square$

Lemma 5.6 (Ingredient 2). *If p is a prime of the form $4k + 3$, and if for integers
a, b, $p \mid a^2 + b^2$, then $p \mid a$ and $p \mid b$.*

Proof. If $p \nmid a$, then $a^2 + b^2 \equiv 0 \bmod p$ implies

$$(ba^{-1})^2 \equiv -1 \quad \bmod p.$$

This means the equation $x^2 \equiv -1 \bmod p$ has a solution u modulo p. By Theorem 2.51 there is a $g \bmod p$ such that $o_p(g) = p - 1$. We write $u \equiv g^i \bmod p$ for some $0 < i < p - 1$. On the other hand, since $g^{(p-1)/2} \not\equiv +1 \bmod p$, and $(g^{(p-1)/2})^2 \equiv +1 \bmod p$, we conclude that

$$g^{(p-1)/2} \equiv -1 \quad \bmod p.$$

Consequently,

$$g^{2i} \equiv g^{(p-1)/2} \quad \bmod p.$$

Lemma 2.47 implies

$$2i \equiv \frac{p-1}{2} \quad \bmod (p-1).$$

But since $2i$ and $p - 1$ are even, and $(p - 1)/2$ is odd, this is a contradiction. $\square$

We will see in Lemma 6.7 that the equation $x^2 \equiv -1 \bmod p$ has a solution modulo p if and only if p is not of the form $4k + 3$.

The next ingredient is a substantial theorem due to Fermat. Here we will give an algebraic proof for the theorem. We will also present a geometric proof in Chapter 10 and another proof using the theory of quadratic forms in §12.2.

Theorem 5.7 (Ingredient 3). *An odd prime number is expressible as a sum of two squares if and only if it is of the form $4k + 1$.*

Proof. First suppose $p = x^2 + y^2$. Look at everything modulo 4. Then $x^2 \equiv 0, 1 \bmod 4$, and as a result $x^2 + y^2 \equiv 0, 1, 2 \bmod 4$. This means $p \equiv 0, 1, 2 \bmod 4$. Obviously if $p \equiv 0 \bmod 4$, it cannot be a prime number. If $p \equiv 2 \bmod 4$, then $p = 2$, and not odd. Consequently, $p \equiv 1 \bmod 4$ is the only possibility, proving the necessity of the condition.

Now we show if p is of the form $4k + 1$, then p is expressible as a sum of two squares. The proof of this statement requires several steps:

Step 1. There exists a such that $a^2 \equiv -1 \bmod p$.

By Wilson's Theorem, Equation (2.8), $(p - 1)! \equiv -1 \bmod p$. Next,

$$x \cdot (p - x) \equiv -x^2 \bmod p.$$

Hence

$$-1 \equiv (p - 1)! \equiv (-1)^{(p-1)/2} \left((\frac{p-1}{2})! \right)^2 \equiv \left((\frac{p-1}{2})! \right)^2 \bmod p$$

as for $p \equiv 1 \bmod 4$, $(p - 1)/2$ is even. Consequently, $a = ((p - 1)/2)!$ satisfies $a^2 \equiv -1 \bmod p$. Note that this means $a^2 + 1 \equiv 0 \bmod p$. It is also clear that if b is another integer such that $b^2 \equiv -1 \bmod p$, then $a \equiv \pm b \bmod p$.

Step 2. Now, with the choice of a as in Step 1, for every integer x we have $x^2(a^2 + 1) \equiv 0 \bmod p$, or $x^2 + (ax)^2 \equiv 0 \bmod p$. This means $p \mid x^2 + (ax)^2$. So if y is an integer such that $y \equiv \pm ax \bmod p$, we have $p \mid x^2 + y^2$. Conversely, suppose we have integers u, v such that $p \mid u^2 + v^2$, but p is not a factor of either u or v. Then $v^2 \equiv -u^2 \bmod p$, and consequently, $(vu^{-1})^2 \equiv -1 \bmod p$, so that $vu^{-1} \equiv \pm a \bmod p$, and $v \equiv \pm au \bmod p$.

Step 3. Assume there are integers x, y such that $p = x^2 + y^2$. By Step 2, $y \equiv \pm ax \bmod p$. Furthermore, since $x^2 < p$ and $y^2 < p$, we have $x \le [\sqrt{p}]$ and $y \le [\sqrt{p}]$. Conversely, suppose we have non-zero integers x, y such that $x, y \le [\sqrt{p}]$ and $y \equiv \pm ax \bmod p$. Then by Step 2, $p \mid x^2 + y^2 \le [\sqrt{p}]^2 + [\sqrt{p}]^2 < \sqrt{p}^2 + \sqrt{p}^2 = p + p = 2p$. (Here we have used the fact that since p is not a square, $\sqrt{p}$ is not an integer, and as such, $[\sqrt{p}] < \sqrt{p}$.) Hence, $x^2 + y^2$ is a positive integer smaller than $2p$ and divisible by p. This means $x^2 + y^2 = p$.

Step 4. So we are reduced to proving the following statement which is known as Thue's Lemma: Suppose a satisfies $a \not\equiv 0 \bmod p$. Then there are integers $x, y \in \{1, \ldots, [\sqrt{p}]\}$ such that $y \equiv \pm ax \bmod p$. We prove this fact using the Pigeon-Hole Principle, Theorem A.7. Look at the following set:

$$A = \{ax - y \mid 0 \le x, y \le [\sqrt{p}]\}.$$

The number of choices for the pairs (x, y) is $(1 + [\sqrt{p}])^2 > p$. By Theorem A.7, there are distinct pairs (x, y) and (x', y') such that $ax - y \equiv ax' - y' \bmod p$, or $y - y' \equiv a(x - x') \bmod p$. Note that $-[\sqrt{p}] \le y - y', x - x' \le [\sqrt{p}]$. By multiplying with appropriate signs we may assume that $y - y' \ge 0$ and $x - x' \ge 0$. The price to pay is an ambiguity of sign which we write as $y - y' \equiv \pm a(x - x') \bmod p$. Since the pairs (x, y) and (x', y') are distinct, at least one of the quantities $y - y'$ or $x - x'$ is non-zero, but whichever is non-zero, it will also be non-zero modulo p, and the relation $y - y' \equiv \pm a(x - x') \bmod p$ implies that the other one is non-zero modulo p as well, and consequently, non-zero. So if we let $X = x - x'$ and $Y = y - y'$, we see that X, Y are non-zero, $0 \le X, Y \le [\sqrt{p}]$ and $Y \equiv \pm aX \bmod p$. This finishes the proof of Thue's Lemma, and hence the proof of the theorem. $\square$

Corollary 5.8. *The equation $x^2 \equiv -1 \bmod p$ has a solution modulo p if and only if p is not of the form $4k + 3$.*

Now we can prove Theorem 5.2:

Proof of Theorem 5.2. It is clear that a number whose square-free part is a sum of two squares is a sum of two squares. By Ingredient 3, every prime of the form $4k + 1$ is a sum of two squares, and also $2 = 1^2 + 1^2$. By Ingredient 1, any product of such is a sum of two squares. This proves one direction of the theorem.

For the other direction, suppose n is a sum of two squares $a^2 + b^2$, and $p^\alpha \| n$. We just need to show that if p is of the form $4k + 3$, then α is even. Write $n = p^\alpha m$ with m coprime to p. By ingredient 2, $p \mid a$ and $p \mid b$, hence we can write $a = pc$ and $b = pd$. Then

$$p^\alpha m = (pc)^2 + (pd)^2 = p^2(c^2 + d^2),$$

so

$$p^{\alpha-2}m = c^2 + d^2.$$

If α is odd, by repeating this process we reach

$$pm = r^2 + s^2$$

for natural numbers r, s. Ingredient 2 again implies $p \mid r$, $p \mid s$, and consequently $p^2 \mid pm$. This last statement means $p \mid m$. This is a contradiction. $\square$

5.4   Irreducible elements in $\mathbb{Z}[i]$

We now determine the collection of irreducible elements in $\mathbb{Z}[i]$. To start, we note that if $N(\varpi)$ is a prime number in $\mathbb{Z}$, then ϖ is irreducible. For example, since $N(1 + i) = 2$, $1 + i$ is irreducible. More interestingly, if p is a rational prime such that $p \equiv 1 \bmod 4$, then by Theorem 5.7, $p = a^2 + b^2$ with $a, b \in \mathbb{Z}$. This implies $N(a + ib) = p$. Consequently, $a + ib$ is irreducible in $\mathbb{Z}[i]$. We now examine primes of the form $4k + 3$. Suppose p is one such prime and that we can write

$$p = z \cdot w$$

for $z, w \in \mathbb{Z}[i]$. Then by taking norms we get

$$p^2 = N(z)N(w). \tag{5.1}$$

This implies $p \mid N(z)N(w)$. Hence, p must divide either $N(z)$ or $N(w)$. Suppose $z = \alpha + i\beta$ and $p \mid N(z) = \alpha^2 + \beta^2$. By Lemma 5.6, $p \mid \alpha$ and $p \mid \beta$. Consequently, $p^2 \mid \alpha^2 + \beta^2$. Equation (5.1) then implies $N(z) = p^2$ and $N(w) = 1$. This means w is a unit. This discussion provides support for the following theorem:

Theorem 5.9.   *The elements $1 \pm i$, $a + ib$ with $a^2 + b^2$ a prime congruent to 1 modulo 4, and primes of the form $4k + 3$, and their associates are all the irreducible elements in $\mathbb{Z}[i]$.*

Proof. If ϖ is an irreducible element in $\mathbb{Z}[i]$, $\varpi \mid \varpi\overline{\varpi} = N(\varpi) \in \mathbb{Z}$. Write the prime factorization of $N(\varpi)$ as $p_1 p_2 \ldots p_k$, with p_j's not necessarily distinct. Now back in $\mathbb{Z}[i]$, $\varpi \mid p_1 p_2 \ldots p_k$. Since ϖ is irreducible, and $\mathbb{Z}[i]$ is a Euclidean domain, it has to be prime, so there is at least one j such that $\varpi \mid p_j$. This means ϖ must occur as a factor of some rational prime p. If $p \equiv 1 \bmod 4$ or $p = 2$, then $p = a^2 + b^2$ for integers a, b, and as we observed in the paragraph preceding the statement of the theorem, $a + ib$ and $a - ib$ are irreducible elements in $\mathbb{Z}[i]$. Since $\varpi \mid p = (a + ib)(a - ib)$, we conclude that either $\varpi \mid a + ib$ or $\varpi \mid a - ib$. If $\varpi \mid a + ib$, since both of these are irreducible, they have to be associates, and similarly for $a - ib$. On the other hand, if $p \equiv 3 \bmod 4$ we proceed as follows. Write $\varpi = m + in$. Since $\varpi \mid p$, then $N(\varpi) \mid N(p) = p^2$. This means $m^2 + n^2 \mid p^2$, from which it follows that $m^2 + n^2 = 1$, or $m^2 + n^2 = p$, or $m^2 + n^2 = p^2$. The case $m^2 + n^2 = 1$ is not possible as that would imply that ϖ is a unit. If $m^2 + n^2 = p$, then Lemma 5.6 shows that $p \mid m$, $p \mid n$, from which it follows $p^2 \mid m^2 + n^2$, a contradiction. So the only possibility is $m^2 + n^2 = p^2$. Again Lemma 5.6 shows that $m = pm_1$ and $n = pn_1$ for $m_1, n_1 \in \mathbb{Z}$. Hence $p^2 = m^2 + n^2 = (pm_1)^2 + (pn_1)^2 = p^2(m_1^2 + n_1^2)$. As a result $m_1^2 + n_1^2 = 1$, implying that $m_1 + in_1$ is a unit in $\mathbb{Z}[i]$. Consequently, $\varpi = m + in = pm_1 + ipn_1 = p(m_1 + in_1)$, showing that ϖ is an associate of p. $\square$

If p is a prime number of the form $4k + 1$, there is a unique representation of the form $p = a^2 + b^2$ with $a > b > 0$. We set

$$\varpi_p = a + ib.$$

We call the irreducibles $1 + i$, ϖ_p and $\overline{\varpi}_p$ for primes of the form $4k + 1$, and primes q of the form $4k + 3$, *standard*. Every other irreducible is an associate of a standard irreducible.

The following theorem follows from general properties of unique factorization domains [25, Ch. 3, §7]:

Theorem 5.10. *Every Gaussian integer can be written as*

$$u m (1 + i)^a \prod_{p \equiv 1 \bmod 4} \varpi_p^{e_p} \overline{\varpi}_p^{f_p}$$

in an essentially unique fashion, i.e., unique up to a permutation of the factors. Here u is one of the four units in $\mathbb{Z}[i]$; m a rational integer which is a product of primes of the form $4k + 3$; and all but finitely many of the nonnegative integers e_p, f_p are zero, meaning the product is finite.

For example, the number 2 considered as an element of $\mathbb{Z}[i]$ has the prime factorization $-i(1 + i)^2$, and

$$12 = -3 \cdot (1 + i)^4, \quad 60 = -3 \cdot (1 + i)^4 \cdot (2 + i) \cdot (2 - i).$$

5.5 Proof of Theorem 5.1

Now we can prove the main theorem of this chapter, Theorem 5.1. By Theorem 3.4 the hypotenuse of a primitive right triangle is an odd number which is the sum of two squares that are coprime to each other. Using Ingredient 2 above we see that no prime factor of such a number can be of the form $4k + 3$. Next we show that every number all of whose prime factors are of the form $4k + 1$ is the hypotenuse of some primitive right triangle.

We proceed in two steps:

Step 1. Suppose $(m, n) = 1$, and assume $m = a^2 + b^2$ and $n = c^2 + d^2$ with $(a, b) = 1$ and $(c, d) = 1$. Then $mn = (ac - bd)^2 + (ad + bc)^2$ with $(ac - bd, ad + bc) = 1$.

We have

$$mn = (ac - bd)^2 + (ad + bc)^2.$$

We claim that $\gcd(ac - bd, ad + bc) = 1$. Suppose for a prime p, $p \mid ac - bd$ and $p \mid ad + bc$. Then

$$p \mid c(ac - bd) + d(ad + bc) = a(c^2 + d^2) = an,$$

and

$$p \mid -d(ac - bd) + c(ad + bc) = b(c^2 + d^2) = bn.$$

Similarly, $p \mid cm$ and $p \mid dm$. Since $\gcd(m, n) = 1$, p cannot divide both m and n, so suppose p does not divide m. Then since $p \mid cm$, $p \mid c$ and similarly $p \mid d$. This contradicts the assumption that $\gcd(c, d) = 1$. The case $p \nmid n$ is similar.

Step 2. Let p be a prime of the form $4k + 1$. Then if $t \in \mathbb{N}$, p^t is the sum of two squares that are coprime to each other.

Note that this is not obvious. It is of course clear that if we write $p = u^2 + v^2$, then $(u, v) = 1$, because if $(u, v) = \delta$, then $\delta^2 \mid p$ which is impossible, unless $\delta = 1$. Next, assuming $p = u^2 + v^2$, we get $p^3 = (pu)^2 + (pv)^2$, so for $t > 1$, there are certainly expressions $p^t = a^2 + b^2$ such that a, b are *not* coprime. The content of this step is that it is possible to find an expression $p^t = a^2 + b^2$ such that a, b *are* coprime, but not that every such expression has the property that $(a, b) = 1$.

Write $p = u^2 + v^2 = N(u + iv)$. Then

$$p^t = N(u + iv)^t = N((u + iv)^t) = N(u_t + iv_t) = u_t^2 + v_t^2,$$

where u_t and v_t are defined to be the real and imaginary parts of $(u + iv)^t$, respectively. We claim $\gcd(u_t, v_t) = 1$. Suppose not, and let q be a prime factor of $\gcd(u_t, v_t)$. Then $q \mid u_t^2 + v_t^2 = p^t$. This implies that $q = p$, meaning every prime factor of $\gcd(u_t, v_t)$ is equal to p, i.e., $\gcd(u_t, v_t) = p^r$ for some r. We wish to show $r = 0$. We do this by induction on t. We already know the statement to be true for $t = 1$. Suppose for some t,

$$\gcd(u_{t-1}, v_{t-1}) = 1.$$

Then we have

$$u_t + iv_t = (u + iv)^t = (u + iv)(u + iv)^{t-1} = (u + iv)(u_{t-1} + iv_{t-1})$$

$$= (uu_{t-1} - vv_{t-1}) + i(uv_{t-1} + vu_{t-1}).$$

As in the first step we conclude $p^r \mid u_{t-1}(u^2 + v^2) = pu_{t-1}$ and $p^r \mid pv_{t-1}$, but since by assumption $\gcd(u_{t-1}, v_{t-1}) = 1$, we conclude that $r \leq 1$. If $r = 0$, we are done. So assume $r = 1$. This means $p \mid u_t$, which, if we use $u_t = uu_{t-1} - vv_{t-1}$, gives

$$uu_{t-1} \equiv vv_{t-1} \quad \mod p. \tag{5.2}$$

Now we write u_{t-1} and v_{t-1} in terms of u, v, u_{t-2}, v_{t-2}:

$$u_{t-1} = uu_{t-2} - vv_{t-2}, \quad v_{t-1} = vu_{t-2} + uv_{t-2}.$$

Using these identities Equation (5.2) reads

$$u(uu_{t-2} - vv_{t-2}) \equiv v(vu_{t-2} + uv_{t-2}) \quad \mod p.$$

Rearranging terms gives

$$(u^2 - v^2)u_{t-2} \equiv 2uvv_{t-2} \quad \mod p.$$

Since $u^2 + v^2 = p$, $-v^2 \equiv u^2 \mod p$, and we obtain

$$2u^2 u_{t-2} \equiv 2uvv_{t-2} \quad \mod p.$$

Canceling out $2u$ gives

$$uu_{t-2} \equiv vv_{t-2} \quad \mod p.$$

As a result, $p \mid uu_{t-2} - vv_{t-2} = u_{t-1}$. Since $v_{t-1}^2 = p^{t-1} - u_{t-1}^2$, and $t \geq 2$, we conclude $p \mid v_{t-1}^2$, and consequently, $p \mid v_{t-1}$. As a result, $p \mid u_{t-1}$, $p \mid v_{t-1}$, and this contradicts $\gcd(u_{t-1}, v_{t-1}) = 1$. □

Exercises

5.1 Write the following numbers as a product of irreducibles of $\mathbb{Z}[i]$:

a. 56;
b. $4 + 6i$;
c. $3 + 5i$;
d. $9 + i$;
e. $7 + 24i$.

5.2 Compute $\gcd(6 - 17i, 18 + i)$.

5.3 Solve the equation $x + y + z = xyz = 1$ in Gaussian integers.

5.4 Determine all irreducible elements with norm less than 100.

5.5 Devise a test to decide whether $x + iy$ is the square of a Gaussian integer.

5.6 Determine all Gaussian integers which are the sum of two squares of Gaussian integers.

5.7 Show that a Gaussian integer $x + iy$ is a sum of the squares of three Gaussian integers if and only if y is even.

5.8 What can you say about right triangles with integral sides such that the legs differ by 1? What if the difference is a fixed number d?

5.9 What can you say about right triangles with integral sides such that the sum of the legs is a fixed number s?

5.10 What can you say about a right triangle with integral sides such that the perimeter and the hypotenuse are squares?

5.11 Write 45305 as a sum of two squares.

5.12 For a natural number n, show that if the equation $n = x^2 + y^2$, $x, y > 0, 2 \mid x$, has more than one solution, then n is not prime.

5.13 Find a formula for the number of primitive right triangles with a leg equal to a number n in terms of the divisors of n.

5.14 Prove the following result of Gauss [16, page 172]: Every hypotenuse composed of k distinct primes belongs to

$$\left[\frac{k}{1}\right] + 2\left[\frac{k}{2}\right] + 2^2\left[\frac{k}{3}\right] + \cdots + 2^{k-1}\left[\frac{k}{k}\right]$$

different right triangles. Of these triangles, 2^{k-1} are primitive.

5.15 (✠) Determine if 31897485916040 is a sum of two squares. If it is, determine in how many ways.

Notes

Primes of special forms

The problem of deciding which polynomials produce prime numbers goes back centuries. Euler made the famously wrong claim that the polynomial $f(x) = x^2 - x + 41$ has the property that $f(n)$ is a prime number for every integer n. The values $f(0), f(1), f(2), f(3), \ldots, f(40)$ are all prime, though $f(41)$ is clearly not. We will see in Exercise 6.2 that there are no non-constant polynomials $f(x)$ such that $f(n)$ is a prime number for every integral value of n. Despite this rather disappointing statement, one could still ask whether there are polynomials that produce infinitely many primes. The answer is a definite yes. For example, every odd prime is either congruent to 1 modulo 4 or congruent to 3 modulo 4. This means that at least one of the polynomials $4x + 1$ or $4x + 3$ produces infinitely many primes. We will see in Chapter 6 that in fact both of these polynomials capture infinitely many primes.

For a general polynomial of degree 1, one can effectively decide whether the polynomial produces infinitely many primes. Suppose $f(x) = ax + b$ with $a, b \in \mathbb{Z}$. If $\gcd(a, b) = d > 1$, then the polynomial has no chance of producing infinitely many primes. It turns out that this is the only obstruction. The following is a celebrated theorem of Dirichlet [2, Theorem 7.3]:

Theorem 5.11. *If a, b are integers with $b > 0$, and $\gcd(a, b) = 1$, then the arithmetic progression*

$$a, a + b, a + 2b, a + 3b, a + 4b, \ldots$$

contains infinitely many prime numbers.

Unfortunately, we do not know an algebraic/elementary proof of this fact. The standard proofs of Dirichlet's Theorem use complex analysis and, though not terribly hard, are beyond the scope of this small volume. We give several examples of this theorem in Chapter 6. We also present the proof of an important special case in Exercise 6.22.

For polynomials of degree larger than 1 the situation is considerably more complicated. For example, in 1912, Landau conjectured that the polynomial $f(x) = x^2 + 1$ produces infinitely many primes. At the time of this writing it is still not known if Landau's Conjecture is true. The best result in this direction is due to Henryk Iwaniec who in 1978 proved that there are infinitely many integers n such that $n^2 + 1$ is the product of at most two prime numbers.

If we consider quadratic polynomials in more than one variable, then the situation is better understood. Theorem 5.7 gives a linear necessary and sufficient condition for the representability of a prime by a quadratic expression—namely, that an odd prime p is representable by the quadratic form $x^2 + y^2$ if and only if p is of the form $4k + 1$, implying that there are infinitely many primes of the form $x^2 + y^2$. There are other results of similar nature for representability of prime numbers by polynomials of the form $x^2 + ny^2$ dating back to, at least, Fermat and Euler. For example, a prime is of the form $x^2 + 2y^2$ for integers x, y if and only if $p \equiv 1, 3 \mod 8$. See Cox [14] for an in-depth study of primes that are representable by quadratic forms in two variables.

Algebraic number theory

We understand the phrase *algebraic number theory* in two different, but related, ways. The first one is *algebraic number/theory*, as in number theory done using algebraic methods, and the second one is *algebraic/number theory* as in the theory of algebraic numbers. In terms of the first interpretation, Chinese Remainder Theorem 2.24 is really a statement about ideals in a general ring; Euler's Theorem 2.31 is a special case of Lagrange's theorem in finite group theory; Lemma 2.49 is a consequence of the statement that every finite subgroup of a field is cyclic. What we did in this chapter with Gaussian integers is part of the second interpretation, and as we saw in this chapter we used our results on Gaussian integers to prove a statement about ordinary integers. Another example is our results from Appendix B which we will use in our proof of the Law of Quadratic Reciprocity in Chapter 7.

An important result in this chapter is Theorem 5.10 which establishes unique factorization in Gaussian integers. Unfortunately, this uniqueness of factorization fails in general number rings. A famous example is the ring $\mathbb{Z}[\sqrt{-5}] = \{x + y\sqrt{-5} \mid x, y \in \mathbb{Z}\}$. We have $6 = 2 \cdot 3 = (1 + \sqrt{-5}) \cdot (1 - \sqrt{-5})$, and it is not hard to see that $2, 3, 1 + \sqrt{-5}, 1 - \sqrt{-5}$ are all irreducible elements. It was Richard Dedekind who discovered that the fix for the failure of unique factorization in this and other number rings was to utilize *ideals*. Let us briefly explain Dedekind's ideas in a slightly more modern language than was available to him. We will use the notion of an *algebraic integer* defined in Appendix B. We define a *number field* to be a field K obtained by adjoining a finite number of algebraic integers to $\mathbb{Q}$. Define *the ring of integers* $\mathcal{O}_K$ of K to be set of all algebraic integers contained in K. Theorem B.4 shows that $\mathcal{O}_K$ is a ring. Dedekind showed that every ideal of $\mathcal{O}_K$ is a product of prime ideals of $\mathcal{O}_K$ in an essentially unique fashion. Since every ideal of $\mathbb{Z}$ and $\mathbb{Z}[i]$ is principal, Dedekind's result implies the unique factorization theorems of these rings.

Algebraic number theory was brought to new heights in the hands of David Hilbert and Emil Artin who early in 20th century found spectacular generalizations of the Law of Quadratic Reciprocity, known as *Reciprocity Laws*. These laws were further generalized by Shimura and Taniyama, who also discovered new connections to the theory of elliptic curves and modular forms. The most general reciprocity laws were conjectured by Robert Langlands in the 60s and 70s. Even though these conjectures remain largely open, they have inspired much progress in the last few decades.

For an elementary introduction to algebraic number theory, see [50]. Samuel [43] is a timeless classic. Murty and Esmonde's book [37] is a much recommended problem-solving-based approach to algebraic number theory. More advanced readers already familiar with basic algebraic number theory, abstract algebra, and measure theory are encouraged to read Weil's *Basic Number Theory* [56]. This book is far from basic, but in the words of Norbert Schappacher, if you learn number theory from this book, you will never forget it. Mazur [86] is an excellent expository article explaining the connections between modular forms and Diophantine equations. The book [17] is an account of the history of class field theory. Gelbart [75] is a not-so-elementary introduction to the Langlands program.

Chapter 6
Primes of the form $4k + 1$

The main goal of this chapter is to prove that there are infinitely many primes of the form $4k + 1$. We model the proof of this fact on Euclid's proof of the infinitude of prime numbers which we explain. We then discuss quadratic residues and study their basic properties. We state, and prove in the next chapter, the Law of Quadratic Reciprocity. At the end of the chapter we use the Law of Quadratic Reciprocity to prove the infinitude of primes of the form $3k + 1$. In the Notes, we discuss Euclid's original writing of his proof of the infinitude of prime numbers, talk about primality testing, and review some recent progress on the Twin Prime Conjecture.

6.1 Euclid's theorem on the infinitude of primes

We saw in Chapter 5 that in order for a prime to divide the side length of a primitive right triangle, it has to be of the form $4k + 1$. It would be extremely surprising, and rather unfortunate, if there were only finitely many such primes. In this chapter we will prove the following theorem:

Theorem 6.1. *There are infinitely many primes of the form $4k + 1$.*

In general it is actually quite hard to prove there are infinitely many primes of a special form. For example, at the time of this writing it is not known if there are infinitely many primes of the form $n^2 + 1$ (Landau's Conjecture, Notes to Ch. 5), or that there are infinitely many primes p such that $p + 2$ is also prime (Twin Prime Conjecture, Notes to this chapter), or that there are infinitely many primes p such that $2p + 1$ is prime (Infinitude of Sophie Germain Primes), etc. Even the proof of the existence of infinitely many primes without any additional restrictions is a non-trivial result that requires a real idea. This goes back to Euclid, and the proof we present here is essentially Euclid's original argument [20, Book IX, Proposition 20].

© Springer Nature Switzerland AG 2018
R. Takloo-Bighash, *A Pythagorean Introduction to Number Theory*,
Undergraduate Texts in Mathematics, https://doi.org/10.1007/978-3-030-02604-2_6

Theorem 6.2 (Euclid). *There are infinitely many prime numbers.*

Proof. Suppose not, and let $\{p_1, \ldots, p_m\}$ be the (finite) set of all prime numbers. Let

$$M = p_1 \cdots p_m + 1.$$

The number M is not divisible by any of the primes p_i, but M is divisible by some prime q, which is necessarily one p_i's. In particular, $q \mid p_1 \cdots p_m$. Consequently,

$$q \mid M - p_1 \cdots p_m = 1.$$

But 1 is not divisible by any primes. This is a contradiction. □

One can try to adapt this argument to prove Theorem 6.1. So let's suppose that $\{p_1, \ldots, p_m\}$ is the finite set of primes of the form $4k + 1$, and set

$$M = p_1 \cdots p_m + 1.$$

We would then ask if this number, for some reason, has to have a new prime factor of the form $4k + 1$. It is wise to do some experiments. Let us start with the first two primes of the form $4k + 1$, namely 5 and 13. Then,

$$M = 5 \times 13 + 1 = 66 = 2 \times 3 \times 11,$$

none of whose factors are of the desired type, as $3 = 0 \times 4 + 3$ and $11 = 2 \times 4 + 3$. One might complain that the issue with this approach was that the resulting number M is not of the form $4k + 1$—in fact, it is always even. So it makes sense to define M this way:

$$M = 4p_1 \cdots p_m + 1.$$

This idea fails too. For example, the numbers 5 and 17 are both primes of the form $4k + 1$. We have

$$M = 4 \times 5 \times 17 + 1 = 341 = 11 \times 31.$$

The primes 11 and 31 are both of the form $4k - 1$. The problem is that when we multiply two primes of the form $4k - 1$, or of the form $4k + 3$, we get a number of the form $4k + 1$:

$$(4m - 1)(4n - 1) = 16mn - 4m - 4n + 1 = 4(4mn - m - n) + 1.$$

But not all is lost! In fact, this last computation suggests that maybe instead of proving the infinitude of primes of the form $4k + 1$, Euclid's idea can be adapted to prove the infinitude of primes of the form $4k - 1$. The key observation is that when we multiply numbers of the form $4k + 1$ the result is always a number of the form $4k + 1$, i.e.,

$$(4m + 1)(4n + 1) = 16mn + 4m + 4n + 1 = 4(4mn + m + n) + 1. \qquad (6.1)$$

Theorem 6.3. *There are infinitely many primes of the form* $4k - 1$.

Proof. Let $p_1, \ldots, p_m$ be a finite set of primes of the form $4k - 1$. Let

$$M = 4p_1 \cdots p_m - 1.$$

The number M is of the form $4k - 1$, and not divisible by any of the primes $p_1, \ldots, p_m$. Also, not all of M's prime factors can be of the form $4k + 1$, because in that case by Equation (6.1) M would be of the form $4k + 1$. As a result M has a prime factor of the form $4k - 1$ and we have found a new prime of the desired form.

$\square$

Going back to Theorem 6.1, the idea is to find an expression M in terms of $p_1, \ldots, p_m$, which is odd, not divisible by any of the p_i's, and provably possessing a new prime factor of the form $4k + 1$. One way to do this to make sure that M has *no* prime factors of the form $4k - 1$. The key to making this happen is Lemma 5.6 of Chapter 5.

Proof of Theorem 6.1. Let $p_1, \ldots, p_m$ be the set of all primes of the form $4k + 1$. Let

$$M = (2p_1 \cdots p_m)^2 + 1.$$

This number is not divisible by any of the p_i. It is odd, and by Lemma 5.6 none of its prime factors can be of the form $4k - 1$. Every prime factor of M is new prime number of the form $4k + 1$. $\square$

For example, the numbers 5, 13, 17, 29, 37 are all primes of the form $4k + 1$. Then

$$(2 \cdot 5 \cdot 13 \cdot 17 \cdot 29 \cdot 37)^2 + 1 = 233 \cdot 593 \cdot 3301 \cdot 12329,$$

with the factors on the right all being prime numbers of the form $4k + 1$.

6.2 Quadratic residues

The main point of the proof of Theorem 6.1 is that a number of the form $n^2 + 1$ cannot have any prime factors of the form $4k - 1$. This suggests that one may be able to prove the infinitude of other sets of prime numbers by exploring prime factors of numbers of the form $n^2 - a$ for integers a. In the argument above, $a = -1$.

Question 6.4. For an integer a, for what primes p, are there no integers n such that $p \mid n^2 - a$?

Gauss systematically studied this question, [21, §IV, Article 95], and proved a number of fundamental results. Let p be an odd prime. Suppose p does not divide a, and define the *Legendre symbol*, or the *quadratic residue symbol*, by

$$\left(\frac{a}{p}\right) = \begin{cases} +1 & \exists n, n^2 \equiv a \bmod p; \\ -1 & \text{otherwise.} \end{cases}$$

If $p \mid a$, we set

$$\left(\frac{a}{p}\right) = 0.$$

We call an integer a a *quadratic residue modulo* p if $p \nmid a$ and the equation $x^2 \equiv a \bmod p$ is solvable, i.e., if $\left(\frac{a}{p}\right) = +1$. It is clear that if $a \equiv b \bmod p$, then

$$\left(\frac{a}{p}\right) = \left(\frac{b}{p}\right),$$

so we often think of $\left(\frac{\cdot}{p}\right)$ as a function on the set of congruence classes modulo p.

Sometimes, when there is no danger of confusion, we write (a/p) instead of $\left(\frac{a}{p}\right)$.

Lemma 6.5 (Euler). *Let p be an odd prime. We have*

$$\left(\frac{a}{p}\right) \equiv a^{\frac{p-1}{2}} \bmod p.$$

Proof. If $a \equiv 0 \bmod p$, the lemma is obvious. So we assume $a \not\equiv 0 \bmod p$. Let g be a primitive root modulo p. Then if $a \equiv g^i \bmod p$, we have

$$\left(\frac{a}{p}\right) = (-1)^i.$$

If i is even,

$$a^{\frac{p-1}{2}} \equiv g^{i\frac{p-1}{2}} \equiv g^{\frac{i}{2}\cdot(p-1)} \equiv 1 \bmod p.$$

On the other hand, if i is odd,

$$a^{\frac{p-1}{2}} \equiv g^{i\frac{p-1}{2}} \equiv g^{\frac{p-1}{2}} \bmod p.$$

As we saw in the proof of Lemma 5.6

$$g^{\frac{p-1}{2}} \equiv -1 \bmod p,$$

and this finishes the proof. $\square$

Lemma 6.6. *Let p be an odd prime. For all integers a, b,*

$$\left(\frac{a}{p}\right)\left(\frac{b}{p}\right) = \left(\frac{ab}{p}\right). \tag{6.2}$$

Proof. By Lemma 6.5 we have

$$\left(\frac{ab}{p}\right) \equiv (ab)^{\frac{p-1}{2}} \equiv a^{\frac{p-1}{2}} b^{\frac{p-1}{2}} \equiv \left(\frac{a}{p}\right)\left(\frac{b}{p}\right) \quad \text{mod } p.$$

This means

$$p \mid \left(\frac{ab}{p}\right) - \left(\frac{a}{p}\right)\left(\frac{b}{p}\right).$$

Since the possible values of the quadratic residue symbol are $+1, -1$, the expression on the right can take values $+2, 0, -2$. Of these numbers, the only one that is divisible by p is 0, and this observation proves the identity. $\square$

The following lemma is a reformulation of Corollary 5.8.

Lemma 6.7. *If p is an odd prime, then*

$$\left(\frac{-1}{p}\right) = (-1)^{\frac{p-1}{2}}. \tag{6.3}$$

Proof. By Lemma 6.5 we have

$$\left(\frac{-1}{p}\right) \equiv (-1)^{\frac{p-1}{2}} \quad \text{mod } p.$$

Now an argument similar to the proof of Lemma 6.6 gives the lemma. $\square$

This last statement means that there is an n such that $n^2 \equiv -1$ mod p precisely when

$$(-1)^{\frac{p-1}{2}} = +1,$$

i.e., when $(p-1)/2$ is even, which is equivalent to p being of the form $4k+1$. For example, 13 and 17 are primes of the form $4k+1$ and $5^2 \equiv -1$ mod 13 and $4^2 \equiv -1$ mod 17.

These facts are enough to compute quadratic residues modulo every prime number. Let us illustrate this by computing $\left(\frac{15}{31}\right)$. By Equation (6.2) we have

$$\left(\frac{15}{31}\right) = \left(\frac{3}{31}\right)\left(\frac{5}{31}\right).$$

To compute $\left(\frac{3}{31}\right)$, we write

$$\left(\frac{3}{31}\right) \equiv 3^{\frac{31-1}{2}} \equiv 3^{15} \equiv 27^5 \equiv (-4)^5 \equiv -4^5$$

$$\equiv -4^3 \cdot 4^2 \equiv -2 \cdot 16 \equiv -1 \quad \text{mod } 31.$$

This means, $\left(\frac{3}{31}\right) = -1$. Next,

$$\left(\frac{5}{31}\right) \equiv 5^{\frac{31-1}{2}} \equiv 5^{15} \equiv (5^3)^5 \equiv 125^5 \equiv 1^5 \equiv 1 \mod 31.$$

Consequently, $\left(\frac{5}{31}\right) = +1$. Putting everything together,

$$\left(\frac{15}{31}\right) = \left(\frac{3}{31}\right)\left(\frac{5}{31}\right) = (-1) \cdot (+1) = -1.$$

To see a slightly more complicated example, we also compute $\left(\frac{17}{31}\right)$. By Lemma 6.5 we have

$$\left(\frac{17}{31}\right) \equiv 17^{\frac{31-1}{2}} \equiv 17^{15} \equiv (17^5)^3 \equiv 26^3 \equiv (-5)^3 \equiv -1 \mod 31.$$

As a result,

$$\left(\frac{17}{31}\right) = -1.$$

Equation (6.2) shows that in order to compute quadratic residues modulo p, we need to know $\left(\frac{q}{p}\right)$ for primes p. At first glance, $\left(\frac{q}{p}\right)$ and $\left(\frac{p}{q}\right)$ should have no relationship with each other—we often think of primes numbers as *independent* of each other, and in many situations they behave as if they are completely unaware of each other's presence. However, in this case, primes p and q, knowing either of $\left(\frac{q}{p}\right)$ and $\left(\frac{p}{q}\right)$, tells us the value of the other one. The exact relationship was conjectured by Euler around 1745, and was proved rigorously for the first time by Gauss in 1796, though Legendre had proved some special cases as early as 1785. By the time he died, Gauss had produced eight different proofs for the theorem, the *Law of Quadratic Reciprocity*.

Theorem 6.8 (Law of Quadratic Reciprocity).

1. *If p, q are distinct odd primes, then*

$$\left(\frac{p}{q}\right) = (-1)^{\frac{p-1}{2} \cdot \frac{q-1}{2}} \left(\frac{q}{p}\right);$$

   Explicitly, $(p/q) = -(q/p)$ only when $p \equiv q \equiv 3 \mod 4$, and in all other situations $(p/q) = (q/p)$.
2. *If p is an odd prime,*

$$\left(\frac{2}{p}\right) = (-1)^{\frac{p^2-1}{8}}.$$

   Explicitly written out, this means if $p \equiv 1, 7 \mod 8$, then 2 is a quadratic residue modulo p, and if $p \equiv 3, 5 \mod 8$, it is not.

Even though at the time of this writing there are literally hundreds of proofs of this fundamental fact available in print, unfortunately, none of them are trivial. In the next chapter we will present one of Gauss's original proofs using quadratic Gauss sums. The Law of Quadratic Reciprocity is a truly impressive theorem. This theorem has now been generalized magnificently through the works of Artin, Hilbert, and Langlands [75], and has inspired an incredible amount of mathematics. In fact, the works of four Fields medalists (V. Drinfeld, L. Lafforgue, B. C. Ngô, and M. Bhargava) have been directly or indirectly inspired by Gauss's work on the Law of Quadratic Reciprocity and its generalizations. This is indeed one of the most important theorems in all of mathematics.

One consequence of Theorem 6.8 is that it allows one to compute (a/p) very quickly. For example, suppose we want to compute $(194/7919)$. By Equation (6.2) we have

$$\left(\frac{194}{7919}\right) = \left(\frac{2}{7919}\right)\left(\frac{97}{7919}\right).$$

By the second part of the theorem,

$$\left(\frac{2}{7919}\right) = (-1)^{(7919^2-1)/8} = (-1)^{7838820} = +1.$$

Next, by the first part,

$$\left(\frac{97}{7919}\right) = (-1)^{(97-1)(7919-1)/4}\left(\frac{7919}{37}\right) = \left(\frac{7919}{97}\right) = \left(\frac{62}{97}\right),$$

as $7919 \equiv 62 \bmod 97$. So far we know

$$\left(\frac{194}{7919}\right) = \left(\frac{62}{97}\right).$$

Now we apply the same procedure to the latter quadratic residue. We have

$$\left(\frac{62}{97}\right) = \left(\frac{2}{97}\right) \cdot \left(\frac{31}{97}\right).$$

By the second part of the theorem

$$\left(\frac{2}{97}\right) = (-1)^{(97^2-1)/8} = +1.$$

Next, by the first part,

$$\left(\frac{31}{97}\right) = (-1)^{(31-1)(97-1)/4}\left(\frac{97}{31}\right) = \left(\frac{97}{31}\right).$$

Since $97 \equiv 4 \bmod 31$, we have

$$\left(\frac{97}{31}\right) = \left(\frac{4}{31}\right) = \left(\frac{2}{31}\right)^2 = +1.$$

Putting everything together, we have

$$\left(\frac{194}{7919}\right) = +1.$$

6.3   An application of the Law of Quadratic Reciprocity

Let us return to Question 6.4. For $a \in \mathbb{Z}$, one can use the Law of Quadratic Reciprocity to characterize p such that

$$\left(\frac{a}{p}\right) = +1.$$

For example, let us study the case where $a = -3$. Suppose $p > 3$. We have

$$\left(\frac{-3}{p}\right) = \left(\frac{-1}{p}\right)\left(\frac{3}{p}\right).$$

Equation (6.3) gives

$$\left(\frac{-1}{p}\right) = (-1)^{\frac{p-1}{2}}.$$

Quadratic Reciprocity implies

$$\left(\frac{3}{p}\right) = (-1)^{(3-1)(p-1)/4}\left(\frac{p}{3}\right) = (-1)^{(p-1)/2}\left(\frac{p}{3}\right).$$

Next,

$$\left(\frac{-3}{p}\right) = \left(\frac{-1}{p}\right)\left(\frac{3}{p}\right) = (-1)^{(p-1)/2}(-1)^{(p-1)/2}\left(\frac{p}{3}\right)$$

$$= \left(\frac{p}{3}\right) = \begin{cases} +1 & p \equiv 1 \quad \bmod 3; \\ -1 & p \equiv 2 \quad \bmod 3. \end{cases}$$

Let's think a moment about what just happened. We are trying to determine for what primes p, -3 is a quadratic residue modulo p. This is a question about quadratic residues modulo p: There are $(p - 1)/2$ of these and that number grows with p, and that's somewhat of a moving target. The Law of Quadratic Reciprocity allows us to turn the problem around and transform it into a problem about quadratic residues modulo 3. The beauty of this idea is that there is only one non-zero quadratic residue modulo 3, the congruence class of 1.

Next, following the argument leading to the proof of the infinitude of primes of the form $4k + 1$, we observe that if we multiply numbers of the form $3k + 1$ we will get another number of the form $3k + 1$:

$$(3m + 1)(3n + 1) = 3(3mn + m + n) + 1. \tag{6.4}$$

This is significant, as every prime number $p \neq 3$ either is of the form $3k + 1$ or of the form $3k + 2$. We can now prove the following theorem:

Theorem 6.9. *There are infinitely many primes of the form $3k + 1$ and infinitely many primes of the form $3k + 2$.*

Proof. The proof for $3k + 2$ is easy. Let $p_1, \ldots, p_m$ be a collection of odd primes of the form $3k + 2$. Set

$$M = 6p_1 \cdots p_m - 1.$$

The number M is not divisible by any of the p_i's, and Equation (6.4) means that not all of its prime factors can be of form $3k + 1$, because then M itself would be of the form $3k + 1$, which it is not. As a result, M must have a new prime factor of the form $3k + 2$, and this proves the second assertion of the theorem.

Next, we prove the first assertion. Again, let $p_1, \ldots, p_m$ be a collection of primes of the form $3k + 1$. Let

$$M = (2p_1 \cdots p_m)^2 + 3.$$

The number M is not divisible by 2, by 3, and by any of the p_i's. But no prime factor of M can be of the form $3k + 2$, because if $q \mid M$, then the equation $n^2 \equiv -3 \bmod q$ will have a solution in n, namely $n = 2p_1 \ldots p_m$. This means M must only consist of primes of the form $3k + 1$, and we have found new primes not among the p_i's. $\square$

So far we have proved that each of the arithmetic progressions $2k + 1$ (Euclid!), $3k + 1, 3k + 2, 4k + 1$, and $4k + 3$ contains infinitely many primes. As we mentioned in the Notes to Chapter 5, a general theorem of Dirichlet, Theorem 5.11, provides a unifying picture for all of these results.

Exercises

6.1 Suppose we have a non-constant polynomial $f(x) \in \mathbb{Z}[x]$. Show that the set of prime numbers p such that $p \mid f(n)$ for some n is infinite.

6.2 Show for every non-constant polynomial $f(x) \in \mathbb{Z}[x]$ there are infinitely many values of n for which $f(n)$ is not prime.

6.3 Show that there are infinitely many primes of the form

    a. $8k + 1$;
    b. $8k + 3$;
    c. $5k + 4$;
    d. $12k + 1$;
    e. $12k + 5$;
    f. $12k + 7$;
    g. $12k + 11$.

6.4 Compute the following Legendre symbols:

    a. $(13/29)$;
    b. $(67/193)$;
    c. $(30/103)$;
    d. $(62/569)$.

6.5 Give a group-theoretic interpretation for the Legendre symbol.

6.6 Suppose p is an odd prime, and $p \nmid a$. Show that the congruence $ax^2 + bx + c \equiv 0 \bmod p$ is solvable if and only if $u^2 \equiv b^2 - 4ac \bmod p$ is solvable.

6.7 Give a characterization for all primes p for which the equation $x^2 + 2x + 3 \equiv 0 \bmod p$ is solvable.

6.8 Determine all primes p that satisfy $(7/p) = +1$.

6.9 Prove that a prime p is of the form $x^2 - 2y^2$ if and only if $p = 2$ or $p \equiv \pm 1 \bmod 8$.

6.10 Prove if $(n/p) = -1$, then

$$\sum_{d \mid n} d^{\frac{p-1}{2}} \equiv 0 \quad \bmod p.$$

6.11 Determine the product of all quadratic residues modulo p.

6.12 Verify the identity

$$x^8 - 16 = (x^2 - 2)(x^2 + 2)((x - 1)^2 + 1)((x + 1)^2 + 1).$$

Use the identity to determine the number of solutions of

$$x^8 \equiv 16 \quad \bmod p.$$

6.13 Determine the number of solutions of the congruence

$$x^6 - 11x^4 + 36x^2 - 36 \equiv 0 \quad \bmod p.$$

6.14 Show that if $p \mid n^4 - n^2 + 1$ for some $n \in \mathbb{Z}$, then $p \equiv 1 \bmod 12$.

6.15 Compute $\sum_{r=1}^{p-2}(r(r+1)/p)$.

6.16 Let $p > 2$ be prime. Determine the number of $1 \leq n \leq p - 2$ such that n and $n + 1$ are both quadratic residues modulo p. To do this, consider

$$\frac{1}{4} \sum_{n=1}^{p-2} \left(1 + \left(\frac{n}{p}\right)\right) \left(1 + \left(\frac{n+1}{p}\right)\right).$$

6.17 Show that if n is not a perfect square, there are infinitely many primes p such that $(n/p) = -1$.

6.18 (✠) We saw in Exercise 2.60 that $p = 2^{17} - 1$ is prime. Compute the quadratic residue symbols (q/p) for q every prime less than 20.

6.19 Prove that there are arbitrarily long non-constant arithmetic progressions such that every two terms of the arithmetic progression are relatively prime.

6.20 Let $k \in \mathbb{N}$. Show that there are integers a, b such that for all $j \in \mathbb{N}$ the number of divisors of $a + bj$ is divisible by k.

6.21 Fix a natural number l. Assuming Theorem 5.11 prove every arithmetic progression $a + bk$, $k \geq 0$, with $\gcd(a, b) = 1$, contains infinitely many terms which are products of l distinct primes.

6.22 The goal of this exercise is to show that if $n \in \mathbb{N}$, then there are infinitely many primes of the form $nk + 1$.

    a. Show that for each $d \in \mathbb{N}$ there is a monic polynomial $\Phi_d(x) \in \mathbb{Z}[x]$, called the d-th cyclotomic polynomial, such that

$$\prod_{d \mid n} \Phi_d(x) = x^n - 1.$$

    b. Show that $\Phi_1(0) = -1$ and for $d > 1$, $\Phi_d(0) = 1$.

    c. (✠) Find the first 100 or so cyclotomic polynomials. Pay close attention to the coefficients of the polynomials.

    d. Suppose $n > 1$ and $a \in \mathbb{Z}$, and let p be a prime divisor of $\Phi_n(a)$. Then show that $\gcd(a, p) = 1$, and if $h = o_p(a)$, $h \mid n$. Furthermore:

    • if $h < n$, then

$$a^n - 1 \equiv (a + p)^n - 1 \equiv 0 \quad \mathrm{mod}\ p^2;$$

    • if $h < n$, then $p \mid n$;

    • if $p \nmid n$, then $h = n$ and $p \equiv 1 \bmod n$.

    e. Conclude there are infinitely many primes of the form $nk + 1$.

Notes

Infinitude of Prime Numbers in **The Elements**

To get a feel for Euclid's style of writing, let us state Euclid's First Theorem, Lemma 2.18:

Theorem 6.10   (Elements, Book VII, Proposition 30). *If two numbers by multiplying one another make some number, and any prime number measures [divides] the product, it will also measure one of the original numbers.*

It may sound like a historical absurdity that Euclid never stated Theorem 2.19—in fact, this particular fact had to wait almost 2000 years to be put in writing by Gauss. However, any rigorous proof of Theorem 2.19 uses mathematical induction which as a tool was not available to Euclid. At any rate, Euclid used this theorem to prove the irrationality of $\sqrt{n}$ for n non-square, which may have been his original goal in writing the number theoretic parts of *The Elements*.

This is Euclid's original formulation of Theorem 6.2:

Theorem 6.11   (Elements, Book IX, Proposition 20). *Prime numbers are more than any assigned multitude of prime numbers.*

Here we will reproduce Euclid's original argument. Note that here Euclid illustrates the idea by working out the proof for a special case:

Let A, B, and C be the assigned prime numbers. I say that there are more prime numbers than A, B, and C.

Take the least number DE measured by A, B, and C. Add the unit DF to DE. Then EF is either prime or not.

First, let it be prime. Then the prime numbers A, B, C, and EF have been found which are more than A, B, and C.

Next, let EF not be prime. Therefore it is measured by some prime number. Let it be measured by the prime number G. I say that G is not the same with any of the numbers A, B, and C. If possible, let it be so.

Now A, B, and C measure DE; therefore G also measures DE. But it also measures EF. Therefore G, being a number, measures the remainder, the unit DF, which is absurd.

Therefore G is not the same with any one of the numbers A, B, and C. And by hypothesis it is prime. Therefore the prime numbers A, B, C, and G have been found which are more than the assigned multitude of A, B, and C.

Therefore, prime numbers are more than any assigned multitude of prime numbers.

At the time of this writing, the largest known prime number is $2^{77,232,917} - 1$ discovered in 2017. This number has $23, 249, 425$ digits. For comparison, the number of atoms in the entire observable universe is a number which is supposed to have about 80 digits. The discovery of this largest prime was part of *The Great Internet Mersenne Prime Search* accessible through

https://www.mersenne.org/

Primality testing

The first *primality test* is due to Eratosthenes (276–194 BCE) who observed that a number n is prime if and only if it is not divisible by any primes up to $\sqrt{n}$; see Exercise 2.18. For n reasonably small this provides a quick way of determining the primality of a number n, but as n gets large this method becomes impractical fairly quickly. Ideally one would like to be able to find a way to tell the primality of a number n in a number of steps that grows like a polynomial in the number of digits of n, and Eratosthenes' algorithm fails this expectation fairly miserably. Such an algorithm was not available until 2004 when the now-famous paper by M. Agrawal, N. Kayal, and N. Saxena [58] came out.

The algorithm presented in this paper is known as the AKS algorithm. Before AKS what was available in literature was an array of probabilistic algorithms, and some of these work quite well. A favorite example is the Miller–Rabin test [53, §6.3] which is based on Fermat's Little Theorem in elementary number theory. The Miller–Rabin test is extremely quick, but the trouble is that it gives *false positives*, in that some composite numbers are marked as primes.

A closely related problem we currently do not know how to solve, which is mentioned in the Notes of Chapter 2, is to factorize a large number as a product of its prime factors with reasonable efficiency. The solution of this problem would have far reaching consequences in terms of cryptography and internet security.

Twin Prime Conjecture

The following conjecture is considered very difficult:

Conjecture 6.12 (de Polignac, 1849). For every even natural number h, there are infinitely many prime numbers p such that $p + h$ is prime.

The case $h = 2$ is known as the *Twin Prime Conjecture* which at the time of this writing is still open. In 1915 Viggo Brun attempted to prove the Twin Prime Conjecture by proving that

$$\sum_{p,\, p+2 \text{ prime}} \frac{1}{p} \tag{6.5}$$

diverges. This idea goes back to Euler who proved the infinitude of prime numbers by showing that the series

$$\sum_{p \text{ prime}} \frac{1}{p}$$

diverges. However, surprisingly, Brun proved that the series (6.5) is convergent! Even more surprisingly, the proof was fairly elementary; see Exercise 9.2.7 of [35] and the exercises leading up to it for a presentation of the argument. The theory of *sieves* that Brun used in his proof has now become a powerful tool in number theory. The next major breakthrough, again involving the theory of sieves, was achieved in 1973 by Jingrun Chen [65] who showed that there are infinitely many primes p such that $p + 2$ is the product of at most two primes. In the same paper Chen also proved an approximation to Goldbach's conjecture; Chen proved every even number is the sum of a prime and a product of at most two primes. In 2005, Goldston, Pintz, and Yıldırım [76] proved a truly remarkable theorem. To state their theorem we will define a piece of notation. For a prime number p, let p_{next} be the smallest prime number larger than p. Using this notation, the Twin Prime Conjecture would assert the existence of infinitely many primes p such that $p_{next} - p = 2$. Goldston, Pintz, and Yıldırım used the theory of sieves in an ingenious way to prove

$$\liminf_{p \to \infty} \frac{p_{next} - p}{\log p} = 0.$$

It is clear that de Polignac's conjecture for any h would imply this result, but knowing this result would not give any information about de Polignac's conjecture. The spectacular work of Yitang Zhang in 2013, building on the techniques of Goldston, Pintz, and Yıldırım, changed the landscape overnight. Zhang [112] showed that there are infinitely many primes p such that

$$p_{next} - p < 7 \times 10^7.$$

This was a major achievement in that it showed the difference between consecutive primes was bounded by a uniform bound. In the last few years the bound of 7×10^7 has been substantially improved by Maynard [85] and the Polymath Project [91]. At the time of this writing we know by [91] that there are infinitely many primes p such that

$$p_{next} - p \leq 246.$$

At this time it is not clear how to reduce the bound 246, and this might require a new idea. The same paper proves that there are infinitely many primes p such that

$$(p_{next})_{next} - p \leq 38130.$$

It would also be of great interest to improve this bound, but, again, this might require an entirely new idea.

Chapter 7
Gauss Sums, Quadratic Reciprocity, and the Jacobi Symbol

Our first goal in this chapter is to present Gauss' sixth proof of his Law of Quadratic Reciprocity. The presentation here follows [32, §3.3] fairly closely, except that our Gauss sums are over the complex numbers, as opposed to *ibid.* where Gauss sums are considered over a finite field. Later in the chapter we introduce the Jacobi symbol and study its basic properties. We will also prove the Law of Quadratic Reciprocity for the Jacobi symbol. At the end of the chapter we will show examples that demonstrate how the Jacobi symbol can be used to compute the Legendre symbol efficiently. The Jacobi symbol will make an appearance in Chapter 10 when we give a proof of the Three Squares Theorem. In the Notes, we give some references for the various proofs of the Law of Quadratic Reciprocity.

7.1 Gauss sums and Quadratic Reciprocity

For an odd prime p, let $\zeta = e^{\frac{2\pi i}{p}}$ and define the pth *Gauss sum* by

$$\tau_p = \sum_{k=1}^{p-1} \left(\frac{k}{p} \right) \zeta^k.$$

We start with the following lemma:

Lemma 7.1. *For all odd primes p,*

$$\tau_p^2 = \left(\frac{-1}{p} \right) p.$$

Proof. We have

$$\tau_p^2 = \sum_{k=1}^{p-1} \sum_{l=1}^{p-1} \left(\frac{k}{p} \right) \left(\frac{l}{p} \right) \zeta^{k+l} = \sum_{k=1}^{p-1} \sum_{l=1}^{p-1} \left(\frac{kl}{p} \right) \zeta^{k+l}.$$

© Springer Nature Switzerland AG 2018
R. Takloo-Bighash, *A Pythagorean Introduction to Number Theory*,
Undergraduate Texts in Mathematics, https://doi.org/10.1007/978-3-030-02604-2_7

We make a change of variables by introducing a new variable m by $l \equiv mk \bmod p$. When k, l range over $\{1, \ldots, p - 1\}$, m varies over the same set. So we get

$$\tau_p^2 = \sum_{k=1}^{p-1} \sum_{m=1}^{p-1} \left(\frac{mk^2}{p}\right) \zeta^{k+mk} = \sum_{k=1}^{p-1} \sum_{m=1}^{p-1} \left(\frac{m}{p}\right) \zeta^{k+mk}$$

$$= \sum_{m=1}^{p-1} \left(\frac{m}{p}\right) \sum_{k=1}^{p-1} \zeta^{k(m+1)}.$$

The innermost sum is a geometric sum, and if $\zeta^{m+1} \neq 1$, we get

$$\sum_{k=1}^{p-1} \zeta^{k(m+1)} = \frac{(\zeta^p)^{m+1} - \zeta^{m+1}}{\zeta^{m+1} - 1} = \frac{1 - \zeta^{m+1}}{\zeta^{m+1} - 1} = -1.$$

If on the other hand $\zeta^{m+1} = 1$, we have

$$e^{\frac{2\pi i (m+1)}{p}} = 1.$$

Consequently, $p \mid m + 1$, and, since $1 \leq m \leq p - 1$, we conclude that $m = p - 1$. In this case,

$$\sum_{k=1}^{p-1} \zeta^{k(m+1)} = p - 1.$$

Putting everything together,

$$\tau_p^2 = \sum_{m=1}^{p-1} \left(\frac{m}{p}\right) \sum_{k=1}^{p-1} \zeta^{k(m+1)}$$

$$= \sum_{m=1}^{p-2} \left(\frac{m}{p}\right) \sum_{k=1}^{p-1} \zeta^{k(m+1)} + (p-1) \left(\frac{p-1}{p}\right)$$

$$= - \sum_{m=1}^{p-2} \left(\frac{m}{p}\right) + (p-1) \left(\frac{p-1}{p}\right)$$

$$= - \sum_{m=1}^{p-1} \left(\frac{m}{p}\right) + \left(\frac{p-1}{p}\right) + (p-1) \left(\frac{p-1}{p}\right)$$

$$= - \sum_{m=1}^{p-1} \left(\frac{m}{p}\right) + p \left(\frac{p-1}{p}\right) = - \sum_{m=1}^{p-1} \left(\frac{m}{p}\right) + p \left(\frac{-1}{p}\right).$$

So in order to prove the lemma it suffices to prove

$$\sum_{m=1}^{p-1} \left(\frac{m}{p}\right) = 0.$$

To see this, let

$$X = \sum_{m=1}^{p-1} \left(\frac{m}{p}\right).$$

Pick an integer b, e.g, a primitive root modulo p, such that $(b/p) = -1$. Then

$$-X = \left(\frac{b}{p}\right) X = \sum_{m=1}^{p-1} \left(\frac{b}{p}\right)\left(\frac{m}{p}\right) = \sum_{m=1}^{p-1} \left(\frac{bm}{p}\right).$$

But when m ranges over the numbers $\{1, \ldots, p-1\}$, the product mb ranges over the same set modulo p. Consequently, the last expression is equal to X as well. Hence

$$-X = X.$$

This implies $X = 0$, and we are done.

Now we can proceed to prove the Quadratic Reciprocity, presenting a variation of Gauss's extremely clever argument. This proof uses Gauss sums. In the course of the proof we will use algebraic integers as introduced in Appendix B.

Proof of Theorem 6.8. For the first part we start with the observation that

$$\tau_p^q = (\tau_p^2)^{\frac{q-1}{2}} \cdot \tau_p = \left(\left(\frac{-1}{p}\right)p\right)^{\frac{q-1}{2}} \cdot \tau_p$$

$$= (-1)^{\frac{p-1}{2} \cdot \frac{q-1}{2}} p^{\frac{q-1}{2}} \tau_p,$$

after using Lemma 6.7. Next,

$$\tau_p^q = \left(\sum_{k=1}^{p-1} \left(\frac{k}{p}\right) \zeta^k\right)^q.$$

By Lemma 2.28 this last expression is equal to

$$\sum_{k=1}^{p-1} \left(\frac{k}{p}\right) \zeta^{kq} + qC \tag{7.1}$$

for some complex number C. It follows from Theorem B.4 and the fact that roots of unity are algebraic integers that the number C is an algebraic integer; see Exercise 7.3. Let q^{-1} be the multiplicative inverse of q modulo p. Then the sum is equal to

$$\sum_{k=1}^{p-1} \left(\frac{kq^{-1}}{p}\right) \zeta^k = \left(\frac{q^{-1}}{p}\right) \sum_{k=1}^{p-1} \left(\frac{k}{p}\right) \zeta^k = \left(\frac{q^{-1}}{p}\right) \tau_p,$$

Since $q \cdot q^{-1} \equiv 1 \bmod p$, we have

$$\left(\frac{q^{-1}}{p}\right) = \left(\frac{q}{p}\right).$$

Putting everything together,

$$\left(\frac{q}{p}\right)\tau_p + qC = (-1)^{\frac{p-1}{2}\cdot\frac{q-1}{2}} p^{\frac{q-1}{2}}\tau_p.$$

Dividing by τ_p gives,

$$\left(\frac{q}{p}\right) + q\frac{C}{\tau_p} = (-1)^{\frac{p-1}{2}\cdot\frac{q-1}{2}} p^{\frac{q-1}{2}}. \tag{7.2}$$

This expression in particular shows that $m := qC/\tau_p$ is an integer. We claim that m is divisible by q. We have

$$m^2 = \frac{q^2 C^2}{\tau_p^2} = \pm\frac{q^2 C^2}{p}. \tag{7.3}$$

Since C is an algebraic integer, by Theorem B.4, C^2 is an algebraic integer. Equation 7.3 shows that C^2 is a rational number. Corollary B.3 shows that $C^2 \in \mathbb{Z}$.

Since $p \mid q^2 C^2$ and $(p, q^2) = 1$, Theorem 2.17 implies that $p \mid C^2$. Consequently, m^2 is divisible by q^2. This means m is divisible by q. Now that we know that qC/τ_p is an integer which is divisible by q, we reduce Equation (7.2) modulo q. We have by Lemma 6.5:

$$\left(\frac{q}{p}\right) \equiv (-1)^{\frac{p-1}{2}\cdot\frac{q-1}{2}} p^{\frac{q-1}{2}} \equiv (-1)^{\frac{p-1}{2}\cdot\frac{q-1}{2}}\left(\frac{p}{q}\right) \quad \mathrm{mod}\ q.$$

So we conclude that

$$\left(\frac{q}{p}\right) \equiv (-1)^{\frac{p-1}{2}\cdot\frac{q-1}{2}}\left(\frac{p}{q}\right) \quad \mathrm{mod}\ q.$$

Since the two sides of the equation are ± 1 and q is odd, an argument similar to the one in the proof of Lemma 6.6 gives

$$\left(\frac{q}{p}\right) = (-1)^{\frac{p-1}{2}\cdot\frac{q-1}{2}}\left(\frac{p}{q}\right),$$

as claimed.

We now proceed to prove the second part of Theorem 6.8. Set

$$\zeta = e^{\frac{\pi i}{4}},$$

an eighth root of unity. We have

$$\zeta^2 = e^{\frac{\pi i}{2}} = \cos\frac{\pi i}{2} + i\sin\frac{\pi i}{2} = i,$$

and

$$\zeta^{-2} = i^{-1} = -i.$$

Now set

$$\rho = \zeta + \zeta^{-1}.$$

We have

$$\rho^2 = (\zeta + \zeta^{-1})^2 = \zeta^2 + \zeta^{-2} + 2 = i + i^{-1} + 2 = 2.$$

Next,

$$\rho^p = (\rho^2)^{\frac{p-1}{2}} \cdot \rho = 2^{\frac{p-1}{2}} \cdot \rho.$$

On the other hand,

$$\rho^p = (\zeta + \zeta^{-1})^p = \zeta^p + \zeta^{-p} + \sum_{k=1}^{p-1} \binom{p}{k}\zeta^k\zeta^{-(p-k)}$$

$$= \zeta^p + \zeta^{-p} + \sum_{k=1}^{(p-1)/2} \binom{p}{k}(\zeta^k\zeta^{-(p-k)} + \zeta^{-k}\zeta^{p-k})$$

$$= \zeta^p + \zeta^{-p} + \sum_{k=1}^{(p-1)/2} \binom{p}{k}(\zeta^{2k-p} + \zeta^{p-2k}).$$

If $8 \mid k$, then $\zeta^k = 1$. For this reason, for an odd number l, the value of $\zeta^l + \zeta^{-l}$ depends only on the residue of l modulo 8. We only need to consider the residue classes $1, 3, 5, 7$:

- If $l \equiv 1 \bmod 8$, then $\zeta^l + \zeta^{-l} = \zeta + \zeta^{-1} = \rho$.
- If $l \equiv 3 \bmod 8$, then

$$\zeta^l = \zeta^3 = e^{\frac{3\pi i}{4}} = \cos\frac{3\pi}{4} + i\sin\frac{3\pi}{4} = -\cos\frac{\pi}{4} + \sin\frac{\pi}{4} = -\zeta^{-1},$$

and similarly, $\zeta^{-l} = -\zeta$. Hence,

$$\zeta^l + \zeta^{-l} = -\zeta^{-1} - \zeta = -\rho.$$

- If $l \equiv 5 \bmod 8$, then

$$\zeta^l = \zeta^5 = \cos\frac{5\pi}{4} + i\sin\frac{5\pi}{4} = -\cos\frac{\pi}{4} - i\sin\frac{\pi}{4} = -\zeta,$$

and also $\zeta^{-l} = -\zeta$. This means that in this case

$$\zeta^l + \zeta^{-l} = -\zeta - \zeta^{-1} = -\rho.$$

- If $l \equiv 7 \bmod 8$, then $\zeta^l = \zeta^{-1}$, $\zeta^{-l} = \zeta$, and $\zeta^l + \zeta^{-l} = \zeta^{-1} + \zeta = \rho$.

These computations mean

$$\zeta^p + \zeta^{-p} = (-1)^{\frac{p^2-1}{8}} \rho,$$

and for all $1 \le k \le p - 1$,

$$\zeta^{2k-p} + \zeta^{p-2k} = (-1)^{\frac{(p-2k)^2-1}{8}} \rho.$$

Consequently,

$$2^{\frac{p-1}{2}} \cdot \rho = (-1)^{\frac{p^2-1}{8}} \rho + \sum_{k=1}^{(p-1)/2} \binom{p}{k}(-1)^{\frac{(p-2k)^2-1}{8}} \rho.$$

Dividing by ρ gives

$$2^{\frac{p-1}{2}} = (-1)^{\frac{p^2-1}{8}} + \sum_{k=1}^{(p-1)/2} \binom{p}{k}(-1)^{\frac{(p-2k)^2-1}{8}}. \tag{7.4}$$

By Lemma 2.27, the binomial coefficient $\binom{p}{k}$ for $1 \le k \le p - 1$ is divisible by p. Reduce Equation (7.4) modulo p to get

$$2^{\frac{p-1}{2}} \equiv (-1)^{\frac{p^2-1}{8}} \mod p.$$

Lemma 6.5 now gives the result. $\square$

7.2 The Jacobi Symbol

In this section we introduce the *Jacobi symbol* which is a generalization of the Legendre symbol.

Definition 7.2. Let b be an odd positive integer, and let a be an integer. We define the Jacobi symbol $\left(\frac{a}{b}\right)$ as follows. If $b = p_1 \dots p_k$, with p_i's not necessarily distinct, we set

$$\left(\frac{a}{b}\right) = \prod_{i=1}^{k} \left(\frac{a}{p_i}\right)$$

For example,

$$\left(\frac{2}{15}\right) = \left(\frac{2}{3}\right)\left(\frac{2}{5}\right).$$

In the case where b is an odd prime number, the Jacobi symbol is identical with the Legendre symbol. There is an important difference between the Legendre symbol and the Jacobi symbol, however. The Legendre symbol (a/p) for a prime number

p determines the solvability of the congruence equation $X^2 \equiv a \bmod p$. In general, the Jacobi symbol (a/b) gives no information about the solvability of the equation $X^2 \equiv a \bmod b$. Suppose, for example, $b = p^2$, with p a prime number. If $X^2 \equiv a \bmod b$ is solvable, then, since $p \mid b$, so is $X^2 \equiv a \bmod p$. So if $(a/p) = -1$, the equation $X^2 \equiv a \bmod b$ will not be solvable. However, $(a/b) = (a/p^2) = (a/p)^2 = (\pm 1)^2 = +1$. The simplest example of this is when $a = -1$ and $b = 9 = 3^2$. In this case, $(-1/9) = (-1/3)^2 = (-1)^2 = +1$, but the equation $X^2 \equiv -1 \bmod 9$ is not solvable. Despite this issue the Jacobi symbol is a useful tool that allows to compute the Legendre symbol without having to factorize integers. We will see some examples at the end of this section.

We have the following theorem:

Theorem 7.3 (Quadratic Reciprocity for the Jacobi symbol).

1. If m is an odd natural number,

$$\left(\frac{-1}{m}\right) = (-1)^{\frac{m-1}{2}}.$$

2. If m is an odd natural number,

$$\left(\frac{2}{m}\right) = (-1)^{\frac{m^2-1}{8}}.$$

3. For odd natural numbers m and n,

$$\left(\frac{m}{n}\right) = \left(\frac{n}{m}\right)(-1)^{\frac{m-1}{2}\cdot\frac{n-1}{2}}.$$

Before we start the proof of the theorem, we note that the above theorem is a generalization of Theorem 6.8.

We start the proof of the theorem with a lemma:

Lemma 7.4. *For an odd natural number q and a natural number α the following identities hold:*

1. $\frac{q^\alpha - 1}{2} \equiv \frac{\alpha(q-1)}{2} \bmod 2$;
2. $\frac{q^{2\alpha} - 1}{8} \equiv \frac{\alpha(q^2-1)}{8} \bmod 2$.

Proof. Proof is by induction. Clearly both identities are true for $\alpha = 1$. So assume that the identities are true for α, and we wish to show their validity for $\alpha + 1$.

We have

$$\frac{q^{\alpha+1} - 1}{2} = \frac{q^{\alpha+1} - q^\alpha + q^\alpha - 1}{2} = q^\alpha \left(\frac{q-1}{2}\right) + \frac{q^\alpha - 1}{2}.$$

This last expression, by the induction assumption, is congruent to

$$\frac{q-1}{2} + \frac{\alpha(q-1)}{2} \equiv \frac{(\alpha+1)(q-1)}{2} \quad \text{mod 2}.$$

The second identity is proved in a completely similar way.

We can now prove the theorem.

Proof of Theorem 7.3. To prove the first part we do induction on the number of distinct prime divisors of m. Write $m = p_1^{\alpha_1} \cdots p_r^{\alpha_r}$, and we do induction on r. We need to prove

$$\frac{p_1^{\alpha_1} \cdots p_r^{\alpha_r} - 1}{2} \equiv \alpha_1 \frac{p_1 - 1}{2} + \cdots + \alpha_r \frac{p_r - 1}{2} \quad \text{mod 2}.$$

The $r = 1$ case is the first part of Lemma 7.4. Now suppose the identity is valid for r, and we wish to prove it for $r + 1$. We have

$$\frac{p_1^{\alpha_1} \cdots p_r^{\alpha_r} p_{r+1}^{\alpha_{r+1}} - 1}{2} = \frac{p_1^{\alpha_1} \cdots p_r^{\alpha_r} p_{r+1}^{\alpha_{r+1}} - p_1^{\alpha_1} \cdots p_r^{\alpha_r} + p_1^{\alpha_1} \cdots p_r^{\alpha_r} - 1}{2}$$

$$= p_1^{\alpha_1} \cdots p_r^{\alpha_r} \frac{p_{r+1}^{\alpha_{r+1}} - 1}{2} + \frac{p_1^{\alpha_1} \cdots p_r^{\alpha_r} - 1}{2}$$

$$\equiv \frac{p_{r+1}^{\alpha_{r+1}} - 1}{2} + \frac{p_1^{\alpha_1} \cdots p_r^{\alpha_r} - 1}{2} \quad \text{mod 2} \quad (\text{as } p_1^{\alpha_1} \cdots p_r^{\alpha_r} \text{ is odd.})$$

$$\equiv \alpha_1 \frac{p_1 - 1}{2} + \cdots + \alpha_r \frac{p_r - 1}{2} + \alpha_{r+1} \frac{p_{r+1} - 1}{2} \quad \text{mod 2}$$

after using the first part of Lemma 7.4 and the induction hypothesis.

The proof of the second part of the theorem is completely similar to our proof of the first part, except that here we need to use the second part of Lemma 7.4 and the computation of $(2/p)$ for an odd prime p from Theorem 6.8.

We now prove the last part of the theorem. Let $m = p_1^{\alpha_1} \cdots p_r^{\alpha_r}$ and $n = q_1^{\beta_1} \cdots q_s^{\beta_s}$ be the prime factorizations of m and n. If m and n are not coprime, both sides of the identity are equal to zero, and there is nothing to prove. So we assume that the p_i's and q_j's are distinct primes. By definition,

$$\left(\frac{m}{n}\right) = \prod_{i=1}^{r} \prod_{j=1}^{s} \left(\frac{p_i}{q_j}\right)^{\alpha_i \beta_j}$$

$$= \prod_{i=1}^{r} \prod_{j=1}^{s} \left(\frac{q_j}{p_i}\right)^{\alpha_i \beta_j} (-1)^{\alpha_i \beta_j \frac{p_i-1}{2} \cdot \frac{q_j-1}{2}} \quad \text{(by Theorem 6.8)}$$

$$= \left(\frac{n}{m}\right) \prod_{i=1}^{r} \prod_{j=1}^{s} (-1)^{\alpha_i \beta_j \frac{p_i-1}{2} \cdot \frac{q_j-1}{2}}.$$

So to prove the third part we just need to show that

$$\frac{m-1}{2} \cdot \frac{n-1}{2} \equiv \sum_{i=1}^{r} \sum_{j=1}^{s} \alpha_i \beta_j \frac{p_i-1}{2} \cdot \frac{q_j-1}{2} \quad \text{mod } 2.$$

To see this, we note that by the proof of the first part of the Theorem,

$$\frac{m-1}{2} \equiv \sum_{i=1}^{r} \alpha_i \frac{p_i-1}{2} \quad \text{mod } 2,$$

and

$$\frac{n-1}{2} \equiv \sum_{j=1}^{s} \beta_j \frac{q_j-1}{2} \quad \text{mod } 2.$$

Multiplying these identities gives the result. □

Now, we will use the Jacobi symbol to compute some Legendre symbols. Let's start with a small example. Suppose we want to compute

$$\left(\frac{37}{89}\right).$$

Since both 37 and 89 are odd primes we can use the Law of Quadratic Reciprocity, Theorem 6.8, to obtain

$$\left(\frac{37}{89}\right) = (-1)^{(37-1)(89-1)/4} \left(\frac{89}{37}\right) = \left(\frac{89}{37}\right).$$

Since $89 \equiv 15 \bmod 37$, the latter quadratic residue symbol is equal to

$$\left(\frac{15}{37}\right).$$

If we were to use the methods of Chapter 6 at this point we would use the fact that $15 = 3 \times 5$ to write $(15/37) = (3/37) \cdot (5/37)$, and then we would apply the Law of Quadratic Reciprocity twice to compute these latter quadratic residue symbols. The problem with this approach is that it requires factorizing 15, and this is something we can do because 15 is a small number. As mentioned in Notes to Chapters 2 and 6, at present we do not know how to factorize a very large natural number in reasonable time. Using the Jacobi symbol allows us to bypass this obstacle. In fact, by Theorem 7.3 we have

$$\left(\frac{15}{37}\right) = (-1)^{(15-1)(37-1)/4} \left(\frac{37}{15}\right) = \left(\frac{37}{15}\right) = \left(\frac{7}{15}\right),$$

as $37 \equiv 7 \bmod 15$. Applying Theorem 7.3 to $(7/15)$ gives

$$\left(\frac{7}{15}\right) = (-1)^{(7-1)(15-1)/2}\left(\frac{15}{7}\right) = -\left(\frac{15}{7}\right) = -\left(\frac{1}{7}\right) = -1,$$

after using $15 \equiv 1 \bmod 7$. Putting everything together, we obtain

$$\left(\frac{37}{89}\right) = -1.$$

Let us now examine an example involving larger numbers. We wish to compute the Legendre symbol

$$\left(\frac{2455927}{36838897}\right).$$

By Theorem 7.3 we have

$$\left(\frac{2455927}{36838897}\right) = (-1)^{(2455927-1)(36838897-1)/4}\left(\frac{36838897}{2455927}\right)$$

$$= \left(\frac{36838897}{2455927}\right) = \left(\frac{2455919}{2455927}\right),$$

as $36838897 \equiv 2455919 \bmod 2455927$. Again using Theorem 7.3 gives

$$\left(\frac{2455919}{2455927}\right) = (-1)^{(2455919-1)(2455927-1)/4}\left(\frac{2455927}{2455919}\right)$$

$$= \left(\frac{2455927}{2455919}\right) = \left(\frac{8}{2455919}\right) = \left(\frac{2}{2455919}\right)^3.$$

Here we have used the fact that $2455927 \equiv 8 \equiv 2^3 \bmod 2455919$, and also the multiplicativity of the Jacobi symbol. Since $(\pm 1)^3 = \pm 1$, the latter Jacobi symbol is equal to $(2/2455919)$. So we have established that

$$\left(\frac{2455927}{36838897}\right) = \left(\frac{2}{2455919}\right).$$

To finish the computation we use the second part of Theorem 7.3 to get

$$\left(\frac{2}{2455919}\right) = (-1)^{(2455919^2-1)/8} = +1.$$

We have proved

$$\left(\frac{2455927}{36838897}\right) = +1.$$

The important point to note here is that we did not have to worry the primality of the numbers that showed up in the computation. In fact, $2455919 = 6841 \times 359$ is not prime.

Exercises

7.1 Compute τ_p for $p = 3, 5$, and verify Lemma 7.1 directly.

7.2 (✠) Compute τ_p for $p = 17$.

7.3 Show that the complex number C defined in Equation (7.1) is an algebraic integer.

7.4 Prove the second part of Lemma 7.4.

7.5 Prove the second part of Theorem 7.3.

7.6 Determine all natural numbers n such that $(n/15) = +1$.

7.7 Determine $(215/997)$ and $(113/1093)$ using the Jacobi symbol.

7.8 Find five pairs of integers (a, b) such that the Jacobi symbol $(a/b) = +1$ but $x^2 \equiv a \bmod b$ is not solvable.

7.9 Show that for all $n > 1$ we have the following identities for Jacobi symbols

$$\left(\frac{n}{4n-1}\right) = -\left(\frac{-n}{4n-1}\right) = 1.$$

7.10 Show that for an integer d with $|d| > 1$ we have

$$\left(\frac{d}{|d|-1}\right) = \begin{cases} 1 & d > 0; \\ -1 & d < 0. \end{cases}$$

7.11 Let $k \in \mathbb{N}$, and let $\gcd(d, k) = 1$. Prove that the number of solutions of $x^2 \equiv d \bmod 4k$ is

$$2 \sum_{\substack{f|k \\ f \text{ squarefree}}} \left(\frac{d}{f}\right).$$

7.12 Show that for an odd prime p, and $a \in \mathbb{N}$ with $p \nmid a$, we have

$$\left(\frac{a}{p}\right) = \left(\frac{a}{p-4a}\right).$$

7.13 This exercise gives another proof of the Law of Quadratic Reciprocity due to Rousseau [94]. The proof uses a bit of group theory. Let p, q be odd primes, and define $G = (\mathbb{Z}/pq\mathbb{Z})^\times/\{\pm1\}$.

a. Show that the set

$$S = \left\{ (x, y) \mid 1 \le x \le p-1, 1 \le y \le \frac{q-1}{2} \right\}$$

is a set of representatives for G. What is the product of elements of S modulo $\{\pm1\}$?

b. Show that the set

$$S' = \left\{ (z \bmod p, z \bmod q) \mid 1 \le z \le \frac{pq-1}{2} \right\}$$

is another set of representatives of G. Determine the product of elements of S' modulo $\{\pm 1\}$.

c. Derive the Law of Quadratic Reciprocity from the first two parts.

Notes

Proofs of quadratic reciprocity

As mentioned in Chapter 6, the Law of Quadratic Reciprocity was conjectured by Euler around 1745, in a paper titled "Theoremata circa divisores numerorum in hac forma $pa^2 \pm qb^2$ contentorum" available from the *Euler Archive* at

http://eulerarchive.maa.org/index.html

though here the conjecture is not explicitly stated as such. The explicit formulation of the conjecture appears in a later paper of Euler's, titled "Observationes circa divisionem quadratorum per numeros primos" available at

http://eulerarchive.maa.org/pages/E552.html

Gauss noted in his notebook that he had found a proof on April 8, 1796. So far over 200 proofs of the Law of Quadratic Reciprocity have been obtained by various mathematicians. Franz Lemmermeyer, the author of [32], maintains a website that keeps track of the various proofs of theorem. The website is available at

http://www.rzuser.uni-heidelberg.de/~hb3/fchrono.html

Generalizations

One can generalize the Law of Quadratic Reciprocity in two different directions, one is by considering higher powers, and the other by considering other number fields, introduced in the Notes to Chapter 5. For introductions to reciprocity laws for higher powers we refer the reader to Lemmermeyer [32] or Cox [14], especially §4. For the generalization of Quadratic Reciprocity to other number fields, known as *Hilbert's Law of Reciprocity*, see the Notes to Chapter 8.

Part II
Advanced Topics

Chapter 8
Counting Pythagorean triples modulo an integer

In this chapter we study the Pythagorean Equation in integers modulo a natural number n and count the number of solutions. In the first section we consider the case where n is a prime number. Later in the chapter we discuss the general case. By using the Chinese Remainder Theorem we show that in order to count the number of solutions modulo a natural number n, it suffices to count the number of solutions modulo prime power divisors of n. We then devise a recursive process to count the number of solutions modulo prime powers. At the end of the chapter we show how the recursive process introduced earlier can be used to find solutions of equations such as $x^2 \equiv 2 \bmod 7^k$ for any natural number k. We will, for example, show that for each k, this equation has precisely two solutions modulo 7^k. The strategy used here is what is usually called Hensel's Lemma. We explore this lemma in Exercises 8.4 and 8.5. In the Notes, we discuss p-adic numbers. We finish with the statement of Hilbert's Law of Reciprocity which is a massive generalization of the Gauss's Law of Quadratic Reciprocity.

8.1 The Pythagorean Equation modulo a prime number p

One interesting feature of the geometric method discussed in §3.2 and explored further in §3.3 is that one does not need to do the geometric constructions presented there just in the real plane. One can repeat the same constructions over every field, provided that one cares that the denominators of the fractions that appear are not zero. This is not an issue over the real numbers, and obviously over the rationals, as for m a real number, $m^2 + 1$ is never zero. But as soon as we start working over the complex numbers, there are in fact choices for m that make $m^2 + 1$ equal to zero. The same problem occurs when considering the Pythagorean Equation modulo a prime number p.

We start by determining $N_p := \#S_p$ with

© Springer Nature Switzerland AG 2018
R. Takloo-Bighash, *A Pythagorean Introduction to Number Theory*,
Undergraduate Texts in Mathematics, https://doi.org/10.1007/978-3-030-02604-2_8

$$S_p = \{(x, y) \mid 1 \le x, y \le p, \; x^2 + y^2 \equiv 1 \bmod p\}.$$

Let's examine a few small primes. We have

$$S_3 = \{(0, 1), (0, 2), (1, 0), (2, 0)\}, \quad N_3 = 4 = 3 + 1;$$

$$S_5 = \{(0, 1), (0, 4), (1, 0), (4, 0)\}, \quad N_5 = 4 = 5 - 1;$$

$$S_7 = \{(0, 1), (0, 6), (1, 0), (6, 0), (2, 2), (2, 5), (5, 2), (5, 5)\}, \quad N_7 = 8 = 7 + 1;$$

$$S_{11} = \{(0, 1), (0, 10), (1, 0), (10, 0), (3, 6), (3, 5), (8, 6), (8, 5),$$

$$(6, 3), (5, 3), (6, 8), (5, 8)\}, \quad N_{11} = 12 = 11 + 1;$$

$$S_{13} = \{(0, 1), (0, 12), (1, 0), (12, 0), (2, 6), (2, 11), (11, 6), (11, 7),$$

$$(6, 2), (11, 2), (6, 11), (7, 11)\}, \quad N_{13} = 12 = 13 - 1.$$

In these examples, $N_p = p - a(p)$ with $a(p) = +1$ whenever p is of the form $4k + 1$, for $p = 5, 13$, and $a(p) = -1$ when p is of the form $4k + 3$, for $p = 3, 7, 11$. Equation 6.3 shows that at least for these primes $a(p) = (-1/p)$. So a reasonable guess is

$$N_p = p - \left(\frac{-1}{p} \right).$$

We will show that this is indeed the case. In order to prove our guess we first find a parametrization for all the solutions of the equation $x^2 + y^2 \equiv 1 \bmod p$. There are several obvious solutions, e.g., $(-1, 0)$ as in the real case. Fix a residue class $m \bmod p$. We consider the intersection of the "line" of "slope" m passing through $(-1, 0)$, i.e., the collection of pairs (x, y) with $1 \le x, y \le p$ such that

$$y \equiv m(x + 1) \quad \bmod p,$$

with the "circle" $x^2 + y^2 \equiv 1 \bmod p$. As before, we obtain

$$m^2(x + 1)^2 + x^2 \equiv 1 \quad \bmod p,$$

or

$$(m^2 + 1)x^2 + 2m^2 x + (m^2 - 1) \equiv 0 \quad \bmod p.$$

If $m^2 + 1 \not\equiv 0 \bmod p$, then it will be invertible, and we obtain the equation

$$x^2 + 2(m^2 + 1)^{-1} m^2 + (m^2 + 1)^{-1}(m^2 - 1) \equiv 0 \quad \bmod p.$$

By construction, $x \equiv -1 \bmod p$ is one of the solutions of this equation. There is a second solution,

$$x \equiv (m^2 + 1)^{-1}(1 - m^2) \quad \bmod p.$$

By using the equation of the "line" we obtain y as

$$y \equiv (m^2 + 1)^{-1} 2m \quad \bmod p.$$

Consequently, the set of solutions of the equation $x^2 + y^2 \equiv 1 \bmod p$ aside from the pair $(-1, 0)$ coincides with the collection of pairs

$$((m^2 + 1)^{-1}(1 - m^2) \bmod p, \ (m^2 + 1)^{-1} 2m \bmod p)$$

for $1 \leq m \leq p$ subject to $m^2 + 1 \not\equiv 0 \bmod p$. If $p \equiv 3 \bmod 4$, there is no m with $p \mid m^2 + 1$. For $p \equiv 1 \bmod 4$, there are two values of m that need to be excluded. If $p = 2$, it is clear that $m = 1$ needs to be omitted. We should also not forget our seed point $(-1, 0)$. These observations mean:

$$N_p = \begin{cases} p + 1 & p \equiv 3 \bmod 4; \\ p - 1 & p \equiv 1 \bmod 4; \\ 2 & p = 2. \end{cases}$$

For p odd this formula can be written alternatively as

$$N_p = p - \left(\frac{-1}{p}\right), \tag{8.1}$$

confirming our observations.

We can also count the number of solutions of the three-variable Pythagorean equation in numbers modulo p. Set

$$N(p) = \#\{(x, y, z) \mid 1 \leq x, y, z \leq p, x^2 + y^2 \equiv z^2 \bmod p\}.$$

The quantity $N(p)$ can easily be computed knowing N_p.

First we account for solutions of $x^2 + y^2 \equiv z^2 \bmod p$ where $z \not\equiv 0 \bmod p$. For every (a, b) satisfying $a^2 + b^2 \equiv 1 \bmod p$, we have $p - 1$ solutions to $x^2 + y^2 \equiv z^2 \bmod p$, namely, the triples

$$(ac, bc, c)$$

for $1 \leq c \leq p - 1$.

Now we count the number of pairs (x, y) with $1 \leq x, y \leq p$ with $x^2 + y^2 \equiv 0 \bmod p$. If $p \equiv 3 \bmod 4$, by Lemma 5.6, there is a unique pair (p, p) that satisfies the equation. If $p \equiv 1 \bmod 4$, then we certainly have the solution (p, p), but we also have solutions (x, y) with x, y not divisible by p. In fact, there are numbers u, v such that $u \not\equiv v \bmod p$ but $u^2 + 1 \equiv v^2 + 1 \equiv 0 \bmod p$. Then we have $2(p - 1)$ additional solutions to the Pythagorean Equation:

$$(x, xu, 0), \quad 1 \leq x \leq p - 1,$$

and

$$(x, xv, 0), \quad 1 \leq x \leq p - 1.$$

This means,

$$N(p) = \begin{cases} (p-1)N_p + 1 & p \equiv 3 \bmod 4; \\ (p-1)N_p + 2(p-1) + 1 & p \equiv 1 \bmod 4. \end{cases}$$

Consequently,

$$N(p) = (p-1)N_p + \left(1 + \left(\frac{-1}{p}\right)\right)(p-1) + 1$$

$$= (p-1)\left(p - \left(\frac{-1}{p}\right)\right) + \left(1 + \left(\frac{-1}{p}\right)\right)(p-1) + 1 = p^2.$$

Let us collect these findings as a proposition:

Proposition 8.1. *If p is a prime number, then*

$$N_p = \begin{cases} p - \left(\frac{-1}{p}\right) & p \ odd; \\ 2 & p = 2, \end{cases}$$

and

$$N(p) = p^2.$$

We also present an alternative evaluation of $N(p)$ using Gauss sums for p odd. For x, y, whether there is a z, $z \not\equiv 0 \bmod p$, such that $x^2 + y^2 \equiv z^2 \bmod p$ is determined by $(x^2 + y^2/p)$. If there is a z, there will be exactly two of them. If on the other hand $x^2 + y^2 \equiv 0 \bmod p$, then there is a unique z, i.e., $z = p$. Hence the total number of solutions is

$$N(p) = \sum_{x,y=1}^{p} \left(1 + \left(\frac{x^2 + y^2}{p}\right)\right) = p^2 + \sum_{x,y=1}^{p} \left(\frac{x^2 + y^2}{p}\right).$$

In order to evaluate the sum, we introduce a variation of the Gauss sum introduced in Chapter 7. Recall the definition of the Gauss sum. We set $\zeta = e^{\frac{2\pi i}{p}}$ and define the pth Gauss sum by

$$\tau_p = \sum_{k=1}^{p-1} \left(\frac{k}{p}\right) \zeta^k.$$

For $1 \le a \le p$, we set

$$\tau_p(a) = \sum_{k=1}^{p-1} \left(\frac{k}{p}\right) \zeta^{ak}$$

Lemma 8.2. *If p is prime, then*

$$\tau_p(a) = \left(\frac{a}{p}\right) \tau_p.$$

Proof. In fact, for $1 \leq a \leq p - 1$ the identity follows from a change of variables in k. When $a = p$ the identity is equivalent to the statement that

$$\sum_{k=1}^{p-1} \left(\frac{k}{p}\right) = 0.$$

We verified this last identity in the proof of Lemma 7.1. $\square$

By the lemma,

$$\sum_{x,y=1}^{p} \left(\frac{x^2 + y^2}{p}\right) = \frac{1}{\tau_p} \sum_{x,y=1}^{p} \tau_p(x^2 + y^2) = \frac{1}{\tau_p} \sum_{x,y=1}^{p} \sum_{k=1}^{p-1} \left(\frac{k}{p}\right) \zeta^{k(x^2+y^2)}$$

$$= \frac{1}{\tau_p} \sum_{k=1}^{p-1} \left(\frac{k}{p}\right) \sum_{x,y=1}^{p} \zeta^{k(x^2+y^2)} = \frac{1}{\tau_p} \sum_{k=1}^{p-1} \left(\frac{k}{p}\right) \left(\sum_{x=1}^{p} \zeta^{kx^2}\right)^2$$

If we write the inner sum $\sum_{1 \leq x \leq p} \zeta^{kx^2}$ as $\sum_{1 \leq t \leq p} a_t \zeta^{kt}$ then we see that

$$a_t = \begin{cases} 2 & (t/p) = 1; \\ 1 & (t/p) = 0; \\ 0 & (t/p) = -1. \end{cases}$$

Consequently,

$$\sum_{1 \leq x \leq p} \zeta^{kx^2} = \sum_{1 \leq t \leq p} \left(1 + \left(\frac{t}{p}\right)\right) \zeta^{kt} = \sum_{1 \leq t \leq p} \zeta^{kt} + \sum_{1 \leq t \leq p} \left(\frac{t}{p}\right) \zeta^{kt}$$

$$= \tau_p(k) = \left(\frac{k}{p}\right) \tau_p,$$

as for $1 \leq k \leq p - 1$

$$\sum_{1 \leq t \leq p} \zeta^{kt} = 0.$$

In particular, for $1 \leq k \leq p - 1$,

$$\left(\sum_{1 \leq x \leq p} \zeta^{kx^2}\right)^2 = \tau_p^2.$$

This means

$$\sum_{x,y=1}^{p} \left(\frac{x^2 + y^2}{p}\right) = \frac{1}{\tau_p} \sum_{k=1}^{p-1} \left(\frac{k}{p}\right) \tau_p^2 = \tau_p \sum_{k=1}^{p-1} \left(\frac{k}{p}\right) = 0,$$

by the computation in the proof of Lemma 7.1. Again, we obtain

$$N(p) = p^2.$$

The Gauss sum method described here is applicable to far more general equations than just the Pythagorean Equation. See, for example, [8, Theorem 3, Ch. 1] and [108].

8.2 Solutions modulo n for a natural number n

In this section we discuss the solutions of the Pythagorean Equation modulo a number n which is not necessarily prime. For a natural number n we set $N_n = \#S_n$ with

$$S_n = \{(x, y) \mid 1 \le x, y \le n, x^2 + y^2 \equiv 1 \bmod n\}.$$

Lemma 8.3. *The function N_n is multiplicative in n, i.e., if $\gcd(m, n) = 1$, then*

$$N_{nm} = N_n \cdot N_m.$$

Proof. We will show there is a bijection $S_{nm} \to S_n \times S_m$. This would then mean $\#S_{nm} = \#S_n \cdot \#S_m$, and that's what we are trying to prove. In order to show the existence of the bijection, we need some preparation. For $n \in \mathbb{N}$, we set

$$A_n = \{1, 2, \ldots, n\}.$$

We also let $A_n^2 = A_n \times A_n$. Observe that for each n, $S_n \subset A_n^2$.

Suppose $n \in \mathbb{N}$ and $d \mid n$. We construct a map

$$\rho_{n/d} : A_n \to A_d,$$

by defining $\rho_{n/d}(x)$, for $1 \le x \le n$, to be the unique $1 \le y \le d$ such that $x \equiv y \bmod d$. We also define a map $\rho_{n/d}^2 : A_n^2 \to A_d^2$ by defining $\rho_{n/d}^2(x, y) = (\rho_{n/d}(x), \rho_{n/d}(y))$ for $x, y \in A_n$.

We start with the observation that if $\gcd(m, n) = 1$, then the map

$$\rho_{n,m} : A_{nm} \to A_n \times A_m$$

defined by

$$\rho_{n,m}(x) = (\rho_{nm/n}(x), \rho_{nm/m}(x))$$

is a bijection. In fact, by the Chinese Remainder Theorem 2.24, if $(y_1, y_2) \in \{1, 2, \ldots, n\} \times \{1, 2, \ldots, m\}$, then there is a unique $1 \le x \le nm$ such that $\rho_{mn/n}(x) = y_1, \rho_{nm/m}(x) = y_2$. Clearly, $\rho_{n,m}(x) = (y_1, y_2)$. The fact that x exists and is unique means that $\rho_{n,m}$ is a bijection.

We can also define a two variable version of $\rho_{n,m}$. We define a map

$$\rho^2_{n,m} : A^2_{nm} \to A^2_n \times A^2_m$$

by defining

$$\rho^2_{n,m}(x, y) = (\rho^2_{nm/n}(x, y), \rho^2_{nm/m}(x, y)),$$

for $(x, y) \in A^2_{nm}$. The map $\rho^2_{n,m}$, too, is a bijection provided that $\gcd(n, m) = 1$.

Now consider the set $\rho^2_{n,m}(S_{nm}) \subset A^2_n \times A^2_m$ for $\gcd(n, m) = 1$. Since $\rho^2_{n,m}$ is a bijection, $\rho^2_{n,m}(S_{nm})$ is in bijection with S_{nm}, and consequently,

$$\#\rho^2_{n,m}(S_{nm}) = \#S_{nm}. \tag{8.2}$$

We claim

$$\rho^2_{n,m}(S_{nm}) = S_n \times S_m. \tag{8.3}$$

Once we establish Equation (8.3) we obtain

$$\#\rho^2_{n,m}(S_{nm}) = \#S_n \cdot \#S_m.$$

Comparing this last statement with Equation (8.2) gives the result.

In order to prove Equation (8.3), as $\rho^2_{n,m}$ is a bijection, it suffices to prove

$$(\rho^2_{n,m})^{-1}(S_n \times S_m) = S_{nm}.$$

We start with a general fact whose proof we leave as an exercise to the reader. Suppose we have we sets X, Y and a map $f : X \to Y$. Also let $A \subset X, B \subset Y$. Then $f^{-1}(B) = A$ if the following statement holds: $x \in A$ if and only if $f(x) \in B$. Because of this general statement we need to prove that $(x, y) \in S_{nm}$ if and only if $\rho^2_{n,m}(x, y) \in S_n \times S_m$. In concrete terms this means that for integers x, y, if $\gcd(n, m) = 1$, $x^2 + y^2 \equiv 1$ mod nm if and only if $x^2 + y^2 \equiv 1$ mod n and $x^2 + y^2 \equiv 1$ mod m. This last statement is completely obvious, and we are done.

$\square$

The lemma implies that in order to determine N_n for all n, we just need to determine N_{p^α} for primes p, because then if $n = p_1^{\alpha_1} \cdots p_k^{\alpha_k}$, we have

$$N_n = N_{p_1^{\alpha_1}} \cdots N_{p_k^{\alpha_k}}.$$

So we proceed to determine N_{p^α}. As a test case let's start with N_{p^2} for an odd prime p. The key observation is that if $x^2 + y^2 \equiv 1$ mod p^2, then $x^2 + y^2 \equiv 1$ mod p. This determines a map

$$\eta_{p^2/p} : S_{p^2} \to S_p,$$

which is simply reduction modulo p.

We now fix an element $(x_0, y_0) \in S_p$, and study the set of $(x, y) \in S_{p^2}$ that reduce to (x_0, y_0), i.e., $\eta^{-1}_{p^2/p}(x_0, y_0) \subset S_{p^2}$. Every such pair (x, y) would have to be of the form

$$(x_0 + kp, y_0 + lp)$$

for some k, l mod p. We need to have

$$(x_0 + pk)^2 + (y_0 + pk)^2 \equiv 1 \quad \text{mod } p^2.$$

Squaring gives

$$(x_0^2 + y_0^2 - 1) + 2p(kx_0 + ly_0) + p^2(k^2 + l^2) \equiv 0 \quad \mathrm{mod}\ p^2,$$

or

$$(x_0^2 + y_0^2 - 1) + 2p(kx_0 + ly_0) \equiv 0 \quad \mathrm{mod}\ p^2.$$

Since $x_0^2 + y_0^2 \equiv 1 \bmod p$, $x_0^2 + y_0^2 - 1$ is divisible by p. Dividing by p gives

$$\frac{x_0^2 + y_0^2 - 1}{p} + 2(kx_0 + ly_0) \equiv 0 \quad \mathrm{mod}\ p.$$

Since $p \neq 2$, 2 will have a multiplicative inverse $2^{-1} \bmod p$. Then this last equation says

$$kx_0 + ly_0 \equiv -2^{-1} \frac{x_0^2 + y_0^2 - 1}{p} \quad \mathrm{mod}\ p.$$

Since $(x_0, y_0) \neq (0, 0)$, there are p choices for (k, l) that satisfy this congruence. Consequently, we see that if $p \neq 2$, then

$$S_{p^2} = pS_p.$$

Indeed this is typical:

Lemma 8.4. *If* $p \neq 2$, *for each* $n \geq 1$,

$$N_{p^{n+1}} = pN_{p^n}.$$

In particular, for $n \geq 1$

$$N_{p^n} = p^n - \left(\frac{-1}{p}\right) p^{n-1}.$$

Proof. We define a map

$$S_{p^{n+1}} \to S_{p^n}$$

by reduction modulo p^n. We will see in a moment that this map is surjective. Let $(x_0, y_0) \in S_{p^n}$. We determine all (x, y) that reduce to (x_0, y_0). Every such pair (x, y) will be of the form

$$(x_0 + kp^n, y_0 + lp^n)$$

for some x, y modulo p. Then $x^2 + y^2 \equiv 1 \bmod p^{n+1}$ is equivalent to saying

$$(x_0 + kp^n)^2 + (y_0 + kp^n)^2 \equiv 1 \quad \mathrm{mod}\ p^{n+1},$$

or

$$(x_0^2 + y_0^2 - 1) + 2p^n(kx_0 + ly_0) + p^{2n}(k^2 + l^2) \equiv 0 \quad \mathrm{mod}\ p^{n+1}.$$

Since $2n \geq n + 1$, this last equation is equivalent to

$$(x_0^2 + y_0^2 - 1) + 2p^n(kx_0 + ly_0) \equiv 0 \quad \mathrm{mod}\ p^{n+1}. \tag{8.4}$$

Dividing by p^n gives

$$kx_0 + ly_0 \equiv -2^{-1}\frac{x_0^2 + y_0^2 - 1}{p^n} \quad \text{mod } p.$$

This equation has p solutions in k, l mod p, and we are done. □

For $p = 2$ the situation is more complicated. For example, the map $S_{2^{n+1}} \to S_{2^n}$ is in general not surjective, i.e., there may be $(x_0, y_0) \in S_{2^n}$ for which there is no $(x, y) \in S_{2^{n+1}}$ satisfying $x \equiv x_0$ mod 2^n and $y \equiv y_0$ mod 2^n. To see this in a concrete situation, let $n = 2$. Then a quick search gives

$$S_4 = \{(4, 1), (4, 3), (2, 1), (2, 3), (1, 4), (3, 4), (1, 2), (3, 2)\}.$$

Similarly we have

$$S_8 = \{(4, 1), (4, 3), (4, 5), (4, 7), (8, 1), (8, 3), (8, 5), (8, 7), (1, 4),$$

$$(3, 4), (5, 4), (7, 4), (1, 8), (3, 8), (5, 8), (7, 8)\}.$$

The image of the reduction modulo 4 map from S_8 to S_4 is

$$\{(4, 1), (4, 3), (1, 4), (3, 4)\},$$

which is visibly not all of S_4.

This in particular means that Lemma 8.4 as written is not valid for $p = 2$. One might of course try to trace the steps of the proof of Lemma 8.4 to see if any of it can be salvaged for $p = 2$. The main issue with the proof of the lemma is that in Equation (8.4) the term $2p^n(kx_0 + ly_0)$ vanishes modulo 2^{n+1} if $p = 2$, so unless $x_0^2 + y_0^2 - 1$ is already divisible by 2^{n+1}, one gets nothing. However, the key to the proof of Lemma 8.4 is that the term $2p^n(kx_0 + ly_0)$ is divisible by p^n and not p^{n+1}. In order to adapt this argument to $p = 2$, we make one small adjustment:

Lemma 8.5. *The following identity holds:*

$$N_{2^n} = \begin{cases} 2 & n = 1; \\ 2^{n+1} & n \geq 2. \end{cases}$$

Proof. That $N_2 = 2$ is obvious. We already determined N_4 and N_8. Our goal is to show that

$$N_{2^{n+1}} = 2N_{2^n}, \tag{8.5}$$

for each $n \geq 2$. Once we know this identity, an easy induction gives the lemma.

We start by obtaining some information about the structure of S_{2^n}. We define an equivalence relation on S_{2^n} by defining $(x, y) \sim (x', y')$ for $(x, y), (x', y') \in S_{2^n}$ if $x \equiv x'$ mod 2^{n-1} and $y \equiv y'$ mod 2^{n-1}. Let $\mathscr{E}_1, \mathscr{E}_2, \ldots, \mathscr{E}_R$ be the equivalence classes.

Our first claim is that each equivalence class $\mathscr{E}_i$ has exactly four elements. In fact, if $(x, y), (x', y') \in S_{2^n}$ and $(x, y) \sim (x', y')$, then

$$\begin{cases} x' \equiv x + k \cdot 2^{n-1} \bmod 2^n, \\ y' \equiv y + l \cdot 2^{n-1} \bmod 2^n, \end{cases} \tag{8.6}$$

for $k, l \in \{0, 1\}$. Now, let $(x, y) \in S_{2^n}$, and for $k, l \in \{0, 1\}$ define x', y' by (8.6). We will prove that $(x', y') \in S_{2^n}$. In order to see this we compute

$$(x')^2 + (y')^2 \equiv (x + k \cdot 2^{n-1})^2 + (y + l \cdot 2^{n-1})^2 \quad \bmod 2^n$$

$$\equiv x^2 + y^2 + 2^n(k + l) + (k^2 + l^2) \cdot 2^{2(n-1)} \quad \bmod 2^n$$

$$\equiv x^2 + y^2 \quad \bmod 2^n$$

$$\equiv 1 \quad \bmod 2^n.$$

This means that every element of the equivalence class of (x, y) is of the form (8.6), and every pair (x', y') of the form (8.6) is equivalent to (x, y). Since there are four choices for the pairs (k, l) we conclude that the equivalence class of (x, y) has four elements, as claimed. Note that this means

$$N_{2^n} = 4R. \tag{8.7}$$

For each i, fix a representative (x_i, y_i) of $\mathscr{E}_i$. The above analysis shows that

$$S_{2^n} = \bigcup_{i=1}^{R} \bigcup_{k,l \in \{0,1\}} \{(x_i + k \cdot 2^{n-1}, y_i + l \cdot 2^{n-1})\}. \tag{8.8}$$

As before we consider the reduction map

$$\eta : S_{2^{n+1}} \to S_{2^n}.$$

Let $(X, Y) \in S_{2^{n+1}}$, and $\eta(X, Y) = (x, y)$. This means that

$$X \equiv x + r \cdot 2^n, \quad Y \equiv y + s \cdot 2^n \quad \bmod 2^{n+1}.$$

Combined with (8.8) we conclude that there are $k, l, r, s \in \{0, 1\}$, and $1 \leq i \leq R$ such that

$$X \equiv x_i + k \cdot 2^{n-1} + r \cdot 2^n, \quad Y \equiv y_i + l \cdot 2^{n-1} + s \cdot 2^n \quad \bmod 2^{n+1}.$$

Now we examine the identity $X^2 + Y^2 \equiv 1 \bmod 2^{n+1}$ to see the types of restrictions we need on k, l, r, s. We have

$$X^2 + Y^2 \equiv (x_i + k \cdot 2^{n-1} + r \cdot 2^n)^2 + (y_i + l \cdot 2^{n-1} + s \cdot 2^n)^2 \quad \bmod 2^{n+1}$$

$$\equiv (x_i^2 + y_i^2) + 2^n(k \cdot x_i + l \cdot y_i) \quad \bmod 2^{n+1}.$$

Since $X^2 + Y^2 \equiv 1 \bmod 2^{n+1}$ we conclude

$$(x_i^2 + y_i^2) + 2^n(k \cdot x_i + l \cdot y_i) \equiv 1 \mod 2^{n+1}.$$

Consequently, we need

$$k \cdot x_i + l \cdot y_i \equiv -\frac{x_i^2 + y_i^2 - 1}{2^n} \mod 2. \tag{8.9}$$

Since $x_i^2 + y_i^2 \equiv 1 \mod 2^n$, not both of x_i, y_i can be divisible by 2. As a result, there will be two pairs (k, l) with $k, l \in \{0, 1\}$ such that (8.9) is satisfied. Furthermore, once appropriate k, l are chosen, any choice of $r, s \in \{0, 1\}$ will work. Finally, $N_{2^{n+1}}$ is equal to the number of possible pairs (x_i, y_i), which is equal to R, multiplied by the number of acceptable pairs (k, l), equal to 2, multiplied by the number of all pairs (r, s), equal to 4, i.e.,

$$N_{2^{n+1}} = 2 \cdot 4 \cdot R = 8R.$$

Comparing this identity with (8.7) proves (8.5). □

Clearly this proof was much more subtle than the proof of Lemma 8.4. As noted above, what prevented us from carrying out the proof of Lemma 8.4 for $p = 2$ was the fact that in the binomial expansion

$$(x + y)^2 = x^2 + 2xy + y^2$$

the middle term is divisible by 2—and this is zero modulo 2. This suggests that if we were to consider an equation of the form

$$x^3 \equiv a \mod p^n$$

then we should run into a problem for $p = 3$, the reason being that the coefficient of $x^2 y$ in the binomial expansion

$$(x + y)^3 = x^3 + 3x^2 y + 3xy^2 + y^3$$

is divisible by 3.

The coefficients that cause trouble in these examples are related to the derivatives of the polynomials x^2 and x^3, respectively. There is an underlying general result, Hensel's Lemma, that explains these examples. See Exercise 8.4 below for Hensel's Lemma, and Exercise 8.5 for a generalization.

Example 8.6. In this example, following the method described above, we will show that for each n, $x^2 = 2 \mod 7^n$ has two solutions. We proceed by induction. If $n = 1$, then $x = 3, 4$ are the two solutions. Now suppose the assertion is true for n, and let x_n be one of the two solutions of $x^2 \equiv 2 \mod 7^n$. We will show that there is a unique $x_{n+1} \mod 7^{n+1}$ such that $x_{n+1} \equiv x_n \mod 7^n$ and $x_{n+1}^2 \equiv 2 \mod 7^{n+1}$. As before, let $x_{n+1} = x_n + k \cdot 7^n$. Since we wish to get $x_{n+1}^2 \equiv 2 \mod 7^{n+1}$ we write

$$(x_n + k \cdot 7^n)^2 \equiv x_n^2 + 2kx_n \cdot 7^n + k^2 \cdot 7^{2n} \equiv x_n^2 + 2kx_n \cdot 7^n \mod 7^{n+1}.$$

In the last step we used the fact that for $n \geq 1$, $2n \geq n + 1$, and hence $7^{2n} \equiv 0 \mod 7^{n+1}$. Then we wish to have

$$x_n^2 + 2kx_n \cdot 7^n \equiv 2 \mod 7^{n+1},$$

or

$$(x_n^2 - 2) + 2kx_n \cdot 7^n \equiv 0 \mod 7^{n+1}.$$

Since $x_n^2 \equiv 2 \mod 7^n$, $7^n \mid x_n^2 - 2$. Dividing the congruence by 7^n gives

$$\frac{x_n^2 - 2}{7^n} + 2kx_n \equiv 0 \mod 7.$$

Since $7 \nmid 2x_n$, $2x_n$ is invertible modulo 7, and we obtain

$$k \equiv -(2x_n)^{-1} \cdot \frac{x_n^2 - 2}{7^n} \mod 7.$$

This means there is a unique choice for k modulo 7, and this is enough to establish the induction step.

Let us illustrate this procedure by computing the first few values of x_n. Suppose we start with $x_1 \equiv 3 \mod 7$. Write $x_2 = 3 + 7k$. We have

$$(3 + 7k)^2 \equiv 2 \mod 7^2.$$

Multiplying out gives $9 + 42k + 7^2k^2 \equiv 2 \mod 7^2$. Consequently,

$$7 + 42k \equiv 0 \mod 7^2.$$

Divide by 7 to obtain,

$$1 + 6k \equiv 0 \mod 7.$$

This gives $k = 1$, and consequently, $x_2 = 10$. We also examine x_3. Write $x_3 = x_2 + l \cdot 7^2 = 10 + l \cdot 7^2$. Then we have

$$(10 + l \cdot 7^2)^2 \equiv 100 + 2 \cdot 10 \cdot 7^2 \cdot l + 7^4 \equiv 100 + 2 \cdot 10 \cdot 7^2 \cdot l \mod 7^3.$$

Since we wish to have $x_3^2 \equiv 2 \mod 7^3$, we get

$$100 + 2 \cdot 10 \cdot 7^2 \cdot l \equiv 2 \mod 7^3.$$

Consequently, $20 \cdot 7^2 \cdot l + 98 \equiv 0 \mod 7^3$. Divide by $2 \cdot 7^2$ to obtain

$$1 + 10l \equiv 0 \mod 7.$$

We obtain $l \equiv 2 \bmod 7$. This gives $x_3 = 10 + 2 \cdot 7^2 = 108$. So, $x_1 = 3$, $x_2 = 3 + 7$, $x_3 = 3 + 7 + 2.7^2$, and the process continues. If we had started with $x_1 = 4$, we would have gotten $x_2 = 39 = 4 + 5 \cdot 7$ and $x_3 = 235 = 4 + 5 \cdot 7 + 4 \cdot 7^2$.

Exercises

8.1 Show that the equation $a_1x_1 + \cdots + a_nx_n = b$ with the a_i's integers is solvable in integers if and only if the congruence equation

$$a_1x_1 + \cdots + a_nx_n \equiv b \quad \bmod m$$

is solvable for all natural numbers m.

8.2 (✠) Numerically verify Equation (8.1) for a few small values of p.

8.3 Find an explicit formula for N_n in terms of the prime factorization of n.

8.4 Prove Hensel's Lemma: Let $f \in \mathbb{Z}[X]$, and suppose $x_1 \in \mathbb{Z}$ is such that $f(x_1) \equiv 0 \bmod p$, but $f'(x_1) \not\equiv 0 \bmod p$. Then for each $n \geq 1$, there is $x_n \in \mathbb{Z}$, uniquely determined modulo p^n, such that $f(x_n) \equiv 0 \bmod p^n$, and $x_n \equiv x_1 \bmod p$.

8.5 Here is a generalization of Hensel's Lemma: Let $f \in \mathbb{Z}[x]$. Suppose for some N and $a \in \mathbb{Z}$, we have $p^{2N+1} \mid f(a)$, $p^N \mid f'(a)$, but $p^{N+1} \nmid f'(a)$. Show that for each $M > N$ there is an $x_M \in \mathbb{Z}$, uniquely determined modulo p^M, such that $f(x_M) \equiv 0 \bmod p^M$ and $x_M \equiv a \bmod p^{N+1}$.

8.6 Show that the equation

$$(x^2 - 13)(x^2 - 17)(x^2 - 221) \equiv 0 \quad \bmod m$$

is solvable for all m. This is [8, Page 3, Problem 4].

8.7 Find a homogeneous cubic polynomial in three variables x, y, z such that

$$f(x, y, z) \equiv 0 \quad \bmod 2$$

has only the zero solution.

8.8 Let ζ be a primitive pth root of unity. Let $f(x_1, \ldots, x_n)$ be a polynomial of n variables with integral coefficients. Show that the number of solutions of the congruence equation

$$f(x_1, \ldots, x_n) \equiv 0 \quad \bmod p$$

is equal to

$$\frac{1}{p} \sum_{x_1, \ldots, x_n} \sum_x \zeta^{xf(x_1, \ldots, x_n)}$$

where all the sums are over the set of integers $\{1, \ldots, p\}$.

8.9 Let $f(x, y) = x^3 + 7y^5$. Use the previous exercise to give an estimate the number of solutions of $f(x, y) \equiv 0 \bmod p$ for a large enough prime number p. For a generalization, see [8, Ch. 1, §2].

8.10 (✠) Let $f(x) = x^2 + 2x + 7$. For each prime p, solve the equation $f(x) \equiv$ 0 mod p, and pick representatives for the roots $0 \leq v_1, v_2 \leq p - 1$, allowing for the possibility that v_1 and v_2 may be equal. Normalize the roots by considering $v_1/p, v_2/p \in [0, 1]$. How are these numbers distributed in the interval $[0, 1]$ as p gets large? Experiment with other polynomials, including quadratic polynomials with or without rational roots, and polynomials of higher degree.

8.11 (✠) Investigate the number of solutions of the equation $x^2 \equiv a$ mod 2^n for several values of a and n.

Notes

p-adic numbers

In the proofs of Lemma 8.4, Lemma 8.5, and in Example 8.6 we encountered sequences $(x_n)_{n \geq 1}$ with the property that

- x_n is a congruence class modulo p^n, represented by an integer, denoted by the same letter x_n, $0 \leq x_n < p^n$;
- $x_{n+1} \equiv x_n$ mod p^n, for each $n \geq 1$.

We define a *p-adic integer* to be a sequence of integers $(x_n)_n$ satisfying these properties. We denote the set of p-adic integers by $\mathbb{Z}_p$. Note that for each $r \in \mathbb{Z}$, the ordinary set of integers, we obtain a constant sequence $\bar{r} := (r \bmod p^n)_{n \geq 1} \in \mathbb{Z}_p$, showing that $\mathbb{Z}$ is naturally a subset of $\mathbb{Z}_p$. (Here r mod p is the remainder of the division of r by p, note that for $p > r$, r mod $p = r$.) The set $\mathbb{Z}_p$ is a commutative ring equipped with the following operations:

$$(x_n)_{n \geq 1} + (y_n)_{n \geq 1} := (x_n + y_n)_{n \geq 1};$$

$$(x_n)_{n \geq 1} \cdot (y_n)_{n \geq 1} := (x_n y_n)_{n \geq 1}.$$

The zero element and the multiplicative identity of $\mathbb{Z}_p$ are given by the constant sequences $\bar{0}$ and $\bar{1}$, respectively. When there is no confusion we drop the line on top of an ordinary integer when thinking of it as a p-adic integer, e.g., we write 0 instead of $\bar{0}$.

It is not hard to see that $\mathbb{Z}_p$ has no zero divisors, i.e., if $xy = 0$, then either $x = 0$ or $y = 0$. We denote by $\mathbb{Q}_p$ the field of fractions of $\mathbb{Z}_p$, and call it *the field of p-adic numbers*. It is clear that $\mathbb{Q}_p$ contains $\mathbb{Q}$.

Let $x = (x_n) \in \mathbb{Z}_p$. Since $p^n \mid x_{n+1} - x_n$, we can write $x_{n+1} = x_n + a_n p^n$ for some $0 \leq a_n < p$, and, if with analogy, we let $x_1 = a_0$, we get $x_1 = a_0, x_2 = a_0 + a_1 \cdot p, x_3 = a_0 + a_1 \cdot p + a_2 \cdot p^2, x_4 = a_0 + a_1 \cdot p + a_2 \cdot p^2 + a_3 \cdot p^3$, etc. We often write the p-adic integer x as a formal sum $\sum_{k=0}^{\infty} a_k \cdot p^k$, with each a_k in the set $\{0, \ldots, p - 1\}$. For example, $-1 = \sum_{k=0}^{\infty}(p - 1) \cdot p^k$. If $a_0 \neq 0$, then $x = \sum_{k=0}^{\infty} a_k \cdot p^k$ is invertible in $\mathbb{Z}_p$. If we denote the set of all invertible elements in $\mathbb{Z}_p$ by $\mathbb{Z}_p^{\times}$, then every non-zero $x \in \mathbb{Z}_p$ can be written as $x = \varepsilon \cdot p^m$ with $\varepsilon \in \mathbb{Z}_p^{\times}$, $m \geq 0$. By considering quotients of such expressions, we see that every non-zero element of $\mathbb{Q}_p$ can be written as $\varepsilon \cdot p^m$ for $\varepsilon \in \mathbb{Z}_p^{\times}, m \in \mathbb{Z}$.

Exercise 8.4 can be interpreted in terms of p-adic integers in the following form, also known as Hensel's Lemma: Let $f \in \mathbb{Z}[X]$, and suppose $x_1 \in \mathbb{Z}$ is such that $f(x_1) \equiv 0 \bmod p$, but $f'(x_1) \not\equiv 0 \bmod p$. Then there is $x \in \mathbb{Z}_p$ such that $f(x) = 0$ in $\mathbb{Z}_p$. Let's examine the equation $x^2 + 1 = 0$. Clearly, this equation has no solutions in $\mathbb{Q}$. If p is an odd prime such that $p \equiv 1 \bmod 4$, then Equation (6.3) implies that the equation $x^2 + 1 \equiv 0 \bmod p$ has a solution x_1. Also if we let $f(x) = x^2 + 1$, $f'(x) = 2x$, and this implies $f'(x_1) \not\equiv 0 \bmod p$. Hensel's Lemma now implies that $x^2 + 1 = 0$ has a solution in $\mathbb{Z}_p$, and consequently in $\mathbb{Q}_p$. If on the other hand, $p \equiv 3 \bmod 4$, then since $x^2 + 1 \equiv 0 \bmod p$ has no solutions, the equation $x^2 + 1 = 0$ will have no solutions in $\mathbb{Q}_p$. It can also be shown that $x^2 + 1 = 0$ has no solutions in $\mathbb{Q}_2$.

The field of p-adic numbers can also be constructed using topology. This method resembles the way $\mathbb{R}$ is constructed from $\mathbb{Q}$ via Cauchy sequences. Recall that a Cauchy sequence of real numbers is a sequence $(x_n)_n$ such that for every $\varepsilon > 0$, there is N such that $|x_n - x_m| < \varepsilon$ for all $n, m > N$. We say Cauchy sequences $(x_n)_n, (y_n)_n$ are *equivalent*, and write $(x_n)_n \sim (y_n)_n$, if for all $\varepsilon > 0$, there is $N > 0$ such that $|x_n - y_m| < \varepsilon$ for all $n, m > N$. Then the field $\mathbb{R}$ can be thought of as the equivalence classes of Cauchy sequences of rational numbers modulo this equivalence relation $\sim$. Note that in this construction we did not have to specify what $|\cdot|$ is because presumably everyone is familiar with the ordinary absolute value. Let us define a new absolute value on $\mathbb{Q}$ which depends on the choice of a prime number p. For a non-zero rational number γ, we can write

$$\gamma = p^r \cdot \frac{a}{b}$$

with $r \in \mathbb{Z}$, $a, b \in \mathbb{Z}$, with $\gcd(p, ab) = 1$. Then we define $|\gamma|_p = p^{-r}$. We also define $|0|_p = 0$. Then for all rational numbers x, $|x|_p \geq 0$, and $|x|_p = 0$ if and only if $x = 0$. Also, we have a triangle inequality, $|x + y|_p \leq |x|_p + |y|_p$. In fact, we have the much stronger *ultrametric* inequality $|x + y|_p \leq \max(|x|_p, |y|_p)$.) This means that if we define $d_p(x, y) = |x - y|_p$, we obtain a metric on $\mathbb{Q}$, and it makes sense to talk about Cauchy sequences. We define a p-*Cauchy sequence* of rational numbers to be a sequence $(x_n)_n$ such that for $\varepsilon > 0$, there is N such that $|x_n - x_m|_p < \varepsilon$ for all $n, m > N$. We say the p-Cauchy sequences $(x_n)_n, (y_n)_n$ are p-*equivalent*, and write $(x_n)_n \sim_p (y_n)_n$, if for all $\varepsilon > 0$, there is $N > 0$ such that $|x_n - y_m|_p < \varepsilon$ for all $n, m > N$. The field $\mathbb{Q}_p$ is nothing but the p-equivalence classes of p-Cauchy sequences of rational numbers.

The beauty of the topological construction of p-adic fields is that it allows us to construct p-adic type field from other number fields. Let K be a number field as in the Notes to Chapter 5, with $\mathcal{O}$ its ring of integers. Let $\mathfrak{p}$ be a prime ideal in $\mathcal{O}$. Then if $\gamma \in K$ is non-zero, then we can let $e_{\mathfrak{p}}(\gamma)$ be the exponent with which the prime ideal $\mathfrak{p}$ occurs in the factorization of $\gamma \mathcal{O}$ as a product of prime ideals. We then define

$$|\gamma|_{\mathfrak{p}} = \#(\mathcal{O}/\mathfrak{p})^{-e_{\mathfrak{p}}(\gamma)}.$$

Here $\#(\mathcal{O}/\mathfrak{p})$ is the number of element of the quotient additive group $\mathcal{O}/\mathfrak{p}$. As before, we define $|0|_{\mathfrak{p}} = 0$. The function $| \cdot | : K \to \mathbb{R}$ gives rise to a metric, and again it makes sense to talk about Cauchy sequences and equivalence classes of Cauchy sequences. The set of equivalence classes of Cauchy sequences with respect to the metric defined by $| \cdot |_{\mathfrak{p}}$ is called the $\mathfrak{p}$-*adic field* and is denoted by $K_{\mathfrak{p}}$.

Fields of p-adic numbers have many applications in modern number theory, through their algebraic, topological, and measure theoretic properties. We refer to [41, Ch. 2, 3] for generalities regarding metric spaces and Cauchy sequences, and [8], especially Ch. 1, 2, and 4 for some applications of p-adic numbers.

Hilbert's Law of Reciprocity

Quadratic Reciprocity for fields other than $\mathbb{Q}$ is known as Hilbert Reciprocity. The formulation of this reciprocity law requires the notion of p-adic numbers introduced above. Let us explain Hilbert's formulation of the Law of Quadratic Reciprocity over $\mathbb{Q}$. Let a, b be non-zero rational numbers. For each prime p, define the *Hilbert Symbol* $(a, b)_p$ to be $+1$ if the equation $ax^2 + by^2 = z^2$ has solutions in p-adic numbers x, y, z, not all of which are zero; otherwise, we define $(a, b)_p$ to be equal to -1. We define $(a, b)_\infty$ to be $+1$ or -1 depending on whether the equation $ax^2 + by^2 = z^2$ has non-trivial solutions in real numbers, i.e., $(a, b)_\infty = -1$ if $a, b < 0$, and $+1$ otherwise. If a, b are non-zero rational numbers, then $(a, b)_p = +1$ for all but finitely many primes p. Hilbert's Law of Reciprocity for $\mathbb{Q}$ says that for all a, b non-zero rational numbers we have

$$(a, b)_\infty \cdot \prod_{p \text{ prime}} (a, b)_p = +1.$$

It is a pleasant exercise to show that this theorem implies Gauss's Law of Quadratic Reciprocity (Hint: Let $a = p, b = q, p, q$ odd prime numbers). For a proof of this theorem over $\mathbb{Q}$, see Serre [44, Ch. III].

For other number fields, we need to define the generalized Hilbert symbols. Let K be a number field. First we define the analogues of $(\cdot, \cdot)_p$. For a prime ideal $\mathfrak{p}$ of K, if a, b are non-zero elements of K, then we define $(a, b)_{\mathfrak{p}} = +1$ if the equation $ax^2 + by^2 = z^2$ has non-trivial solutions in $K_{\mathfrak{p}}$, otherwise we define it to be -1. To define the analogue of $(a, b)_\infty$, we need the concept of a real embedding. A *real embedding* of K is a non-zero function $\sigma : K \to \mathbb{R}$ such that $\sigma(xy) = \sigma(x)\sigma(y)$ and

$\sigma(x + y) = \sigma(x) + \sigma(y)$ for all $x, y \in K$. For the number field K, there are only finitely many real embeddings, $\sigma_1, \sigma_2, \ldots, \sigma_r$. For example, if $K = \mathbb{Q}(\sqrt{2})$, then every element of K can be written as $u + v\sqrt{2}$ with $u, v \in \mathbb{Q}$, and the real embeddings are $\sigma_1 : u + v\sqrt{2} \mapsto u + v\sqrt{2}$ and $\sigma_2 : u + v\sqrt{2} \mapsto u - v\sqrt{2}$. For $1 \leq j \leq r$ and a, b as above, we define $(a, b)_j$ to be $+1$ if at least one of $\sigma_j(a), \sigma_j(b)$ is a positive number, otherwise we define it to be -1. If $K = \mathbb{Q}$, since $\mathbb{Q}$ has only one real embedding, $(a, b)_1 = (a, b)_\infty$ defined earlier. Hilbert's Law of Reciprocity for K is the statement that if $a, b \in K$ are non-zero, then

$$\prod_{j=1}^{r} (a, b)_j \cdot \prod_{\mathfrak{p} \text{ prime ideal}} (a, b)_\mathfrak{p} = +1.$$

Again, all but finitely many of the factors in this product are equal to 1, so the product makes sense. Hilbert's Law of Reciprocity for an arbitrary number field is a hard theorem. Nowadays, it is most convenient to derive this theorem from the general Artin's Law of Reciprocity which includes all the reciprocity theorems we have mentioned in this chapter. Cox [14, Ch. Two] provides a nice introduction to Artin's Law of Reciprocity. We refer to Lemmermeyer [32], especially the preface, and the references therein, for a history of these ideas.

Chapter 9
How many lattice points are there on a circle or a sphere?

A point in $\mathbb{R}^n$ with integral coordinates is called a *lattice point*. In this chapter we study the distribution of lattice points on circles and spheres in $\mathbb{R}^n$. We start by finding a formula for the number $r(n)$ of points with integral coordinates on the circle $x^2 + y^2 = n$ for a natural number n. We then prove a famous theorem of Gauss that gives an expression for the sum $\sum_{n=1}^{k} r(n)$. We then state similar theorems for the number of points on higher dimensional spheres. At the end we state and prove a theorem of Jarnik (Theorem 9.9), and a recent generalization due to Cilleruelo and Córdoba (Theorem 9.10), about integral points on arcs. In the Note, we discuss the error term in Gauss' theorem mentioned above.

9.1 The case of two squares

For a natural number n, we let $r(n)$ be the number of representations of n as a sum of two integral squares, i.e., the number of integral points on the circle $x^2 + y^2 = n$. By Theorem 5.7 we know that if we write

$$n = m \cdot 2^{\alpha} \prod_{p \equiv 1 \bmod 4} p^{\beta_p}$$

with m a product of primes of the form $4k + 3$, then $r(n) = 0$ unless m is a square.

Theorem 9.1. *If m is a square,*

$$r(n) = 4 \prod_{p} (1 + \beta_p).$$

Proof. If $n = x^2 + y^2$, then $n = N(x + iy)$. So we need to determine the number of Gaussian integers z such that $n = N(z)$. By Theorem 5.10 any such z is a product

$$z = uk(1 + i)^a \prod_{p \equiv 1 \bmod 4} \varpi_p^{e_p} \overline{\varpi}_p^{f_p}.$$

© Springer Nature Switzerland AG 2018
R. Takloo-Bighash, *A Pythagorean Introduction to Number Theory*,
Undergraduate Texts in Mathematics, https://doi.org/10.1007/978-3-030-02604-2_9

Here u is one of the four units in $\mathbb{Z}[i]$; $k \in \mathbb{N}$ is a product of primes of the form $4k+3$; and all but finitely many of the non-negative integers e_p, f_p are zero, meaning the product is finite. Then we have

$$N(z) = N(k)N(1+i)^a \prod_{p \equiv 1 \bmod 4} N(\varpi_p)^{e_p} N(\overline{\varpi}_p)^{f_p}$$

$$= k^2 2^a \prod_{p \equiv 1 \bmod 4} p^{e_p} p^{f_p} = k^2 2^a \prod_{p \equiv 1 \bmod 4} p^{e_p + f_p}.$$

Consequently, since $N(z) = n$ we get

$$m 2^\alpha \prod_j p_j^{\beta_j} = k^2 2^a \prod_{p \equiv 1 \bmod 4} p^{e_p + f_p}.$$

This implies $m = k^2$, $a = \alpha$, and for each p of the form $4k+1$, $e_p + f_p = \beta_p$. The number of such e_p, f_p is $1 + \beta_p$. Since there are four possibilities for the unit u, i.e., ± 1, $\pm i$, we get a total of

$$4 \prod_p (1 + \beta_p)$$

choices for z. This finishes the proof.   □

For example, we have

$$180 = 3^2 \cdot 2^2 \cdot (2+i) \cdot (2-i).$$

So we get the following numbers as the list of numbers z that have the property that $N(z) = 180$:

$$u \cdot 3 \cdot (1+i)^2 \cdot (2+i) = u(-6+12i)$$

and

$$u \cdot 3 \cdot (1+i)^2 \cdot (2-i) = u(6+12i)$$

for $u \in \{+1, -1, i, -i\}$. This means that the possibilities for the ordered pairs (a, b) such that $180 = a^2 + b^2$ are:

$$(\pm 6, \pm 12), \quad (\pm 12, \pm 6),$$

a total of eight possibilities.

Now that we have a formula for $r(n)$ one could ask natural statistical questions about it. For example, one could ask what the average behavior of $r(n)$ is like. Let us make this notion precise.

Definition 9.2. Suppose $f : \mathbb{N} \to \mathbb{C}$ is a function. We say f has *average value* equal to c if the limit

$$\lim_{N \to \infty} \frac{1}{N} \sum_{n=1}^{N} f(n)$$

exists and is equal to c.

It should be clear that not every function has an average value, see Exercise 9.3. There is a more general concept which is the following:

Definition 9.3. For functions $f, g : \mathbb{N} \to \mathbb{C}$, we say f, g have the same average order, or that g is an average order of f, if

$$\lim_{X \to \infty} \frac{\sum_{n \leq X} f(n)}{\sum_{n \leq X} g(n)} = 1.$$

In applications, one of the functions, say f, is the one we are interested in, and the idea is to find a nice function g which imitates the function f on average.

In the case of $r(n)$, the sum $\sum_{n=1}^{N} r(n)$ that appears in the definition of the average value has a neat geometric interpretation. Indeed, we have

$$\sum_{n=1}^{N} r(n) = \sum_{n=1}^{N} \#\{(x, y) \in \mathbb{Z}^2 \mid x^2 + y^2 = n\}$$

$$= \#\{(x, y) \in \mathbb{Z}^2 \mid x^2 + y^2 \leq N\}.$$

This means that $\sum_{n=1}^{N} r(n)$ is the number of integral points inside the circle of radius $\sqrt{N}$. Intuitively, the number of integral points inside the circle of radius $\sqrt{N}$ should be about the area of the circle. One way to see this is to associate a unit square to each integral point as shown in Figure 9.1 for the point $(3, 2)$.

Fig. 9.1 The grey square is completely within the circle of radius 6. The point $(5, 5)$ is outside the circle of radius 6 but the blue square to its lower left intersects the circle. The point $(-2, -5)$ is within the circle of radius 6 but the red square to its lower left is not contained in the circle of radius 6

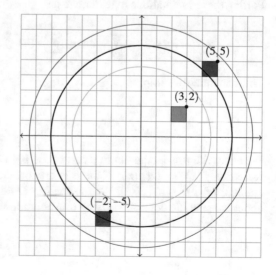

The trouble here is that not every square based on a point inside the circle will be completely within the circle, e.g., the red square in Figure 9.1 whose upper right corner is the point $(-2, -5)$ is not entirely within the circle of radius 6; and also some integral points outside the circle of radius 6 shown in the picture will have squares associated with them that intersect the circle, e.g., the blue square to the lower left of the point $(5, 5)$. The key point, however, is that the troublesome squares cannot stray too much from the boundary of the circle with radius $\sqrt{N}$. In fact, since the diagonal of a unit square is $\sqrt{2}$, each square to the lower left of an integral point within the circle of radius $\sqrt{N}$ will be fully contained in a circle of radius $\sqrt{N} + \sqrt{2}$. For $\sqrt{N} = 6$, the purple circle in the figure has radius $6 + \sqrt{2}$. Consequently, the total area of all unit squares, which is equal to $\sum_{n=1}^{N} r(n)$, is at most the area of the circle with radius $\sqrt{N} + \sqrt{2}$. Hence,

$$\sum_{n=1}^{N} r(n) \leq \pi(\sqrt{N} + \sqrt{2})^2 = \pi N + 2\pi\sqrt{2}\sqrt{N} + 2\pi.$$

Similarly, the entire area of the circle with radius $\sqrt{N} - \sqrt{2}$ is covered by unit squares to the lower left of integral points within the circle of radius $\sqrt{N}$. In the figure above the green circle is the one that has radius $6 - \sqrt{2}$. This means,

$$\sum_{n=1}^{N} r(n) \geq \pi(\sqrt{N} - \sqrt{2})^2 = \pi N - 2\pi\sqrt{2}\sqrt{N} + 2\pi.$$

Putting these inequalities together, we get

$$\pi N - 2\pi\sqrt{2} \cdot \sqrt{N} + 2\pi \leq \sum_{n=1}^{N} r(n) \leq \pi N + 2\pi\sqrt{2}\sqrt{N} + 2\pi.$$

These inequalities imply

$$\left| \sum_{n=1}^{N} r(n) - \pi N - 2\pi \right| \leq 2\pi\sqrt{2}\sqrt{N}.$$

We can write this inequality in terms of the *big O* notation. For real functions f, g, we write $f(x) = O(g(x))$ if there is a constant $C > 0$ such that for all x large enough, $|f(x)| \leq C|g(x)|$. We use the big O notation if we do not have to worry about the specific constants. Using this notation we can write

$$\sum_{n=1}^{N} r(n) = \pi N + 2\pi + O(\sqrt{N}) = \pi N + O(\sqrt{N}).$$

This last identity is a famous result of Gauss which for ease of reference we record as a theorem:

Theorem 9.4 (Gauss). *As $N \to \infty$,*

$$\sum_{n=1}^{N} r(n) = \pi N + O(\sqrt{N}).$$

This theorem has the following rather curious corollary:

Corollary 9.5. *The average value of $r(n)$ is π.*

Remark 9.6. We will prove a variation of Theorem 9.4 in §13.1.

9.2 More than two squares

It is clear that the geometric argument of the proof of Theorem 9.4 can be adapted to every dimension. For $k \geq 2$ and $n \in \mathbb{N}$, we set

$$r_k(n) = \#\left\{ (x_1, \ldots, x_k) \in \mathbb{Z}^k \mid \sum_{i=1}^{k} x_i^2 = n \right\},$$

the number of integral points on the sphere in the k-dimensional space. We have $r_2(n) = r(n)$. Then we have:

Theorem 9.7. *As $N \to \infty$,*

$$\sum_{n=1}^{N} r_k(n) = \frac{\pi^{\frac{k}{2}}}{\Gamma\left(\frac{k}{2}+1\right)} N^{\frac{k}{2}} + O(N^{\frac{k-1}{2}}).$$

For the definition and basic properties of the Gamma function Γ see [4] or [41, Chapter 8]. We review some basic properties in Exercise 9.2. The proof of Theorem 9.7 is sketched in Exercises 9.4–9.6 below.

Note that Theorem 9.7 shows that for $k > 2$, the limit

$$\lim_{N \to \infty} \frac{1}{N} \sum_{n=1}^{N} r_k(N)$$

is not finite, and consequently $r_k(N)$ does not have an average value.

Now we state an extension of Theorem 5.7 for $k > 2$.

Theorem 9.8. *For $n \in \mathbb{N}$, $r_3(n) \neq 0$ if and only if n is not of the form $4^a(8n + 7)$. If $k > 3$, for all n, $r_k(n) \neq 0$, i.e., every natural number is the sum of four integral squares.*

We will give a proof of this fact in Chapter 10 using a theorem of Minkowski. We will give other proofs in Chapters 11 and 12. The most challenging part of the theorem is the statement for sums of three squares. We present two proofs for this theorem in §10.5 and §12.4, but unfortunately, both of these proofs rely in substantial ways on Dirichlet's Arithmetic Progression Theorem, Theorem 5.11 in Notes to Chapter 5.

In Chapter 5 we referred to Theorem 5.2 as the *Two Squares Theorem*. Throughout the text we refer to the portion of Theorem 9.8 that deals with sums of three squares as the *Three Squares Theorem*, and to the part about the expressibility of every natural number as the sum of four squares as the *Four Squares Theorem* often without explicit reference to Theorem 9.8.

Generalizing Theorem 9.1 for $k > 2$ is far more problematic. Computing $r_3(n)$ already poses a serious challenge, [113]. Erdös [73] claims that there is a constant $c > 0$ such that

$$r_3(n) \geq c\sqrt{n} \log \log n$$

but does not provide a proof. For $k = 4$ there is a beautiful explicit formula, due to Jacobi (1834), that says

$$r_4(n) = 8 \sum_{d\mid n, 4\nmid d} d.$$

The short paper [80] contains an elementary proof of this fact. For $k > 5$, the question of determining r_k has a long history. We refer the reader to [79] and [74] for some early works.

9.3   Integral points on arcs

In §3.2 we studied the rational points on the unit circle. If we have a rational point $(a/c, b/c)$ on the unit circle, $a, b, c \in \mathbb{Z}$, we obtain an integral point (a, b) on the circle $x^2 + y^2 = c^2$ of radius $|c|$, an integer. In general, if we have an integral point (a, b) on some circle with equation $x^2 + y^2 = R^2$, R need not be an integer, e.g., the point $(2, 1)$ is on the circle with radius $\sqrt{5}$. As we noted above, Theorem 9.1 counts the number of integral points on the circle $x^2 + y^2 = n$ for a natural number n. As we will see below, it is in general difficult to gain a complete understanding of the distribution of these integral points on the circle, and there are still open problems that we do not know to solve. We learned the material of this section and Theorems 9.9 and 9.10 from Lillian Pierce (private communication).

Suppose we have three integral points A, B, C on a circle of radius R and let L be an arc containing the three points, e.g., $\overset{\frown}{ACB}$, as in Figure 9.2 . Let a, b, c be the lengths of the three sides of the triangle ABC and S the area of the triangle formed by the points.

By Exercise 9.8 we have $abc = 4SR$. Then since $a, b, c \leq \max\{a, b, c\}$ we have

Fig. 9.2 Triangle ABC with area S whose vertices are on a circle of radius R

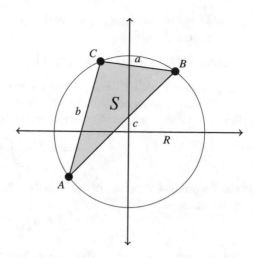

$$4SR = abc \leq \max\{a, b, c\}^3.$$

But since any triangle with three vertices that are integral points has area at least $1/2$ we have

$$\max\{a, b, c\}^3 \geq 2R$$

and consequently,

$$\max\{a, b, c\} \geq (2R)^{1/3}.$$

But the maximum of a, b, c is less than the length of the arc L. This means

$$L \geq (2R)^{1/3}.$$

We state this as the following important theorem:

Theorem 9.9 (Jarnik). *An arc of length less than $(2R)^{1/3}$ in a circle of radius R contains at most two integral points.*

In the case where we have more than three points the situation becomes complicated very quickly. The following is a fairly recent result that gives a non-trivial bound for any number of points.

Theorem 9.10 (Cilleruelo and Córdoba, [67]). *On a circle of radius R centered at the origin, an arc of length*

$$\sqrt{2}R^{\frac{1}{2} - \frac{1}{4\lfloor m/2 \rfloor + 2}}$$

contains at most m integral points.

At the time of this writing the bound obtained in the theorem seems to be the best available in literature; see [68, §5] for several comments on this theorem. The bound is sharp for $m = 3$, but it is not clear whether for $m \geq 4$ it provides the best bound possible. For $m = 4$ the theorem gives the exponent $2/5$, and *ibid.* lists the following as a non-trivial problem:

Question 9.11. Can the exponent $2/5$ be improved?

Proof of Theorem 9.10. We use the notations of the proof of Theorem 9.1. If the circle of radius R contains no lattice points, there is nothing to prove. So we assume that $R = \sqrt{n}$ for some natural number n, and by Theorem 9.1, we may further assume

$$n = k^2 2^\alpha \prod_{p \equiv 1 \bmod 4} p_j^{\beta_p}$$

with k a product of primes of the form $4k + 3$. Then the total number of lattice points on the circle is

$$r(n) = 4 \prod_p (1 + \beta_p).$$

This number $r(n)$ corresponds to the various representations $N(a + ib) = n$, and in fact one can write any such $a + ib$ in the form

$$uk(1 + i)^\alpha \prod_{p \equiv 1 \bmod 4} \varpi_p^{e_p} \overline{\varpi}_p^{f_p}.$$

for a unit $u \in \{\pm 1, \pm i\}$ and $e_p + f_p = \beta_p$ with $e_p, f_p \geq 0$. Here for each prime $p \equiv 1 \bmod 4$, write

$$\varpi_p = \sqrt{p} e^{2\pi i \phi_p}$$

and

$$\overline{\varpi}_p = \sqrt{p} e^{-2\pi i \phi_p}.$$

Then

$$\varpi_p^{e_p} \overline{\varpi}_p^{f_p} = p^{\frac{\beta_p}{2}} e^{2\pi i (e_p - f_p)\phi_p} = p^{\frac{\beta_p}{2}} e^{2\pi i (\beta_p - 2 f_p)\phi_p}.$$

Also each unit in $\mathbb{Z}[i]$ can be written as

$$e^{2\pi i \frac{t}{4}}, \quad t \in \{0, 1, 2, 3\}.$$

Consequently, every $a + ib$ with $N(a + ib) = n$ can be written as

$$\sqrt{n} e^{2\pi i (\phi_2 + \sum_p \gamma_p \phi_p + \frac{t}{4})} \tag{9.1}$$

for $t \in \{0, 1, 2, 3\}$, $|\gamma_p| \leq \beta_p$ and $\gamma_p \equiv \beta_p \bmod 2$, and the sum in the exponent is over primes p with $p \equiv 1 \bmod 4$, and $\phi_2 = \begin{cases} 0 & \alpha \text{ even;} \\ 1 & \alpha \text{ odd.} \end{cases}$

We divide the remainder of the proof into three steps.

Step One. Suppose we have $m + 1$ lattice points on an arc of length $\sqrt{2} R^\theta$. We write these points as

$$a_s + ib_s = \sqrt{n} e^{2\pi i (\phi_2 + \sum_p \gamma_p^s \phi_p + \frac{t^s}{4})},$$

$s \in \{1, \ldots, m + 1\}$, with γ_p^s, t^s integers as above. For $s, s' \in \{1, \ldots, m + 1\}$, $\gamma_p^s \equiv \gamma_p^{s'} \bmod 2$. Define

$$\psi^{s,s'} = \sum_p \frac{\gamma_p^s - \gamma_p^{s'}}{2} \phi_p + \frac{t^s - t^{s'}}{8}.$$

Note that $2\pi |||\psi^{s,s'}|||$ is one half of the central angle between $a_s + ib_s$ and $a_{s'} + ib_{s'}$ in radians (Here and elsewhere, for a real number x, $|||x|||$ is the distance from x to the closest integer). If the length of the arc connecting $a_s + ib_s$ and $a_{s'} + ib_{s'}$ is η, then we have

$$2\pi |||\psi^{s,s'}||| = \frac{\eta}{2R} \leq \frac{\sqrt{2}R^\theta}{2R} = \frac{1}{\sqrt{2}} R^{\theta-1}.$$

We obtain the first main inequality of this proof:

$$|||\psi^{s,s'}||| \leq \frac{1}{2\pi\sqrt{2}} R^{\theta-1}. \tag{9.2}$$

Step two. Now we proceed to obtain a lower bound for $|||\psi^{s,s'}|||$. Comparing this lower bound with Equation (9.2) gives the result. We recognize two cases:

- If $t^s \equiv t^{s'} \bmod 2$, then $(t^s - t^{s'})/8 = t^{s,s'}/4$ for some integer $t^{s,s'}$. In this case, $2\pi\psi^{s,s'}$ is the angle corresponding to a representation of the number

$$\prod_p p^{\frac{|\gamma_p^s - \gamma_p^{s'}|}{2}}$$

as a sum of two squares;

- if $t^s \not\equiv t^{s'} \bmod 2$, then $(t^s - t^{s'}) = 1/8 + t^{s,s'}/4$ for some integer $t^{s,s'}$. In this case, $2\pi\psi^{s,s'}$ is the angle corresponding to a representation of the number

$$2\prod_p p^{\frac{|\gamma_p^s - \gamma_p^{s'}|}{2}}$$

as a sum of two squares.

Note that if $\psi^{s,s'}$ is an integer, then the linear independence of

$$1, \phi_2, \phi_3, \phi_5, \ldots$$

over the rationals (Exercise 9.16) implies that $t^s = t^{s'}$ and $\gamma_p^s = \gamma_p^{s'}$ for every p. This means $a_s + ib_s = a_{s'} + ib_{s'}$. Consequently, if $s \neq s'$, then $|||\psi^{s,s'}||| > 0$. By the above discussion, $\psi := 2\pi |||\psi^{s,s'}|||$ is the angle of a lattice point $P(x_0, y_0)$ not on the x axis and on a circle of radius

$$R_{s,s'} := 2^{\nu/2} \prod_p p^{\frac{|\gamma_p^s - \gamma_p^{s'}|}{4}}$$

Fig. 9.3 In this diagram, η' is the length of the arc connecting $P(x_0, y_0)$ to $(R_{s,s'}, 0)$

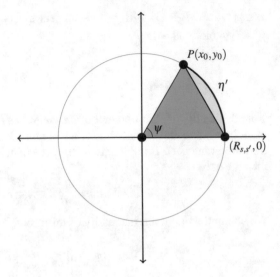

with $\nu = 0$ or 1, depending on whether $t_s \equiv t_{s'}$ mod 2 or not. Let η' be the length of the arc connecting the P to the point $(R_{s,s'}, 0)$ as in Figure 9.3. Then η' is longer than the straight line distance between P and $(R_{s,s'}, 0)$.

This means,

$$\eta' > \sqrt{(x_0 - R_{s,s'})^2 + y_0^2} \geq \sqrt{y_0^2} \geq 1.$$

Then, in the circle of radius $R_{s,s'}$ we have

$$2\pi |||\psi^{s,s'}||| = \frac{\eta'}{R_{s,s}} > \frac{1}{R_{s,s'}} \geq \frac{1}{\sqrt{2}\prod_p p^{\frac{|\gamma_p^s - \gamma_p^{s'}|}{4}}}.$$

We have then obtained our second main inequality:

$$|||\psi^{s,s'}||| > \frac{1}{2\pi\sqrt{2}\prod_p p^{\frac{|\gamma_p^s - \gamma_p^{s'}|}{4}}} \qquad (9.3)$$

for $s \neq s'$.

Step three. Comparing (9.2) and (9.3) gives the following inequality: For all $s \neq s'$ we have

$$\frac{1}{\prod_p p^{\frac{|\gamma_p^s - \gamma_p^{s'}|}{4}}} < R^{\theta-1}. \qquad (9.4)$$

Step four. There are $m(m + 1)/2$ choices for the unordered pairs of numbers s, s'. Multiplying inequalities (9.4) over all of these choices gives

$$\frac{1}{\prod_{s,s'} \prod_p p^{\frac{|\gamma_p^s - \gamma_p^{s'}|}{4}}} < R^{(\theta-1)m(m+1)/2}.$$

We would like to find a lower bound for the left hand side of the above inequality. In order to do this we need to maximize

$$\prod_{s,s'} \prod_p p^{\frac{|\gamma_p^s - \gamma_p^{s'}|}{4}} = \left(\prod_p p^{\sum_{s,s'} |\gamma_p^s - \gamma_p^{s'}|} \right)^{\frac{1}{4}}.$$

In order to do this, we need to maximize

$$\sum_{s,s'} |\gamma_p^s - \gamma_p^{s'}|$$

subject to the following conditions: for each s, $|\gamma_p^s| \le \beta_p$ and $\gamma_p^s \equiv \beta_p$ mod 2. By Exercise 9.17, the maximum value of this expression is

$$\frac{(m + 1)^2 - \delta(m + 1)}{2} \beta_p,$$

with the function δ being given by

$$\delta(n) = \frac{1 - (-1)^n}{2} = \begin{cases} 0 & n \text{ even}; \\ 1 & n \text{ odd}. \end{cases}$$

Putting everything together, we obtain

$$R^{(\theta-1)m(m+1)/2} > \left(\prod_p p^{\frac{(m+1)^2 - \delta(m+1)}{2} \beta_p} \right)^{-\frac{1}{4}} \ge R^{-\frac{(m+1)^2 + \delta(m+1)}{4}}.$$

This inequality implies

$$\theta > 1 - \frac{(m + 1)^2 - \delta(m + 1)}{2m(m + 1)} = \frac{1}{2} - \frac{1}{4[m/2] + 2}. \tag{9.5}$$

This finishes the proof of the theorem. $\square$

We finish this chapter with the following conjecture:

Conjecture 9.12 ([68], Conjecture 14). The number of lattice points on an arc of length $R^{1-\theta}$ on the circle with equation $x^2 + y^2 = R^2$ is bounded uniformly in R.

Exercises

9.1 (✠) Investigate the error term in the asymptotic formula of Theorem 9.4.

9.2 In this exercise we assume the reader is familiar with basic complex analysis.

    a. Show that for each $s \in \mathbb{C}$ with $\Re s > 0$, the integral

$$\Gamma(s) := \int_0^\infty t^{s-1} e^{-t} \, dt$$

    is absolutely convergent.

    b. Show that the for all s with $\Re s > 0$ we have

$$\Gamma(s+1) = s\Gamma(s).$$

    c. Conclude that the function $\Gamma(s)$, originally defined on $\Re s > 0$, has an analytic continuation to a meromorphic function on all of $\mathbb{C}$ with simple poles at $s = 0, -1, -2, -3, \ldots$. Compute the residues at the poles.

    d. Show that $1/\Gamma(s)$ is entire.

    e. Show that for each natural number n, $\Gamma(n) = (n-1)!$.

    f. Show that for all s_1, s_2 with $\Re s_1, \Re s_2 > 0$, we have

$$\int_0^1 t^{s_1-1}(1-t)^{s_2-1} \, dt = \frac{\Gamma(s_1)\Gamma(s_2)}{\Gamma(s_1+s_2)}.$$

9.3 Find an easy function $f : \mathbb{N} \to \mathbb{C}$ which does not have an average value.

9.4 Compute the volume of the sphere of radius R in $\mathbb{R}^k$.

9.5 Compute the diameter of the unit hypercube in $\mathbb{R}^k$.

9.6 Prove Theorem 9.7.

9.7 Show that the function r_k for $k > 2$ does not have an average value. Find a continuous function $f : \mathbb{R} \to \mathbb{R}$ with the same average order as r_k.

9.8 Prove that for a triangle with side lengths a, b, c with area S which is inscribed in a circle of radius R we have

$$abc = 4RS.$$

9.9 Show that if a circle of radius r in $\mathbb{R}^2$ has three points A, B, C such that the distances AB, AC, BC are rational numbers, then r is a rational number.

9.10 Show that every circle in $\mathbb{R}^2$ with rational radius contains infinitely many points every two of which have rational distance.

9.11 Justify Equation (9.1).

9.12 Show that for all real numbers ξ, $|||\xi||| = |\xi + [\xi] - [2\xi]|$.

9.13 Show that for all real numbers ξ, η,

$$|||\xi + \eta||| \le |||\xi||| + |||\eta|||.$$

9.14 Show that for all $\xi \in \mathbb{R}$ and $n \in \mathbb{Z}$, $|||n\xi||| \le |n| \cdot |||\xi|||$.

9.15 Show that for all natural numbers n,

$$n \cdot |||n\sqrt{2}||| \geq 2 \cdot |||2\sqrt{2}||| = 6 - 4\sqrt{2}.$$

9.16 Show that the real numbers $1, \phi_2, \phi_3, \phi_5, \ldots$ appearing in the proof of Theorem 9.10 are linearly independent over the rational numbers.

9.17 Suppose β is a positive integer, and k a natural number. Show that for each choice of $\gamma_1, \ldots, \gamma_k$ such that for i, $|\gamma_i| \leq \beta$ and $\gamma_i \equiv \beta \bmod 2$, we have

$$\sum_{1 \leq i < j \leq k} |\gamma_i - \gamma_j| \leq \frac{k^2 - \delta(k)}{2}\beta$$

where $\delta(k) = \begin{cases} 0 & k \text{ even}; \\ 1 & k \text{ odd}. \end{cases}$. Show that equality is attained if

a. k even: $k/2$ of the γ_i's are equal to β and the other $k/2$ are equal to $-\beta$;
b. k odd: $(k+1)/2$ of the γ_i's are equal to β and the remaining $(k-1)/2$ are equal to $-\beta$.

9.18 Prove inequality (9.5).

9.19 Show that for every natural number m there are infinitely many circles centered at the origin with precisely m integral points on their perimeters.

9.20 Show that for each natural number n, there are infinitely many circles in $\mathbb{R}^2$ which contain exactly n lattice points.

9.21 This problem is about the celebrated theorem of Georg Pick (1859–1942, Theresienstadt Concentration Camp). A simple proof of this theorem appears in [103].

a. Suppose T is a triangle in the plane all of whose vertices are lattice points. Let S be the area of the triangle, E the number of lattice points on the edges, and I the number of lattice points inside the triangle. Show that

$$S = I + \frac{1}{2}E - 1.$$

b. Prove Pick's theorem: Let P be a closed non self-intersecting polygon in $\mathbb{R}^2$ whose vertices are lattice points. Let S be the area, E the number of lattice points on the edges, and I the number of lattice points inside P. Then we have

$$S = I + \frac{1}{2}E - 1.$$

9.22 (✠) Investigate Question 9.11.

9.23 (✠) Do you believe Conjecture 9.12?

9.24 (✠) For each natural number n, consider the sphere S_n defined by

$$x^2 + y^2 + z^2 = n$$

in $\mathbb{R}^3$, and define $S_n(\mathbb{Z})$ to be the collection of points on S_n that have integral coordinates. If $(x, y, z) \in S_n(\mathbb{Z})$, then

$$(\frac{x}{\sqrt{n}}, \frac{y}{\sqrt{n}}, \frac{z}{\sqrt{n}}) \in S_1.$$

Investigate the distribution of the resulting points on the sphere S_1. Experiment with restricting the sequence of n's, e.g., squares, primes, etc.

Notes

Gauss' Circle Theorem

In Theorem 9.4 we showed that if we have a circle of radius r, then the number of lattice points inside the circle is $\pi r^2 + O(r)$. There is a famous conjecture [23, Section F1] asserting that the error term in Gauss' Circle Theorem is $O(r^{1/2+\epsilon})$ for any $\varepsilon > 0$. Richard Guy describes the problem of proving this conjecture as *very difficult*. The best result in this direction is due to Martin Huxley who around the year 2000 proved that the error is $O(r^{131/208})$ improving his own earlier result of $O(r^{46/73})$. Note that $46/73 - 131/208 = 0.000329....$

Chapter 10
What about geometry?

In this chapter we present a geometric theorem of Minkowski, and use it to prove Theorem 9.8. We start with the basic theory of lattices in $\mathbb{R}^n$, discuss the volume of a lattice, and explain the connection of the volume of a lattice to the determinants of certain matrices. We then prove two foundational results of Minkowski, Proposition 10.8 and Theorem 10.10. The remainder of the chapter is devoted to studying sums of squares using the results of Minkowski just mentioned. The Two and Four Square Theorems are relatively easy to prove using Theorem 10.10, but the Three Square Theorem is hard. The proof of the Three Square Theorem occupies §10.5. In the Notes, we discuss Waring's Problem, introduce the functions $g(k)$ and $G(k)$, and explain some recent results obtained using the Circle Method (we also include an explanation of the Circle Method). At the end of the Notes, we say a few words about geometry of numbers.

10.1 Lattices in $\mathbb{R}^n$

Definition 10.1. Let $\mathscr{B} = \{v_1, \ldots, v_n\}$ be a basis of $\mathbb{R}^n$, i.e., a set of n $\mathbb{R}$-linearly independent vectors in $\mathbb{R}^n$. The lattice generated by $\mathscr{B}$, denoted by $\Lambda_{\mathscr{B}}$, is the set of all linear combinations

$$c_1 v_1 + \cdots + c_n v_n$$

with $c_i \in \mathbb{Z}$. We define the *fundamental parallelogram* $\mathscr{P}_{\mathscr{B}}$ by

$$\mathscr{P}_{\mathscr{B}} = \{\alpha_1 v_1 + \cdots + \alpha_n v_n \mid 0 \leq \alpha_1, \ldots, \alpha_n < 1\}.$$

We define Vol $\mathscr{P}_{\mathscr{B}}$ to be the n-dimensional volume of the fundamental parallelogram $\mathscr{P}_{\mathscr{B}}$.

In Figure 10.1, $n = 2$, $\mathscr{B} = \{(2, 1), (1, 3)\}$, and the marked points are the elements of the lattice $\Lambda_{\mathscr{B}}$. Here, the fundamental parallelogram is painted yellow. In this case, Vol $\Lambda_{\mathscr{B}} = 5$.

© Springer Nature Switzerland AG 2018
R. Takloo-Bighash, *A Pythagorean Introduction to Number Theory*,
Undergraduate Texts in Mathematics, https://doi.org/10.1007/978-3-030-02604-2_10

Fig. 10.1 A lattice in $\mathbb{R}^2$.
The fundamental
parallelogram is painted
yellow

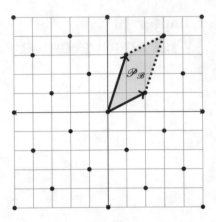

Note that the set $\Lambda_{\mathscr{B}}$ does not uniquely identify the basis $\mathscr{B}$. In fact it is easy to construct distinct bases $\mathscr{B}$ and $\mathscr{B}'$ of $\mathbb{R}^n$ such that

$$\Lambda_{\mathscr{B}} = \Lambda_{\mathscr{B}'};$$

see Exercise 10.1.

Definition 10.2. By a lattice in $\mathbb{R}^n$, we understand a set of the form $\Lambda_{\mathscr{B}}$ for some basis $\mathscr{B}$ of $\mathbb{R}^n$.

The quintessential example of a lattice in $\mathbb{R}^n$ is $\mathbb{Z}^n$ built from the standard basis:

$$e_1 = (1, 0, \ldots, 0),$$

$$e_2 = (0, 1, \ldots, 0),$$

$$\cdots$$

$$e_n = (0, 0, \ldots, 1).$$

The fundamental parallelogram associated to this basis is the unit cube in $\mathbb{R}^n$ whose volume is 1.

Proposition 10.3. *Suppose*

$$v_1 = (a_{11}, a_{12}, \ldots, a_{1n}),$$

$$v_2 = (a_{21}, a_{22}, \ldots, a_{2n}),$$

$$\cdots$$

$$v_n = (a_{n1}, a_{n2}, \ldots, a_{nn})$$

are n linearly independent vectors in $\mathbb{R}^n$. Set $\mathscr{B} = \{v_1, \ldots, v_n\}$. Then

$$\text{Vol } \mathscr{P}_{\mathscr{B}} = |\det(a_{ij})_{ij}| \neq 0$$

Proof. It is well known, e.g., [25, Ch. 6, §9], that the determinant $\det(a_{ij})_{ij}$ is non-zero if and only if the vectors $v_1, \ldots, v_n$ are linearly independent. For the volume statement, see Exercise 10.2. $\square$

Example 10.4. Define a set $\Lambda \subset \mathbb{Z}^2$ as follows:

$$\Lambda = \{(x, y) \in \mathbb{Z}^2 \mid x \equiv y \bmod 2\}.$$

Let $v_1 = (2, 0)$, $v_2 = (1, 1)$, and set $\mathscr{B} = \{v_1, v_2\}$. We will show that $\Lambda = \Lambda_{\mathscr{B}}$. It is clear that $v_1, v_2 \in \Lambda$, and consequently, $\Lambda_{\mathscr{B}} \subset \Lambda$. Now we show the opposite inclusion. Observe that $2v_2 - v_1 = (2, 2) - (2, 0) = (0, 2)$. Now suppose $(x, y) \in \Lambda$. Since $x \equiv y \bmod 2$, there are two possibilities:

- x, y are even. In this case, $(x, y) = (2k, 2l)$ for integers k, l. Then

$$(x, y) = k(2, 0) + l(0, 2) = kv_1 + l(2v_2 - v_1) = (k - l)v_1 + 2lv_2 \in \Lambda_{\mathscr{B}};$$

- x, y are odd. In this case, $(x, y) = (2k + 1, 2l + 1)$ for integers k, l. Then

$$(x, y) = k(2, 0) + l(0, 2) + (1, 1) = (k - l)v_1 + (2l + 1)v_2 \in \Lambda_{\mathscr{B}}.$$

The fundamental domain $\mathscr{P}_{\mathscr{B}}$ is painted yellow in Figure 10.2. Proposition 10.3 shows that

$$\text{Vol } \mathscr{P}_{\mathscr{B}} = \left| \det \begin{pmatrix} 2 & 0 \\ 1 & 1 \end{pmatrix} \right| = 2.$$

Another relevant basis here is $\mathscr{B}' = \{v_1', v_2\}$ with $v_1' = (1, 3)$ and v_2 as above. The associated fundamental domain $\mathscr{P}_{\mathscr{B}'}$ is painted green in Figure 10.2. One easily

Fig. 10.2 The regions painted yellow and green are both fundamental domains

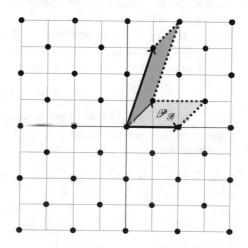

checks that $\Lambda = \Lambda_{\mathscr{B}'}$. Then we have

$$\text{Vol } \mathscr{P}_{\mathscr{B}'} = \left| \det \begin{pmatrix} 0 & 2 \\ 1 & 1 \end{pmatrix} \right| = |-2| = 2.$$

Even though the bases $\mathscr{B}$ and $\mathscr{B}'$ are different, the fundamental parallelograms $\mathscr{P}_{\mathscr{B}}$ and $\mathscr{P}_{\mathscr{B}'}$ have the same volume.

In general, for a lattice $\Lambda \subset \mathbb{R}^n$ there are infinitely many bases $\mathscr{B}$ such that $\Lambda = \Lambda_{\mathscr{B}}$. We will see in Exercise 10.3 that even though the set $\mathscr{P}_{\mathscr{B}}$ depends on the choice of $\mathscr{B}$, its volume, $\text{Vol } \mathscr{P}_{\mathscr{B}}$, is independent of it, and that it depends only on the lattice Λ. This statement inspires the following definition:

Definition 10.5. If $\Lambda \subset \mathbb{R}^n$ is a lattice, then we define $\text{Vol } \Lambda$ to be $\text{Vol } \mathscr{P}_{\mathscr{B}}$ for any basis $\mathscr{B}$ such that $\Lambda = \Lambda_{\mathscr{B}}$.

10.2   Minkowski's Theorem

Let's start with a question:

Question 10.6. Suppose $\Lambda \subset \mathbb{R}^n$ is a lattice and let $S \subset \mathbb{R}^n$ be some subset . Under what conditions on S and Λ does S contain a non-zero point of Λ?

It is impossible to give exact necessary and sufficient conditions in this generality. In this section we state and prove an important theorem of Minkowski from 1896, Theorem 10.10, that gives necessary conditions for the existence of a point as asked in the question in some fairly narrow special cases. The surprising thing is that this theorem has some powerful applications in number theory. Our discussion of Minkowski's Theorem, while not particularly complicated, is, unfortunately, fairly abstract. It is only in the later parts of this chapter, starting with §10.3, that the relevance of what we do in this section to our concrete problems becomes clear.

First some preparation. If x is a vector in $\mathbb{R}^n$ and $S \subset \mathbb{R}^n$, we define

$$x + S = \{x + s \mid s \in S\}.$$

Hence $x + S$ is obtained from shifting the whole set S by the vector x. For example, if S is the disk of radius r centered at the origin in $\mathbb{R}^2$, $x + S$ will be the disk of radius r centered at x.

Lemma 10.7. *Let Λ be a lattice in $\mathbb{R}^n$, and let $\mathscr{B} = \{v_1, \ldots, v_n\}$ be a basis of $\mathbb{R}^n$ such that $\Lambda = \Lambda_{\mathscr{B}}$. Then,*

$$\mathbb{R}^n = \bigcup_{\lambda \in \Lambda} (\lambda + \mathscr{P}_{\mathscr{B}}), \tag{10.1}$$

a disjoint union.

Proof. Let $v \in R^n$. Since $\mathscr{B}$ is a basis, we can write

$$v = r_1 v_1 + \ldots r_n v_n,$$

for $r_i \in \mathbb{R}$. Next, for each i, write $r_i = [r_i] + \{r_i\}$. We obtain

$$v = \sum_{i=1}^{n} [r_i] v_i + \sum_{i=1}^{n} \{r_i\} v_i.$$

It is clear that $\sum_{i=1}^{n} [r_i] v_i \in \Lambda$ and $\sum_{i=1}^{n} \{r_i\} v_i \in \mathscr{P}_{\mathscr{B}}$. Now we show the union in (10.1) is disjoint. Suppose for vectors $\lambda, \lambda' \in \Lambda$, we have

$$(\lambda + \mathscr{P}_{\mathscr{B}}) \cap (\lambda' + \mathscr{P}_{\mathscr{B}}) \neq \varnothing. \tag{10.2}$$

Write

$$\lambda = \sum_{i=1}^{n} c_i v_i, \quad \lambda' = \sum_{i=1}^{n} c'_i v_i$$

for integers $c_1, c'_1, \ldots, c_n, c'_n$. Equation (10.2) means that there are real numbers $\alpha_1, \alpha'_1, \ldots, \alpha_n, \alpha'_n$ with $0 \leq \alpha_i, \alpha'_i < 1$ for each i such that

$$\sum_{i=1}^{n} c_i v_i + \sum_{i=1}^{n} \alpha_i v_i = \sum_{i=1}^{n} c'_i v_i + \sum_{i=1}^{n} \alpha'_i v_i.$$

Consequently,

$$\sum_{i=1}^{n} (c_i + \alpha_i) v_i = \sum_{i=1}^{n} (c'_i + \alpha'_i) v_i.$$

Since $\mathscr{B}$ is a basis of $\mathbb{R}^n$, this last identity implies that for all i we have

$$c_i + \alpha_i = c'_i + \alpha'_i,$$

from which it immediately follows that $c_i = c'_i$ for all i. Hence, $\lambda = \lambda'$, and we are done. $\square$

If we consider the example considered earlier where $n = 2$, $\mathscr{B} = \{(2, 1), (1, 3)\}$, we get the partition of $\mathbb{R}^2$ as a union of parallelograms as in Figure 10.3. (Care is needed about the boundary of each parallelogram!)

The following simple proposition is fundamental:

Proposition 10.8 (Minkowski). *Let Λ be a lattice in $\mathbb{R}^n$. Suppose U is an open set in $\mathbb{R}^n$ such that* Vol $U >$ Vol Λ. *Then there are distinct vectors $u_1, u_2 \in U$ such that $u_1 - u_2 \in \Lambda$.*

Proof. The starting point is Equation (10.1). Intersecting with the open set U gives

$$U = \bigcup_{\lambda \in \Lambda} \{(\lambda + \mathscr{P}_{\mathscr{B}}) \cap U\},$$

Fig. 10.3 The partition of
$\mathbb{R}^2$ as a union of the
translates of the fundamental
domain as in Lemma 10.7

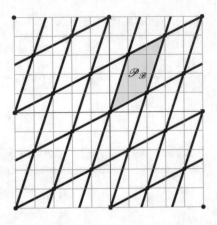

a disjoint union. Now we consider the volume of the set U. Since the right-hand side
of the above equation is a disjoint union we have

$$\text{Vol}\, U = \sum_{\lambda \in \Lambda} \text{Vol}\, \{(\lambda + \mathscr{P}_{\mathscr{B}}) \cap U\}. \tag{10.3}$$

Next, since volume in $\mathbb{R}^n$ is translation invariant, we have

$$\text{Vol}\, \{(\lambda + \mathscr{P}_{\mathscr{B}}) \cap U\} = \text{Vol}\, \{\mathscr{P}_{\mathscr{B}} \cap (-\lambda + U)\}.$$

Denote the set $\mathscr{P}_{\mathscr{B}} \cap (-\lambda + U)$ by $\mathscr{P}_\lambda$. Note that $\mathscr{P}_\lambda \subset \mathscr{P}_{\mathscr{B}}$. Since by assumption
$\text{Vol}\, U > \text{Vol}\, \Lambda = \text{Vol}\, \mathscr{P}_{\mathscr{B}}$, Equation (10.3) gives

$$\text{Vol}\, \mathscr{P}_{\mathscr{B}} < \sum_{\lambda \in \Lambda} \text{Vol}\, \mathscr{P}_\lambda.$$

This equation implies that there are distinct elements $\lambda_1, \lambda_2 \in \Lambda$ such that $\mathscr{P}_{\lambda_1} \cap$
$\mathscr{P}_{\lambda_2} \neq \varnothing$. This means that there are $u_1, u_2 \in U$ such that $-\lambda_1 + u_1 = -\lambda_2 + u_2$ with.
This last equation implies the statement of the proposition with $\lambda = \lambda_1 - \lambda_2$. □

Before we state the main theorem of this chapter we need a couple of definitions.

Definition 10.9. Let S be a set in $\mathbb{R}^n$.

- We call S *symmetric* if $x \in S$ implies $-x \in S$.
- We call S *convex* if for $x, y \in S$ and $0 < \alpha < 1$ we have

$$\alpha x + (1 - \alpha)y \in S,$$

  i.e., if $x, y \in S$, the line segment connecting x and y lies in S.

The quintessential example of a convex symmetric set in $\mathbb{R}^2$ is a filled ellipse of the
form

$$\frac{x^2}{a^2} + \frac{y^2}{b^2} < 1.$$

A particularly important example that makes an appearance later in this chapter is a disk

$$x^2 + y^2 < r^2.$$

The set

$$1 < x^2 + y^2 < 4$$

is symmetric but not convex, and the disk

$$(x-2)^2 + y^2 < 3$$

is convex but not symmetric.

We can now state and prove the following important theorem:

Theorem 10.10 (Minkowski). *Let Λ be a lattice in $\mathbb{R}^n$. Suppose S is a convex symmetric open set in $\mathbb{R}^n$ such that $\mathrm{Vol}\, S > 2^n \mathrm{Vol}\, \Lambda$. Then there is a non-zero vector in the intersection $\Lambda \cap S$.*

Proof. Let $S' = (1/2)S$ be the scaled down version of S. Then S' is open and $\mathrm{Vol}\, S' > \mathrm{Vol}\, \Lambda$. By Proposition 10.8, there are distinct elements $u_1, u_2 \in S$ such that

$$u := \frac{u_1}{2} - \frac{u_2}{2} \in \Lambda - \{0\}.$$

Since $u_2 \in S$ and S is symmetric, $-u_2 \in S$. Also $(u_1 - u_2)/2 \in S$ as S is convex and $(u_1 - u_2)/2$ is the middle point of the line segment connecting $u_1, -u_2 \in S$. The theorem is proved. $\square$

Example 10.11. A lattice Λ is called *unimodular* if $\mathrm{Vol}\, \Lambda = 1$. Let Λ be a unimodular lattice in $\mathbb{R}^2$. Define the set $S_r \subset \mathbb{R}^2$ to be the disk

$$x^2 + y^2 < r^2.$$

For each $r > 0$, S_r is convex, symmetric, and open, and has area πr^2. If $r > 2/\sqrt{\pi} = 1.12837916709551...$, then $\mathrm{Vol}\, S_r > 4 = 2^2 \cdot \mathrm{Vol}\, \Lambda$. Theorem 10.10 implies that the set $A_r := S_r \cap \Lambda - \{(0,0)\}$ is a finite non-empty set. Basic properties of compact sets, e.g., [41, Theorem 2.36], imply that

$$A := \bigcap_{r > 2/\sqrt{\pi}} (S_r \cap \Lambda - \{(0,0)\}) = \overline{S_{2/\sqrt{\pi}}} \cap \Lambda - \{(0,0)\}$$

is non-empty. This means that every unimodular lattice $\Lambda \subset \mathbb{R}^2$ contains at least one non-zero vector v whose length is less than or equal to $2/\sqrt{\pi}$. This result can be generalized to every dimension; see Exercise 10.13.

Despite its innocent abstract appearance, Theorem 10.10 is a powerful result with many applications. In the remainder of this chapter we give three applications of this theorem to questions involving sums of squares.

10.3   Sums of two squares

In our first application we give a second proof of Theorem 5.7.

The second proof of Theorem 5.7. Recall that the non-trivial part of Theorem 5.7 is the statement that every prime p of the form $4k + 1$ is a sum of two squares. As we observed in our proof of Theorem 5.7, it suffices to find a pair of integers (u, v) with the following properties:

1. $u^2 + v^2 < 2p$;
2. $p | u^2 + v^2$;
3. $(u, v) \neq (0, 0)$.

Consider the set

$$S = \{(x, y) \in \mathbb{R}^2 \mid x^2 + y^2 < 2p\}.$$

The set S is convex, symmetric, and open. Also, $\text{Vol } S = 2\pi p$. In order to apply Theorem 10.10, we need to find a lattice Λ such that

(i) $4\text{Vol } \Lambda < 2\pi p$;
(ii) for all $(a, b) \in \Lambda$, $p \mid a^2 + b^2$.

Note that since p is of the form $4k + 1$, by (6.3), there is z such that $z^2 + 1 \equiv 0 \bmod p$. Clearly, the sensible thing to do is to use z to construct the lattice. Consider the vectors:

$$v_1 = (p, 0), \quad v_2 = (z, 1).$$

We have

$$\det \begin{pmatrix} p & 0 \\ z & 1 \end{pmatrix} = p \neq 0,$$

hence the vectors v_1, v_2 are linear independent. Let Λ be the lattice generated by v_1, v_2. By Proposition 10.3, $\text{Vol } \Lambda = p$. Since $4p < 2\pi p$, condition (ii) is satisfied. Next, we verify (i). A typical vector in Λ can be written as $c_1 v_1 + c_2 v_2$ with $c_1, c_2 \in \mathbb{Z}$. We compute the coordinates of the vector as

$$c_1 v_1 + c_2 v_2 = (c_1 p + c_2 z, c_2).$$

We compute the sum of the squares of the coordinates to obtain

$$(c_1 p + c_2 z)^2 + c_2^2 \equiv c_2^2 (z^2 + 1) \equiv 0 \pmod p.$$

This finishes the proof. $\square$

10.4 Sums of four squares

In this section and the next we give a proof of Theorem 9.8. Our goal in this section is to show that every natural number is a sum of four squares following an idea of Davenport [71]. Davenport gives credit to Hermite (1830) for this proof, though Hermite did not have Theorem 10.10 at his disposal, so he had to use other methods. Davenport notes this is a very non-trivial result. According to [71], Euler tried to prove the result unsuccessfully many times between 1730 and 1750, see [90]. This is a testimony to the effectiveness of Minkowski's innocuous looking theorem. We will present another proof of the Four Squares Theorem in Chapter 11 where we will use quaternions.

We start with an identity discovered by Euler [90]. This identity is the analogue of Lemma 5.3 in this setting.

Lemma 10.12 (Euler's identity). *For all complex numbers a, b, c, d, e, f, g, h,*

$$(a^2 + b^2 + c^2 + d^2)(e^2 + f^2 + g^2 + h^2) =$$
$$(ae - bf - cg - dh)^2 + (af + be + ch - dg)^2$$
$$+(ag - bh + ce + df)^2 + (ah + bg - cf + de)^2$$

Proof. See Exercise 10.14 or Lemma 11.4 for a conceptual proof. □

Lemma 10.12 implies that in order to prove that every natural number is the sum of four squares, it suffices to prove that every prime number is a sum of four squares. Since

$$2 = 1^2 + 1^2 + 0^2 + 0^2,$$

we just need to prove the assertion for an odd prime p. As in the case of the Two Squares Theorem, we need to show that there are integers u, v, w, t such that

1. $u^2 + v^2 + w^2 + t^2 < 2p$;
2. $p \mid u^2 + w^2 + v^2 + t^2$;
3. $(u, v, w, t) \neq (0, 0, 0, 0)$.

By Exercise 9.4 the volume of the set S in $\mathbb{R}^4$ defined by

$$S = \{(u, v, w, t) \in \mathbb{R}^4 \mid u^2 + v^2 + w^2 + t^2 < 2p\}$$

is

$$\frac{\pi^2}{2}(\sqrt{2p})^4 = 2\pi^2 p^2.$$

Also, we note that the set S is convex, symmetric, and open. In order to apply Theorem 10.10 we need to construct a lattice Λ such that

(i) $16 \text{Vol } \Lambda < \text{Vol } S$;
(ii) for all $(a, b, c, d) \in \Lambda$, $p \mid a^2 + b^2 + c^2 + d^2$.

We need a lemma:

Lemma 10.13. *If p is an odd prime number, there are integers r, s such that*

$$r^2 + s^2 + 1 \equiv 0 \mod p.$$

Proof. We define functions f, g from $\mathbb{Z}$ to $\mathbb{Z}/p\mathbb{Z}$, the set of congruence classes modulo p, by setting $f(x) = x^2 \mod p$ and $g(x) = -1 - x^2 \mod p$. The assertion of the lemma is equivalent to the existence of integers r, s such that $f(r) = g(s)$. We claim

$$\#f(\mathbb{Z}) = \#g(\mathbb{Z}) = \frac{p+1}{2},$$

where here, for example, $f(\mathbb{Z})$ is the image of the function f and $\#f(\mathbb{Z})$ is the number of elements in the image. We will prove the statement involving f; the one for g follows similarly. The first point to note is that if $x \equiv y \mod p$, then $f(x) = f(y)$ as an element of $\mathbb{Z}/p\mathbb{Z}$. This means that we may in fact think of f as a function from $\mathbb{Z}/p\mathbb{Z}$ to $\mathbb{Z}/p\mathbb{Z}$. Next, $f(x) = f(y)$ if and only if $x \equiv \pm y \mod p$. Now, if $x \not\equiv 0 \mod p$, then $x \not\equiv -x \mod p$. This means that f is 2-to-1 for non-zero congruence classes. Since there are $p - 1$ non-zero congruence classes, there will be $(p - 1)/2$ elements in the images of these classes. We also need to account for $f(0) = 0$. Consequently, the total number of elements in the image of f is $(p - 1)/2 + 1 = (p + 1)/2$. This finishes the proof of $\#f(\mathbb{Z}) = (p + 1)/2$. As mentioned above, the proof of $\#g(\mathbb{Z}) = (p + 1)/2$ is similar. Now, we notice

$$\#f(\mathbb{Z}) + \#g(\mathbb{Z}) = \frac{p+1}{2} + \frac{p+1}{2} = p + 1 > p,$$

hence by the Pigeon-Hole Principle, Theorem A.7, there has to be an overlap between the images of the functions f, g. $\square$

Fix r, s as in the lemma, and consider the four vectors

$$v_1 = (p, 0, 0, 0), \quad v_2 = (0, p, 0, 0), \quad v_3 = (r, s, 1, 0), \quad v_4 = (s, -r, 0, 1).$$

We have

$$\det \begin{pmatrix} p & 0 & 0 & 0 \\ 0 & p & 0 & 0 \\ r & s & 1 & 0 \\ s & -r & 0 & 1 \end{pmatrix} = p^2 \neq 0,$$

and consequently the vectors $\{v_1, v_2, v_3, v_4\}$ are linearly independent. If Λ is the lattice generated by these vectors, Proposition 10.3 implies that $\mathrm{Vol}\,\Lambda = p^2$. Note that since $2\pi^2 > 16$,

$$16\mathrm{Vol}\,\Lambda < \mathrm{Vol}\,S.$$

Now we can apply Theorem 10.10 to conclude that there is $(a, b, c, d) \in S \cap \Lambda$ such that $(a, b, c, d) \neq (0, 0, 0, 0)$. Next, we show that if $(a, b, c, d) \in \Lambda$, then

$$p \mid a^2 + b^2 + c^2 + d^2.$$

In order to see this, we write

$$(a, b, c, d) = c_1 v_1 + c_2 v_2 + c_3 v_3 + c_4 v_4$$

with $c_1, c_2, c_3, c_4 \in \mathbb{Z}$. Then

$$\begin{cases} a = c_1 p + c_3 r + c_4 s, \\ b = c_2 p + c_3 s - c_4 r, \\ c = c_3, \\ d = c_4. \end{cases}$$

Finally,

$$a^2 + b^2 + c^2 + d^2 \equiv (c_1 p + c_3 r + c_4 s)^2 + (c_2 p + c_3 s - c_4 r)^2 + c_3^2 + c_4^2$$

$$\equiv (c_3 r + c_4 s)^2 + (c_3 s - c_4 r)^2 + c_3^2 + c_4^2$$

$$\equiv c_3^2 (r^2 + s^2 + 1) + c_4^2 (r^2 + s^2 + 1) \equiv 0 \mod p$$

by the choices of r, s. This finishes the proof of the Four Square Theorem.

Remark 10.14. Davenport's original proof [71] differs slightly from the above argument. Davenport notes that if

$$m = x^2 + y^2 + z^2 + t^2,$$

then

$$2m = (x + y)^2 + (x - y)^2 + (z + t)^2 + (z - t)^2.$$

So it suffices to prove the theorem for odd m. So we assume that m is an odd natural number. There are integers r, s such that

$$r^2 + s^2 + 1 \equiv 0 \mod m.$$

Then consider the four vectors

$$v_1 = (m, 0, 0, 0), \quad v_2 = (0, m, 0, 0), \quad v_3 = (r, s, 1, 0), \quad v_4 = (s, -r, 0, 1).$$

and form the lattice Λ generated by them. The remainder of the argument is identical to what was presented above. Davenport's clever idea of reducing the general case to the odd m case should be compared to the division-by-$(1 + i)$ step in the proof of the Four Square Theorem presented in §11.3.

10.5   Sums of three squares

We now give a proof of the only remaining statement of Theorem 9.8 that a positive integer m is expressible as a sum of three squares if and only if m is not of the form $4^a(8n + 7)$. We will give one more proof of this fact using the theory of quadratic forms in §12.4.

The fact that numbers of the form $4^a(8n + 7)$ are not expressible as a sum of three squares is not hard; see Exercise 10.17; however, the fact that a number m not of the form $4^a(8n + 7)$ is expressible as a sum of three squares is a much harder theorem. There are several proofs of this result available in literature. We will present a beautiful proof due to Dirichlet in Chapter 12 following the exposition of the classical text by Landau [31]. The remarkable proof we give in this chapter is due to Ankeny [61].

It is clear that we may assume that m is square-free. Following [61], we present the detailed proof for the case where $m \equiv 3 \bmod 8$ to illustrate the method, and refer the reader to the exercises for the remaining cases.

Suppose $m = p_1 \cdots p_r$ is a square-free integer such that $m \equiv 3 \bmod 8$.

Step 1. There is a prime number q such that

- For each i, $1 \le i \le r$,

$$\left(\frac{-2q}{p_i}\right) = +1;$$

- $q \equiv 1 \bmod 4$.

To see this, we note that each condition $(-2q/p_i) = +1$ means that q belongs to some congruence classes modulo p_i. The Chinese Remainder Theorem 2.24 then implies that there is a congruence condition of the form $q \equiv a \bmod 4m$ such that all of these conditions are satisfied. Dirichlet's Arithmetic Progression Theorem, Theorem 5.11 in Notes to Chapter 5, implies the existence of infinitely many primes q with this property.

Step 2. There is an odd integer b and an integer h such that

$$b^2 - 4hq = -m.$$

To see this, we first have to show

$$\left(\frac{-m}{q}\right) = +1.$$

In fact,

$$+1 = \prod_{i=1}^{r}\left(\frac{-2q}{p_i}\right) = \prod_{i=1}^{r}\left(\frac{-2}{p_i}\right)\left(\frac{q}{p_i}\right)$$

$$= \left(\frac{-2}{m}\right)\prod_{i=1}^{r}\left(\frac{q}{p_i}\right).$$

Here $(-2/m)$ is the Jacobi symbol of §7.2. By Quadratic Reciprocity,

$$\left(\frac{q}{p_i}\right) = (-1)^{\frac{q-1}{2}\cdot\frac{p_i-1}{2}}\left(\frac{p_i}{q}\right) = \left(\frac{p_i}{q}\right)$$

as $q \equiv 1 \bmod 4$. Hence,

$$+1 = \left(\frac{-2}{m}\right)\prod_i\left(\frac{p_i}{q}\right) = \left(\frac{-2}{m}\right)\left(\frac{m}{q}\right).$$

This means

$$\left(\frac{-2}{m}\right) = \left(\frac{m}{q}\right).$$

Next, since $q \equiv 1 \bmod 4$, $(-1/q) = +1$, we have,

$$\left(\frac{-m}{q}\right) = \left(\frac{-1}{q}\right)\left(\frac{m}{q}\right) = \left(\frac{m}{q}\right).$$

Putting these identities together gives

$$\left(\frac{-m}{q}\right) = \left(\frac{-2}{m}\right).$$

Next, by Theorem 7.3,

$$\left(\frac{-2}{m}\right) = \left(\frac{-1}{m}\right)\left(\frac{2}{m}\right) = (-1)^{\frac{m-1}{2}}(-1)^{\frac{m^2-1}{8}} = (-1).(-1) = +1,$$

as $m \equiv 3 \bmod 8$. We finally obtain

$$\left(\frac{-m}{q}\right) = +1. \tag{10.4}$$

This means there is an integer b such that $b^2 \equiv -m \bmod q$. By adding q to b if necessary, we assume b is odd. Consequently, there is an integer h_1 such that

$$b^2 - qh_1 = -m.$$

Since b and q are odd, and $m \equiv 3 \bmod 8$, viewing this equation modulo 4 gives $1 - h_1 \equiv +1 \bmod 4$, or $h_1 \equiv 0 \bmod 4$. Write $h_1 = 4h$ to get

$$b^2 - 4qh = -m$$

as claimed.

Step 3. There is an integer t such that

$$t^2 \equiv -1/(2q) \quad \bmod m.$$

In fact, by the choice of q, for each i, there is an integer s_i such that

$$s_i^2 \equiv -2q \pmod{p_i}.$$

If we set $t_i \equiv s_i^{-1} \bmod p_i$, then $t_i^2 \equiv -1/(2q) \bmod p_i$. By the Chinese Remainder Theorem, there is a t modulo m such that for each i, $t \equiv t_i \bmod p_i$. This means $t^2 \equiv t_i^2 \equiv -1/(2q) \bmod p_i$. Consequently, as $m = p_1 \ldots p_r$, $t^2 \equiv -1/(2q) \bmod m$.

Step 4. Define
$$S = \{(u, v, w) \in \mathbb{R}^3 \mid u^2 + v^2 + w^2 < 2m\}.$$

Then S is an open ball in the three-dimensional space, and as such it is convex, symmetric, and open. The volume of S is

$$\frac{4}{3}\pi (2m)^{\frac{3}{2}}.$$

Step 5. We now define a lattice. Set

$$v_1 = (2tq, (2q)^{1/2}, 0), \quad v_2 = (tb, b/(2q)^{1/2}, m^{1/2}/(2q)^{1/2}), \quad v_3 = (m, 0, 0).$$

Since

$$\det \begin{pmatrix} tb & b/(2q)^{1/2} & m^{1/2}/(2q)^{1/2} \\ 2tq & (2q)^{1/2} & 0 \\ m & 0 & 0 \end{pmatrix} = -m^{3/2} \neq 0,$$

$\{v_1, v_2, v_3\}$ is a linearly independent set in $\mathbb{R}^3$. Let Λ be the lattice generated by these vectors. Then Vol $\Lambda = m^{3/2}$.

Step 6. If $(u, v, w) \in \Lambda$, then v, w are not integers. However, we show that $u^2 + v^2 + w^2$ is an integer, and that

$$u^2 + v^2 + w^2 \equiv 0 \pmod{m}.$$

We have for three integers x, y, z,

$$(u, v, w) = xv_1 + yv_2 + zv_3$$

$$= (2tqx + tby + mz, (2q)^{1/2}x + b/(2q)^{1/2}y, m^{1/2}/(2q)^{1/2}y).$$

Consequently,

$$u^2 + v^2 + w^2 = (2tqx + tby + mz)^2 + ((2q)^{1/2}x + b/(2q)^{1/2}y)^2 + (m^{1/2}/(2q)^{1/2}y)^2$$

$$= (2tqx + tby + mz)^2 + \frac{1}{2q}(2qx + by)^2 + \frac{my^2}{2q} \tag{10.5}$$

$$= (2tqx + tby + mz)^2 + 2(qx^2 + bxy + hy^2). \tag{10.6}$$

This shows that $u^2 + v^2 + w^2$ is an integer. We now that it is divisible by m. From (10.5) we have

$$u^2 + v^2 + w^2 \equiv t^2(2qx + by)^2 + (2qx + by)^2/2q \equiv 0 \pmod{m}$$

by the choice of t.

Step 7. Recall Vol $\Lambda = m^{2/3}$ and Vol $S = \frac{4}{3}\pi(2m)^{\frac{3}{2}}$. Since

$$\frac{4}{3}\pi(2m)^{\frac{3}{2}} > 8m^{2/3},$$

we see that Vol $S > 2^3$ Vol Λ. Theorem 10.10 implies that there is a non-zero triple of integers (x_1, y_1, z_1) such that

$$(u_1, v_1, w_1) := x_1v_1 + y_1v_2 + z_1v_3 \in S.$$

Since $(u_1, v_1, w_1) \in S$, we have $u_1^2 + v_1^2 + w_1^2 < 2m$. Step 6 says $u_1^2 + v_1^2 + w_1^2$ is a non-zero integer, and that $m \mid u_1^2 + v_1^2 + w_1^2$. This means

$$u_1^2 + v_1^2 + w_1^2 = m. \qquad (10.7)$$

Now let

$$R_1 = 2tqx + tby + mz, \quad v = qx^2 + bxy + hy^2. \qquad (10.8)$$

The identity (10.6) combined with (10.7) shows

$$m = R_1^2 + 2v. \qquad (10.9)$$

Step 8. Finally, we show that $2v$ is a sum of two squares, and this fact combined with Equation (10.9) finishes the proof of the theorem for $m \equiv 3 \bmod 8$.

It suffices to show that v is a sum of two squares. Indeed, $2 = 1^2 + 1^2$, and by Lemma 5.5 if v is a sum of two squares, $2v$ will be a sum of two squares.

To show that v is a sum of two squares, by Theorem 5.2, we need to show that if $p^{2k+1} \mid v$ but $p^{2k+2} \nmid v$, then $p \equiv 1 \bmod 4$.

There are two cases to consider: $p \nmid m$, and $p \mid m$. We treat each case separately.

If $p \nmid m$, then reducing Equation (10.9) modulo p implies $\left(\frac{m}{p}\right) = +1$. If $p = q$, then by Equation (10.4), $(-1/p) = +1$, and Lemma 6.7 implies $p \equiv 1 \bmod 4$.

Now suppose $p \neq q$. The definition of v from (10.8) shows

$$4qv = (2qx_1 + by_1)^2 + my_1^2.$$

This equation implies that p^{2k+1} divides an expression of the form $e^2 + mf^2$, but p^{2k+2} does not. Consequently, again,

$$\left(\frac{-m}{p}\right) = +1.$$

Again, as before, $(-1/p) = 1$, and $p \equiv 1 \bmod 4$. This settles the case where $p \nmid m$.

Now we consider the case where $p \mid m$. Recall that we have

$$R_1^2 + 2v = m.$$

This identity implies that $p \mid R_1$. We can now rewrite this equation in the following form

$$R_1^2 + \frac{1}{2q}((2qx_1 + by_1)^2 + my_1^2) = m,$$

and this identity implies $p|(2qx_1 + by_1)$. Since by assumption m is square-free, these statements show

$$\frac{1}{2q}\frac{m}{p}y_1^2 \equiv \frac{m}{p} \quad \text{mod } p,$$

or what is the same

$$y_1^2 \equiv 2q \quad \text{mod } p.$$

Consequently, $(2q/p) = +1$. Recall from Step 1 that since $p \mid m$, we have $(-2q/p) = +1$. Hence, $(-1/p) = +1$, and again we arrive at the conclusion that $p \equiv 1 \bmod 4$.

For the cases where $m \equiv 1, 2, 5, 6 \bmod 8$, see Exercise 10.19.

Exercises

10.1 Find bases $\mathcal{B}$ and $\mathcal{B}'$ of $\mathbb{R}^2$ which generate the same lattice but $\mathcal{P}_{\mathcal{B}} \neq \mathcal{P}_{\mathcal{B}'}$.

10.2 Prove Proposition 10.3.

10.3 Show that the volume of $\mathcal{P}_{\mathcal{B}}$ is independent of the choice of the basis of $\mathcal{B}$ that generates the lattice Λ.

10.4 Show that for $n > 3$ we have

$$\det((ij - 1)^2)_{1 \leq i, j \leq n} = 0.$$

10.5 Show that for all $n > 4$

$$\det((ij - 1)^3)_{1 \leq i, j \leq n} = 0.$$

10.6 Generalize the previous two problems by showing that for natural numbers n, k, if $n > k + 1$, we have

$$\det((ij - 1)^k)_{1 \leq i, j \leq n} = 0.$$

10.7 Define a matrix $D = (a_{ij})_{1 \leq i, j \leq n}$ by setting

$$a_{ij} = \begin{cases} 0 & i = j; \\ 1 & i < j; \\ -1 & i > j. \end{cases}$$

Show that

$$\det D = \begin{cases} 0 & n \text{ odd}; \\ 1 & \text{otherwise}. \end{cases}$$

10.8 Define a matrix $E = (b_{ij})_{1 \leq i, j \leq n}$ by setting

$$b_{ij} = \begin{cases} 1 + x^2 & i = j; \\ x & |i - j| = 1; \\ 0 & \text{otherwise.} \end{cases}$$

Compute $\det E$. Hint: Let $D_n = \det E$. Show that for $n \geq 3$ we have $D_n - D_{n-1} = x^2(D_{n-1} - D_{n-2})$.

10.9 For three complex numbers α, β, γ, and $r \in \mathbb{N}$, set $\sigma_r = \alpha^r + \beta^r + \gamma^r$. Show that for $n \in \mathbb{N}$,

$$\det(\sigma_{n+i+j-2})_{1 \leq i,j \leq 3} = (\alpha\beta\gamma)^n \{(\alpha - \beta)(\beta - \gamma)(\gamma - \alpha)\}^2.$$

10.10 Define a matrix $F_n = (c_{ij})_{1 \leq i,j \leq n}$ by setting $c_{ij} = 1 + \delta_{ij}x$, where δ_{ij} is Kronecker's delta. Compute $f_n(x) := \det F_n$ by showing that $f_n'(x) = n f_{n-1}(x)$.

10.11 Define a subset $\Lambda \subset \mathbb{Z}^2$ by setting

$$\Lambda = \{(x, y) \in \mathbb{Z}^2 \mid x + y \equiv 0 \bmod 3\}.$$

Show that Λ is a lattice by finding a basis $\mathscr{B}$ of $\mathbb{R}^2$ such that $\Lambda = \Lambda_{\mathscr{B}}$. Compute Vol Λ.

10.12 Fix a prime p. Define a subset $\Lambda_{p,n} \subset \mathbb{Z}^n$ by setting

$$\Lambda_{p,n} = \left\{ (x_1, \ldots, x_n) \in \mathbb{Z}^n \mid \sum_{i=1}^{n} x_i \equiv 0 \bmod p \right\}.$$

Prove that $\Lambda_{p,n}$ is a lattice. Compute Vol $\Lambda_{p,n}$.

10.13 Verify the details of the argument in Example 10.11. Generalize to all $\mathbb{R}^n$.

10.14 Prove Lemma 10.12 by direct computation.

10.15 Supply the details of Davenport's proof of the Four Square Theorem.

10.16 (✠) Write 4594043492117928 as a sum of four squares. In how many ways is it possible to do this?

10.17 Show that a number of the form $4^a(8n + 7)$ is not expressible as a sum of three squares.

10.18 (✠) Can you write 4594043492117928 as a sum of three squares? In how many ways?

10.19 Complete Ankeny's proof of the remaining cases of the Three Square Theorem, i.e., for the cases where $m \equiv 1, 2, 5, 6 \bmod 8$. Let q be prime, $(-q/p_j) = +1$ for all odd prime divisors of m, and $q \equiv 1 \bmod 4$, and if m is even, $m = 2m_1$, $(-2/q) = (-1)^{(m_1-1)/2}$, $t^2 \equiv -1/q \bmod p_j$, t odd, $b^2 - qh = -m$, and $v_1 = (tq, q^{1/2}, 0)$, $v_2 = (tb, b/q^{1/2}, m^{1/2}/q^{1/2})$, $v_3 = (m, 0, 0)$.

10.20 Show that every integer can be represented as a sum of five cubes of integers in infinitely many ways. Show that 3 can be written as a sum of four cubes not equal to 0 or 1 in infinitely many ways.

Notes

Waring's Problem

In 1770 Edward Waring asked whether for a natural number k, there is an integer s, depending on k, such that every natural number could be written as the sum of at most s natural numbers everyone of which is a k-th power. If the answer is yes, then the smallest possible value of s is denoted by $g(k)$. For example, in this chapter we showed that every natural number is the sum of at most four perfect squares. We also saw that there are many integers that are not sums of three squares. This means that $g(2) = 4$. David Hilbert showed in 1909 that the answer to Waring's question was yes. The first few values of $g(k)$ are as follows: $g(2) = 4$, $g(3) = 9$, $g(4) = 19$, $g(5) = 37$, $g(6) = 73$, etc. The sequence of integers $g(k)$ appears as sequence A002804 in the *Online Encyclopedia of Integer Sequences* available at

$$\text{https://oeis.org/A002804}$$

The following conjecture dates back to the 19th century:

Conjecture 10.15 *(The Ideal Waring's Theorem).* For all k we have

$$g(k) = 2^k + [(3/2)^k] - 2,$$

where in this formula $[x]$ denotes the integer part of a real number x.

It is a theorem of L. E. Dickson and S. S. Pillai from 1936 that this formula for $g(k)$ holds if

$$2^k\{(3/2)^k\} + [(3/2)^k] \le 2^k. \tag{10.10}$$

This last inequality is known to be true for $k \le 471,600,000$, and for k large enough by a result of K. Mahler from 1957. Equation (10.10) is expected to hold for all k; see [106]. In fact, it is a result of David, Waldschmidt, and Laishram [82, 107] that an explicit form of the *abc* Conjecture implies the Ideal Waring's Theorem. For the statement of the *abc* Conjecture and its explicit form see Notes to Chapter 3.

   A related sequence which is considerably more difficult to study is the sequence $G(k)$ defined as the smallest positive integer s such that every *sufficiently large* positive integer can be written as the sum of s k-th powers. It is clear that $G(k) \le g(k)$. One could, however, imagine that there may exist some rogue integers early on that require a lot of k-powers, but past a certain point the situation would stabilize. The only values of $G(k)$ that are currently known are $G(2) = 4$ and $G(4) = 16$ obtained in 1939 by Davenport. It appears that the best available upper bound for $G(k)$ is provided by Trevor Wooley in 1995:

$$G(k) \le k(\log k + \log\log k + 2 + O(\log\log k/\log k)).$$

This should be compared with the conjectured formula for $g(k)$ mentioned earlier. The conjectured value of $g(k)$ grows exponentially with k, whereas the inequality proved by Wooley shows that $G(k)$ grows essentially in a linear fashion. (Professor Ram Murty often jokes that analytic number theorists say "log, log, log" when they drown.)

A powerful method that has been employed to prove many of the results related to Waring's Problem, and other additive questions in number theory, is the *Circle Method* originally invented by Hardy and Ramanujan around 1916. The idea is to define a function

$$f_k(x) = \sum_{n=0}^{\infty} e^{2\pi i n^k x}$$

on the interval $[0, 1]$. Fix a natural number s, and suppose we wish to show that every natural number is the sum of s k-th powers. We have

$$f_k(x)^s = \sum_{n_1} \sum_{n_2} \cdots \sum_{n_s} e^{2\pi i x \sum_{j=1}^{s} n_j^k} = \sum_{n=0}^{\infty} R_s(n) e^{2\pi i n x}$$

with $R_s(n)$ being the number of representations of n as a sum of s k-th powers. Theorem A.3 now implies that for each l,

$$\int_0^1 f_k(x)^s e^{-2\pi i l x} \, dx = \sum_{n=0}^{\infty} R_s(n) \int_0^{\infty} e^{2\pi i (k-l) x} \, dx = \sum_{n=0}^{\infty} R_s(n) \delta_{nl} = R_s(l),$$

with δ_{nl} being Kronecker's delta. So in order to show that $R_s(l) \neq 0$ to gain information about $g(k)$, or $R_s(l) \neq 0$ for l large enough to gain information about $G(k)$ one needs to show that

$$\int_0^1 f_k(x)^s e^{-2\pi i l x} \, dx \neq 0.$$

In order to see that this integral is non-zero the idea is to concentrate on those x's for which the value of $f(x)$ is large. Note that if x is a rational number, then $e^{2\pi i k^2 x} = 1$ for infinitely many k. So for such x, the function $f(x)$ *blows up*. The *art* of the Circle Method is to partition $[0, 1]$ to two pieces: $\mathfrak{M}$, called the *major arcs*, consisting of those x's which are *close* to a rational number with small denominator, and $\mathfrak{m}$, the *minor arcs*, the complement of $\mathfrak{M}$. Then we have

$$\int_0^1 f_k(x)^s e^{-2\pi i l x} \, dx = \int_{\mathfrak{M}} f_k(x)^s e^{-2\pi i l x} \, dx + \int_{\mathfrak{m}} f_k(x)^s e^{-2\pi i l x} \, dx.$$

In most applications, including Waring's Problem, the major arcs integral is not too hard to analyze to obtain a fairly explicit asymptotic formula. The real problem is to show that the contribution of the major arcs is not canceled out by the integral over the minor arcs. To see how this is done in a series of instructive examples, see Vaughan [54]. To see the analysis of the major arcs in a situation where we do not know how to handle the minor arcs see [33, Ch. 14]. A major new development in applications of the Circle Method is Harald Helfgott's recent proof of the Ternary Goldbach Conjecture which asserts that every odd integer larger than 5 is the sum of three prime numbers, available at

https://arxiv.org/abs/1312.7748

Geometry of numbers

Minkowski's Theorem 10.10 and Gauss' Circle Theorem 9.4 belong to an area of mathematics called *geometry of numbers*. Minkowski proved Theorem 10.10 in the course of his work on quadratic forms in relation to Diophantine approximation. To get a feel for the sort of problem Minkowski was interested in, suppose we want to study minimal values of positive definite quadratic forms on integral points. Note if $f(x_1, \ldots, x_n)$ is a positive definite quadratic form with real coefficients, then the set of real points $(x_1, \ldots, x_n)$ such that $f(x_1, \ldots, x_n) < \lambda$ is a bounded convex symmetric domain of the sort considered in this chapter. The volume of this set is $\mathrm{Vol}\{f < 1\} \cdot \lambda^{\frac{n}{2}}$. Since $\mathrm{Vol}\,\mathbb{Z}^n = 1$, Theorem 10.10 implies that if

$$\mathrm{Vol}\{f < 1\} \cdot \lambda^{\frac{n}{2}} > 2^n, \text{ i.e., } \lambda > \frac{4}{\mathrm{Vol}\{f < 1\}^{2/n}},$$

then there is at least one non-zero integral point $\underline{x} \in \mathbb{Z}^n$ such that $f(\underline{x}) < \lambda$. This means that the minimal value of f on non-zero points in $\mathbb{Z}^n$ is at most

$$\frac{4}{\mathrm{Vol}\{f < 1\}^{2/n}}.$$

Suppose, for example, that $n = 2$, and $f(x_1, x_2) = ax_1^2 + bx_1x_2 + cx_2^2$. The positive-definiteness of f means $a, c > 0$ and $b^2 < 4ac$. In this case, it is a nice exercise to show that

$$\mathrm{Vol}\{f < 1\} = \frac{2\pi}{\sqrt{4ac - b^2}}.$$

Putting everything together, we see that the minimum value of a positive definite quadratic form $f(x_1, x_2) = ax_1^2 + bx_1x_2 + cx_2^2$ on non-zero integral points (x_1, x_2) is at most

$$\frac{2}{\pi}\sqrt{4ac - b^2}.$$

Analogues of the Gauss's Circle Theorem 9.4 appear often in contemporary research papers. In many number theoretic problems one needs to count integral points in a certain domain. Ideally one should be able to replace the number of integral points by the area of the region, which is often not too hard to compute, plus an error term contributed by the points along, or close to, the boundary. However, in order to bound the error one needs to show that there are not too many points on the boundary of the regions, or *tucked away in corners*. This can be a real challenge, as anyone studying the works of Manjul Bhargava (Fields medal, 2014) might notice. Geometry of numbers methods have featured prominently in Bhargava's groundbreaking works. To see an expository article on the work of Bhargava and the role played by the geometry of numbers, see Gross' article [78].

Davenport's little article [71] is a nice entry way to the subject of geometry of numbers. The classical texts by Cassels [11] and Siegel [45] are wonderful introductions to this exciting area.

Chapter 11
Another proof of the four squares theorem

The goal of this chapter is to give a second proof of the Four Squares Theorem. This proof uses the theory of quaternions, which we will briefly discuss. The proof of the Four Squares Theorem in this chapter is in the spirit of the argument for the Two Squares Theorem we presented in Chapter 5 using Gaussian integers. Recall that if we have a complex number $z = x + iy$, then if we define $N(z) = x^2 + y^2$, for complex numbers z, w, we have $N(z \cdot w) = N(z) \cdot N(w)$. We used this identity to reduce the Two Squares Theorem to determining which prime numbers are expressible as a sum of two squares. In this chapter we develop a similar method for sums of four squares. Among other things we provide an "explanation" for why Lemma 10.12 is true, though historically speaking, the theory of quaternions was developed because of Lemma 10.12. In the Notes at the end of this chapter, we introduce *Octonions* that provide a framework for identities involving eight squares.

11.1   Quaternions

We typically think of the set of complex numbers as the two-dimensional real vector space consisting of all expressions of the form

$$a + bi$$

with a, b real numbers, and i a formal symbol satisfying $i^2 = -1$. We define the space of the *quaternions* similarly.

Definition 11.1. We define $\mathbb{H}$, the *space of Hamilton quaternions*, to be the four-dimensional real vector space consisting of all elements of the form

$$x = a + bi + cj + dk$$

with a, b, c, d real numbers, and i, j, k formal symbols commuting with real numbers and satisfying

© Springer Nature Switzerland AG 2018

R. Takloo-Bighash, *A Pythagorean Introduction to Number Theory*,
Undergraduate Texts in Mathematics, https://doi.org/10.1007/978-3-030-02604-2_11

$$i^2 = j^2 = k^2 = ijk = -1.$$

We call a, b, c, d the *coordinates* of x.

Theorem 11.2. *The vector space $\mathbb{H}$ is an associative algebra with an identity element.*

The direct proof of this theorem is an excruciating exercise in endurance, and we omit it here. However, we will show in Lemma 11.5 that quaternions can be realized as a set of 2×2 complex matrices. We will use this lemma to give a reasonable proof of the theorem.

It is not hard to check (Exercise 11.1) that

$$ij = -ji = k, \quad jk = -kj = i, \quad ki = -ik = j. \tag{11.1}$$

For example,

$$ij = -ijk^2 = -(ijk)k = -(-1)k = k.$$

One can use these identities to write down the explicit multiplication formula for quaternionic multiplication:

$$(a + bi + cj + dk)(e + fi + gj + hk) =$$

$$(ae - bf - cg - dh) + (af + be + ch - dg)i$$

$$+(ag - bh + ce + df)j + (ah + bg - cf + de)k.$$

The proof of this identity is tedious but completely straightforward. In practice, when multiplying quaternions, we do not use this formula. Instead, we just use standard distribution laws. For example,

$$(1 + 2i) \cdot (2j + 5k) = 1 \cdot 2j + 1 \cdot 5k + 2i \cdot 2j + 2i \cdot 5k = 2j + 5k + 4i \cdot j + 10i \cdot k$$

$$= 2j + 5k + 4k - 10j = -8j + 9k,$$

after using $i \cdot j = k$ and $i \cdot k = -j$.

For a quaternion

$$\tau = a + bi + cj + dk,$$

we define the conjugate of τ, usually denoted by $\overline{\tau}$ in analogy with complex conjugation, by

$$\overline{\tau} = a - bi - cj - dk.$$

Lemma 11.3. *If $\tau_1, \tau_2 \in \mathbb{H}$, then*

$$\overline{\tau_1 \cdot \tau_2} = \overline{\tau_2} \cdot \overline{\tau_1}.$$

Proof. Computation.   □

A straightforward computation shows

$$\tau \cdot \overline{\tau} = a^2 + b^2 + c^2 + d^2 \in \mathbb{R}.$$

The square root of the latter is usually denoted by $|\tau|$, i.e.,

$$|\tau|^2 = \tau \cdot \overline{\tau}.$$

In particular, if $\tau \neq 0$, then $|\tau| \neq 0$. We usually call $|\tau|$ the *length* of τ, and $|\tau|^2$ its *norm*. This has the following interesting consequence:

$$\tau \cdot \frac{1}{|\tau|^2}\overline{\tau} = 1, \tag{11.2}$$

i.e., non-zero quaternions are invertible. In particular, $\mathbb{H}$ is a division ring.

Lemma 10.12 has the following beautiful interpretation:

Lemma 11.4. *If $\tau_1, \tau_2 \in \mathbb{H}$, then*

$$|\tau_1 \cdot \tau_2| = |\tau_1| \cdot |\tau_2|.$$

Proof. We have by Lemma 11.3,

$$|\tau_1 \cdot \tau_2|^2 = \tau_1 \cdot \tau_2 \cdot \overline{\tau_1 \cdot \tau_2} = \tau_1 \cdot \tau_2 \cdot \overline{\tau_2} \cdot \overline{\tau_1} = \tau_1 \cdot |\tau_2|^2 \cdot \overline{\tau_1} = |\tau_2|^2 \cdot \tau_1 \cdot \overline{\tau_1} = |\tau_1|^2 \cdot |\tau_2|^2.$$

$\square$

11.2 Matrix representation

In this section we discuss a method to represent quaternions as 2×2 matrices with complex entries, called *matrix representation*. We will use the matrix representation to prove Theorem 11.2. The representation also clarifies the meaning of Lemma 11.3. Since this representation of quaternions is similar to the matrix representations of complex numbers, we start by recalling the latter as a means to motivate the matrix representation of quaternions.

For a complex number $z = a + ib$, with $a, b \in \mathbb{R}$, we define

$$m_{\mathbb{C}}(z) = \begin{pmatrix} a & b \\ -b & a \end{pmatrix}.$$

Then direct computation shows that for all $z_1, z_2 \in \mathbb{C}$,

$$m_{\mathbb{C}}(z_1 + z_2) = m_{\mathbb{C}}(z_1) + m_{\mathbb{C}}(z_2), \quad m_{\mathbb{C}}(z_1 z_2) = m_{\mathbb{C}}(z_1) m_{\mathbb{C}}(z_2).$$

Furthermore, for $z \in \mathbb{C}$,

$$m_{\mathbb{C}}(\overline{z}) = m_{\mathbb{C}}(z)^T,$$

where for every matrix A, A^T is the transpose of the matrix. And finally,

$$|z|^2 = \det(m(z)).$$

We now explain the matrix representation for quaternions. It is clear that $\mathbb{C} \subset \mathbb{H}$, and in fact every element τ of $\mathbb{H}$ can be written as

$$\tau = x + yj$$

with $x, y \in \mathbb{C}$. We define

$$m_{\mathbb{H}}(\tau) = \begin{pmatrix} x & y \\ -\overline{y} & \overline{x} \end{pmatrix}.$$

Note that if $y = 0$, i.e., $\tau \in \mathbb{C}$, then

$$m_{\mathbb{C}}(\tau) = m_{\mathbb{H}}(\tau).$$

Lemma 11.5.   *1. For $\tau_1, \tau_2 \in \mathbb{H}$,*

$$m_{\mathbb{H}}(\tau_1 + \tau_2) = m_{\mathbb{H}}(\tau_1) + m_{\mathbb{H}}(\tau_2), \quad m_{\mathbb{H}}(\tau_1 \tau_2) = m_{\mathbb{H}}(\tau_1) m_{\mathbb{H}}(\tau_2).$$

2. For $\tau \in \mathbb{H}$,

$$m_{\mathbb{H}}(\overline{\tau}) = \overline{m_{\mathbb{H}}(\tau)}^T.$$

Here, for a complex matrix $A = \begin{pmatrix} a & b \\ c & d \end{pmatrix}$, we define $\overline{A} = \begin{pmatrix} \overline{a} & \overline{b} \\ \overline{c} & \overline{d} \end{pmatrix}$.

3. For $\tau \in \mathbb{H}$,

$$|\tau|^2 = \det m_{\mathbb{H}}(\tau).$$

Proof. This is a computation; see Exercise 11.4.   □

Proof of Theorem 11.2. The only non-trivial part is the associativity of multiplication. Lemma 11.5 shows that $\mathbb{H}$ is a subalgebra of $M_2(\mathbb{C})$ considered as an 8-dimensional algebra over $\mathbb{R}$. Consequently, the associativity follows from the associativity of multiplication of 2×2 matrices.   □

11.3   Four squares

In this section we will explain a proof of the Four Squares Theorem which we learned from Lior Silberman (based on Notes by Matilde Lalin). To see another proof of the Four Squares Theorem using quaternions, see Herstein [25, Ch. 7, §4].

As in §10.4 it is sufficient to show that every odd prime is a sum of four squares. By Lemma 10.13 there are integers r, s such that

$$r^2 + s^2 \equiv -1 \mod p.$$

This means $r^2 + s^2 + 1 = zp$ for some integer z. Since the set $\{-(p-1)/2, -(p-1)/2+1, \ldots, (p-1)/2 - 1, (p-1)/2\}$ is a complete system of residues modulo p, we may assume that $|r|, |s| \le (p-1)/2$, and as a result $zp \le 2(p-1)^2/4 + 1 < p^2$, i.e., $z < p$. We let Q be the set of all quaternions $z = a + bi + cj + dk$ with $a, b, c, d \in \mathbb{Z}$, and Q_p the set of all elements $x \in Q$ with $|x|^2 = mp$ for some integer m with $0 < m < p$. Then $1 + ri + sj \in Q_p$, and in particular $Q_p \ne \varnothing$.

Now, pick an element $\tau = a + bi + cj + dk \in Q_p$ with minimal length among the elements of Q_p. We write $|\tau|^2 = m_0 p$, with $0 < m_0 < p$. Our goal is to show that $m_0 = 1$, i.e., $|\tau|^2 = p$.

Our first observation is that m_0 is odd. Suppose not. Then $|\tau|^2 = a^2 + b^2 + c^2 + d^2 = m_0 p$ is even. Consequently, either $a \equiv b, c \equiv d \mod 2$, or $a \equiv c, b \equiv d \mod 2$, or $a \equiv d, b \equiv c \mod 2$. Without loss of generality, suppose we are in the first situation where $a \equiv b, c \equiv d \mod 2$. Consider, $\tau/(1+i)$. We have

$$\left| \frac{\tau}{1+i} \right|^2 = \frac{|\tau|^2}{|1+i|^2} = \frac{m_0 p}{1^2 + 1^2} = \frac{m_0}{2} \cdot p.$$

So $p \mid |\tau/(1+i)|^2$, but $p^2 \nmid |\tau/(1+i)|^2$. Now we compute $\tau/(1+i)$ explicitly. By Equation (11.2),

$$\frac{\tau}{1+i} = \frac{\tau \cdot (1-i)}{2} = \frac{(a + bi + cj + dk)(1-i)}{2}$$

$$= \left(\frac{a+b}{2} \right) + \left(\frac{-a+b}{2} \right) \cdot i + \left(\frac{c-d}{2} \right) \cdot j + \left(\frac{c+d}{2} \right) \cdot k.$$

Since we assumed $a \equiv b, c \equiv d \mod 2$, it followed that $\tau/(1+i) \in Q_p$, but since $|\tau/(1+i)| < |\tau|$, we reach a contradiction as we had picked τ to have minimal length among the elements of Q_p. If instead of the congruences $a \equiv b, c \equiv d \mod 2$ we had assumed $a \equiv c, b \equiv d \mod 2$, then we would need to consider $\tau/(1+j)$; if $a \equiv d, b \equiv c \mod 2$, then we would consider $\tau/(1+k)$.

Before moving on, let us introduce a piece of notation. For $x_l = a_l + b_l i + c_l j + d_l k$, $l = 1, 2$, elements of Q, and m an integer, write $x_1 \equiv x_2 \mod m$ if $a_1 \equiv a_2, b_1 \equiv b_2, c_1 \equiv c_2, d_1 \equiv d_2 \mod m$. It is easy to check that if $x_1 \equiv x_2 \mod m$, then $\bar{x}_1 \equiv \bar{x}_2 \mod m$. Also if $x_1 \equiv x_2, y_1 \equiv y_2 \mod m$, then $x_1 \pm y_1 \equiv x_2 \pm y_2 \mod m$ and $x_1 \cdot y_1 \equiv x_2 \cdot y_2 \mod m$.

Suppose $m_0 \ne 1$. Pick an element $\sigma \in Q$ with minimal length such that $\sigma \equiv \tau \mod m_0$. Note that $\sigma \ne 0$, as otherwise this would mean that $\tau \equiv 0 \mod m_0$, and its norm would be divisible by m_0^2 which we assume not to be the case, unless of course $m_0 = 1$. Since m_0 is odd, the set of integers $S = \{-\frac{m_0-1}{2}, -\frac{m_0-1}{2} + 1, \ldots, \frac{m_0-1}{2} - 1, \frac{m_0-1}{2}\}$ is a complete system of residues modulo m_0. In particular, as we are making the coordinates as small as possible in their congruence class, we see that if we write $\sigma = u + vi + wj + tk$, then $u, v, w, t \in S$. Then

$$|\sigma|^2 = u^2 + v^2 + w^2 + t^2 \le 4 \left(\frac{m_0 - 1}{2} \right)^2 < m_0^2.$$

Next, since $\sigma \equiv \tau \bmod m_0$, $\overline{\sigma} \equiv \overline{\tau} \bmod m_0$, and as a result $|\sigma|^2 \equiv \sigma \cdot \overline{\sigma} \equiv \tau \cdot \overline{\tau} \equiv m_0 p \equiv 0 \bmod m_0$. So $0 < |\sigma|^2 < m_0^2$ and $|\sigma|^2$ is divisible by m_0, and this implies that $|\sigma|^2 = r m_0$ for some $0 < r < m_0 < p$.

Now let $\xi = \tau \cdot \overline{\sigma}$. Then

$$|\xi|^2 = |\tau|^2 \cdot |\overline{\sigma}|^2 = |\tau|^2 \cdot |\sigma|^2 = m_0^2 r p.$$

Next,

$$\xi \equiv \tau \cdot \overline{\sigma} \equiv \tau \cdot \overline{\tau} \equiv m_0 p \equiv 0 \quad \bmod m_0.$$

This last congruence means that the coordinates of ξ are divisible by m_0. Now, set $\tilde{\xi} = \xi / m_0$. Clearly, $\tilde{\xi} \in Q$, and if $m_0 > 1$, $|\tilde{\xi}|^2 = r p < |\tau|^2$, contradicting the choice of τ as the element with minimal length among the elements of Q_p.

The contradiction shows that $m_0 = 1$. This means that $|\tau|^2 = p$ and we are done.

Exercises

11.1 Verify the statements in (11.1).

11.2 Show that if z is a quaternion such that $z \cdot \tau = \tau \cdot z$ for all $\tau \in \mathbb{H}$, then $z \in \mathbb{R}$.

11.3 Prove Lemma 11.3.

11.4 Prove Lemma 11.5.

11.5 Determine all quaternions $z \in \mathbb{H}$ such that $z^2 + 1 = 0$.

11.6 Let $Q \subset \mathbb{H}$ be the set of all quaternions $z = a + bi + cj + dk$ with $a, b, c, d \in \mathbb{Z}$. Let $z_1, z_2 \in Q$, with $z_2 \ne 0$. Show that it is not always possible to find quaternions $q, r \in Q$ such that $z_1 = q z_2 + r$ and $|r| < |z_2|$.

11.7 Find all $z \in Q$ such that $z^{-1} \in Q$.

11.8 Define the set of *integral quaternions* $\mathbb{H}_{\mathbb{Z}}$ to be

$$\{ a\zeta + bi + cj + dk \mid a, b, c, d \in \mathbb{Z} \}$$

with $\zeta = \frac{1}{2}(1 + i + j + k)$. Show that $\mathbb{H}_{\mathbb{Z}}$ is a subring of $\mathbb{H}$ which is closed under conjugation.

11.9 Show that for all $z \in \mathbb{H}_{\mathbb{Z}}$, $|z|^2 \in \mathbb{Z}$.

11.10 Determine the group of units in $\mathbb{H}_{\mathbb{Z}}$.

11.11 Show that for all $z \in \mathbb{H}_{\mathbb{Z}}$, there are integers b, c such that $z^2 + bz + c = 0$, i.e., α is integral over $\mathbb{Z}$. Is $\mathbb{H}_{\mathbb{Z}}$ the integral closure of $\mathbb{Z}$ in $\mathbb{H}$?

11.12 Let $z_1, z_2 \in \mathbb{H}_{\mathbb{Z}}$, with $z_2 \ne 0$. Show that there are quaternions $q, r \in Q$ such that $z_1 = q z_2 + r$ and $|r| < |z_2|$.

11.13 (✠) Investigate the solutions of the equation $x^2 + 2x + 7 = 0$ in $\mathbb{H}$.

Notes

Octonions

A division algebra over a field k is a ring with identity containing a copy of k such that $ab = 0$ implies that either $a = 0$ or $b = 0$. For example, $\mathbb{C}$ is a division algebra over $\mathbb{R}$. A classical theorem of Frobenius asserts that $\mathbb{R}$, $\mathbb{C}$, and $\mathbb{H}$, respectively of dimension 1, 2, and 4 as vector spaces over $\mathbb{R}$, are the only division algebras over $\mathbb{R}$. Note that in going from $\mathbb{R}$ to $\mathbb{C}$ we had to give up the order relation, and from $\mathbb{C}$ to $\mathbb{H}$ we gave up commutativity. A question that arises is whether we can further relax the definition of a ring by removing associativity to obtain larger rings. The answer is yes. Here we briefly explain the construction of a ring called *Octonions*, denoted by $\mathbb{O}$, which is a non-associative, non-commutative, division algebra of dimension 8 over $\mathbb{R}$. The ring $\mathbb{O}$ has the interesting property that the subalgebra generated by any two elements is associative. It is a theorem going back to 1958, due independently to Kervaire and Bott–Milnor that $\mathbb{R}$, $\mathbb{C}$, $\mathbb{H}$, and $\mathbb{O}$ are the only division algebras over $\mathbb{R}$. We warn the reader that when dealing with non-associative algebras there are many subtleties that one needs to worry about. For example, an associative algebra is a division algebra if and only if every non-zero element has a multiplicative inverse, but this statement may not hold in a non-associative algebra; see [62, §2] for an example. Our reference here is Baez [62]. Another great reference for quaternions and Octonions is the charming book [13].

We define $\mathbb{O}$ to be the 8-dimensional real vector space consisting of all vectors of the form

$$x_0 + x_1 e_1 + \cdots + x_7 e_7$$

with $x_0, \ldots, x_7 \in \mathbb{R}$. We make $\mathbb{O}$ into an algebra by requiring that the e_i's have the following multiplication properties:

- $e_1 e_2 = e_4$;
- $e_1^2 = e_2^2 = \cdots = e_7^2 = -1$;
- For all $i \neq j$, $e_i e_j = -e_j e_i$;
- $e_i e_j = e_k$ implies $e_{i+1} e_{j+1} = e_{k+1}$;
- $e_i e_j = e_k$ implies $e_{2i} e_{2j} = e_{2k}$.

All indices are computed modulo 7 and we take as a complete system of residues modulo 7 the set $\{1, 2, \ldots, 7\}$. For example, since $e_1 e_2 = e_4$, we conclude that $e_2 e_4 = e_8 = e_1$. This latter equality in turn implies $e_3 e_5 = e_2$, etc. This is of course not easy to remember, and [62] contains a couple of different mnemonic devices to remember the multiplication table for Octonions, but since we will not be doing any computations with them in this book we will not review them here. Let us just note here that it follows from the multiplication table that for all i, j, k distinct, we have $(e_i e_j)e_k = -e_i(e_j e_k)$ which shows that the algebra $\mathbb{O}$ is not associative.

A conceptually pleasant method to build the Octonions is the Cayley–Dickson construction which we now explain. We often view complex numbers as pairs of real numbers (a, b), to represent the complex number $a + bi$, with addition done componentwise and multiplication given by

$$(a, b)(c, d) = (ac - db, ad + cb).$$

Complex conjugation is defined by $\overline{(a, b)} = (a, -b)$. We can similarly construct quaternions from complex numbers. Since $ij = k$, we have $a + bi + cj + dk = (a + bi) + (c + di)j$. In this expression, $a + bi, c + di \in \mathbb{C}$, and as a result $\mathbb{H}$ can be identified pairs of complex numbers. Clearly, addition is done componentwise, and a computation shows that

$$(a, b)(c, d) = (ac - d\bar{b}, \bar{a}d + cb), \tag{11.3}$$

and

$$\overline{(a, b)} = (\bar{a}, -b). \tag{11.4}$$

Finally, we define $\mathbb{O}$ to be the collection of pairs of quaternions (a, b) with addition defined componentwise, and multiplication and conjugation defined by Equation (11.3) and Equation (11.4), respectively. We can certainly continue this process, known as the *Cayley–Dickson construction* and build more algebras, but the 16-dimensional algebra constructed from $\mathbb{O}$ will no longer be a division algebra.

We can now establish some basic properties of the algebra $\mathbb{O}$. It is not hard to see that if $(a, b), (c, d) \in \mathbb{O}$ then $\overline{(a, b)} = (c, d)$ if and only if $(a, b) = \overline{(c, d)}$. Also if $(a, b), (c, d) \in \mathbb{O}$, then

$$\overline{(a, b)(c, d)} = \overline{(c, d)} \cdot \overline{(a, b)}.$$

If $(a, b) \in \mathbb{O}$, then

$$(a, b) \cdot \overline{(a, b)} = \overline{(a, b)} \cdot (a, b) = (|a|^2 + |b|^2)(1, 0).$$

Consequently, if $(a, b) \neq (0, 0)$, (a, b) is invertible, and

$$(a, b)^{-1} = \frac{1}{|a|^2 + |b|^2}\overline{(a, b)}.$$

Also, if we set

$$|(a, b)| = \sqrt{(a, b) \cdot \overline{(a, b)}},$$

then

$$|(a, b)(c, d)|^2 = |(a, b)|^2 \cdot |(c, d)|^2.$$

Note that the expression on the right is the product of two sums of eight squares, and the formula expresses this massive product as a sum of eight squares. Compare this identity with Euler's identity, Lemma 10.12. Finally, if $x, y \in \mathbb{O}$, then

$$(xx)y = x(xy), \quad (xy)x = x(yx), \quad (yx)x = y(xx).$$

Octonions have many applications in number theory, algebra, and geometry. We refer the reader to [62] and [13] for a survey of these applications.

Chapter 12
Quadratic forms and sums of squares

Our goal in this chapter is to develop the theory of quadratic forms so we can give another proof of Theorem 9.8, especially in the three square case. Our exposition follows [31, Part 3, Chap IV] closely. We start with the basic theory of quadratic forms and explain the notion of equivalence. We then discuss the concept of representability of an integer by a quadratic form. Since the goal of the chapter is to give a proof of the Three Square Theorem we set the stage by giving a proof of the Two Squares Theorem in §12.2. In this section we develop the theory of binary quadratic forms with integral coefficients, determine representatives for the equivalence classes of positive definite binary quadratic forms of a given discriminant, and use this knowledge to prove the Two Squares Theorem. In the next two sections we develop the analogous theory for ternary quadratic forms and prove the Three Squares Theorem. In the Notes to this chapter, we explain Gauss's beautiful composition law for binary quadratic forms.

12.1 Quadratic forms with integral coefficients

In Chapters 5 and 9 we determined what numbers can be represented as a sum of two, three, or four squares. One way to view these results is to think of them as theorems about the numbers that are represented by certain quadratic forms. For example, if we let

$$f(x, y) = x^2 + y^2,$$

then Theorem 5.2 tells us what $f(\mathbb{Z}^2)$ is. This is an example of a *quadratic form with integral coefficients*.

Definition 12.1. Let $A = (a_{ij})_{1 \le i, j \le n}$ be an $n \times n$ symmetric matrix with integer entries. We call a function $f : \mathbb{Z}^n \to \mathbb{Z}$ defined by

$$f(x_1, \ldots, x_n) = \sum_{\substack{1 \le i \le n \\ 1 \le j \le n}} a_{ij} x_i x_j$$

© Springer Nature Switzerland AG 2018
R. Takloo-Bighash, *A Pythagorean Introduction to Number Theory*,
Undergraduate Texts in Mathematics, https://doi.org/10.1007/978-3-030-02604-2_12

a *quadratic form with integral coefficients associated to the matrix A*. We define the *discriminant of the form* f, denoted disc f, to be the determinant of the matrix A. We call a quadratic form f with integral coefficients *primitive* if it is not an integral multiple of another quadratic form with integral coefficients.

For example, if $n = 2$ and

$$A = \begin{pmatrix} a & b \\ b & c \end{pmatrix}$$

with $a, b, c \in \mathbb{Z}$, then the quadratic form associated to A is

$$f(x, y) = ax^2 + 2bxy + cy^2.$$

It is easy to check that

$$f(x, y) = (x \ y) \begin{pmatrix} a & b \\ b & c \end{pmatrix} \begin{pmatrix} x \\ y \end{pmatrix} = v^T A v$$

with $v = \begin{pmatrix} x \\ y \end{pmatrix}$. This is of course a completely general fact: If f is the quadratic form associated to the matrix A, then

$$f(x_1, \ldots, x_n) = v^T A v \tag{12.1}$$

with $v = \begin{pmatrix} x_1 \\ x_2 \\ \vdots \\ x_n \end{pmatrix}$ the column vector with entries $x_1, \ldots, x_n$.

Lemma 12.2. *The quadratic form f uniquely determines the matrix A.*

Proof. Suppose f is associated to matrices $A = (a_{ij})_{1 \leq i, j \leq n}$ and $A' = (a'_{ij})_{1 \leq i, j \leq n}$. Then we have

$$v^T A v = v^T A' v \tag{12.2}$$

for all v. We will prove $A = A'$ by induction on n. If $n = 1$, then

$$a_{11} x_1^2 = a'_{11} x_1^2$$

for all $x_1 \in \mathbb{Z}$ immediately implies $a_{11} = a'_{11}$. Now suppose the lemma is true for $n - 1$. Let $w = \begin{pmatrix} w_1 \\ w_2 \\ \vdots \\ w_{n-1} \end{pmatrix} \in \mathbb{Z}^{n-1}$ be a column vector, and for each $1 \leq j \leq n$ let

$$w(j) = \begin{pmatrix} w_1(j) \\ w_2(j) \\ \vdots \\ w_n(j) \end{pmatrix}$$ be the vector in $\mathbb{Z}^n$ which is defined as follows:

$$w_i(j) = \begin{cases} w_i & i < j; \\ 0 & i = j; \\ w_{i-1} & i \geq j. \end{cases}$$

For example, if $n = 3$ and $w = \begin{pmatrix} x \\ y \\ z \end{pmatrix}$, then

$$w(3) = \begin{pmatrix} x \\ y \\ 0 \\ z \end{pmatrix}.$$

Next, for each $n \times n$ matrix $B = (b_{kl})_{1 \leq k, l \leq n}$ and each $1 \leq j \leq n$ define a matrix $B(j)$ to be the $(n-1) \times (n-1)$ matrix which is obtained from B by deleting the jth rows and jth column of B, i.e., if we write $B(j) = (b_{kl}(j))_{1 \leq k, l \leq n-1}$, then

$$b_{kl}(j) = \begin{cases} b_{kl} & k, l < j; \\ b_{k,l+1} & k < j, l > j; \\ b_{k+1,l} & k > j, l < j; \\ b_{k+1,l+1} & k > j, l > j. \end{cases}$$

For example, if

$$B = \begin{pmatrix} a & b & c & d \\ e & f & g & h \\ i & j & k & l \\ m & n & o & p \end{pmatrix},$$

then

$$B(1) = \begin{pmatrix} f & g & h \\ j & k & l \\ n & o & p \end{pmatrix}, \quad B(3) = \begin{pmatrix} a & b & d \\ e & f & h \\ m & n & p \end{pmatrix}.$$

The importance of the matrix $B(j)$ lies in the fact that for each $w \in \mathbb{Z}^{n-1}$ and each $B \in M_n(\mathbb{Z})$ we have

$$w(j)^T B w(j) = w^T B(j) w. \tag{12.3}$$

Now we go back to Equation (12.2), and apply it to column vectors of the form $w(j)$, $1 \leq j \leq n$. For each j we have

$$w^T A(j) w = w(j)^T A w(j) = w(j)^T A' w(j) = w^T A'(j) w.$$

Since we are assuming the lemma is true for $n - 1$, this last equation implies that for each j,

$$A(j) = A'(j).$$

The assertion now follows from Exercise 12.2. $\square$

Since the matrix A is symmetric and $x_i x_j = x_j x_i$ for all i, j, we have

$$f(x_1, \ldots, x_n) = \sum_{i=1}^{n} a_{ii} x_i^2 + 2 \sum_{1 \le i < j \le n} a_{ij} x_i x_j,$$

This points to a caveat in our theory, namely that the quadratic forms that we consider have *even* coefficients for their "mixed" terms, i.e., the terms of the form $x_i x_j$ with $i \neq j$. This means that our theory does not include quadratic forms like

$$x^2 + xy + y^2, \quad x^2 + y^2 + z^2 + 3xy + 4xz.$$

One way to avoid this problem is to consider matrices that are not symmetric, or by allowing the off diagonal terms in A be half integers, but either of these ideas brings about complications that we do not want to deal with in this book. We refer the reader to Cassels [12] for a more thorough treatment of quadratic forms over the field of rational numbers.

Definition 12.3. For quadratic forms f and g with integral coefficients, we say f is *equivalent* to g, and write $f \sim g$, if there is a matrix $P = (p_{ij}) \in \mathrm{SL}_n(\mathbb{Z})$ such that

$$f(\sum_{j=1}^{n} p_{1j} x_j, \sum_{j=1}^{n} p_{2j} x_j, \ldots, \sum_{j=1}^{n} p_{nj} x_j) = g(x_1, x_2, \ldots, x_n).$$

For example if $f(x, y) = x^2 + y^2$ and $g(x, y) = x^2 + 2xy + 2y^2$, then $f \sim g$. The reason is that $f(x + y, y) = g(x, y)$, i.e., the definition holds with $P = \begin{pmatrix} 1 & 1 \\ & 1 \end{pmatrix} \in \mathrm{SL}_2(\mathbb{Z})$.

In the above notation, note that

$$\begin{pmatrix} \sum_{j=1}^{n} p_{1j} x_j \\ \sum_{j=1}^{n} p_{2j} x_j \\ \vdots \\ \sum_{j=1}^{n} p_{nj} x_j \end{pmatrix} = P \cdot \begin{pmatrix} x_1 \\ x_2 \\ \vdots \\ x_n \end{pmatrix}.$$

Now if we suppose f and g are associated to the matrices A and B, respectively, then $f \sim g$ means

$$(Pv)^T A(Pv) = v^T Bv$$

for all $v = \begin{pmatrix} x_1 \\ x_2 \\ \vdots \\ x_n \end{pmatrix} \in \mathbb{Z}^n$. Since transposition is order reversing, $(XY)^T = Y^T X^T$, this equation now implies

$$v^T (P^T AP)v = v^T Bv.$$

Lemma 12.2 says

$$P^T A P = B. \tag{12.4}$$

It is clear that this process can be reversed, meaning if there is $P \in \mathrm{SL}_n(\mathbb{Z})$ such that Equation (12.4) holds, then $f \sim g$. We summarize this discussion as the following lemma:

Lemma 12.4. *Suppose f, g are quadratic forms associated to matrices A, B. Then $f \sim g$ if and only if there is $P \in \mathrm{SL}_n(\mathbb{Z})$ such that*

$$P^T A P = B.$$

This lemma has the following important consequence:

Proposition 12.5. *The relation $\sim$ on quadratic forms is an equivalence relation that preserves the discriminant.*

Proof. We need to show that $\sim$ is symmetric, reflexive, and transitive, and that if $f \sim g$, then $\det f = \det g$. We use Lemma 12.4 repeatedly.

Reflexive. We need: $f \sim f$. Clearly I_n, the $n \times n$ identity matrix, is in $\mathrm{SL}_n(\mathbb{Z})$, and $A = I_n^T A I_n$.

Symmetry. We need: $f \sim g$ implies $g \sim f$. Suppose f and g are associated to A, B, respectively. If there is a matrix $P \in \mathrm{SL}_n(\mathbb{Z})$ such that $P^T A P = B$, then since $(P^T)^{-1} = (P^{-1})^T$, $(P^{-1})^T B (P^{-1}) = A$, and $P^{-1} \in \mathrm{SL}_n(\mathbb{Z})$. This means $g \sim f$.

Transitive. We need: $f \sim g$ and $g \sim h$ implies $f \sim h$. Suppose f, g, h are associated to A, B, C, respectively, and that there are $P, Q \in \mathrm{SL}_n(\mathbb{Z})$ such that $P^T A P = B$ and $Q^T B Q = C$. Then

$$C = Q^T B Q = Q^T P^T A P Q = (PQ)^T A (PQ).$$

Determinant preservation. We need: $f \sim g$ implies $\det f = \det g$. Suppose f, g are associated to A, B, respectively, and that there is $P \in \mathrm{SL}_n(\mathbb{Z})$ such that $B = P^T A P$. We have

$$\mathrm{disc}\, g = \det B = \det(P^T A P) = \det(P^T) \det P \det A$$

by multiplicativity of determinant. Then we note that $\det P^T = \det P = 1$ as transposition does not change the value of determinant. This means that

$$\mathrm{disc}\, g = \det A = \mathrm{disc}\, f.$$

$\square$

Definition 12.6. For a quadratic form f and an integer m, we say f *represents* m if there are integers $x_1, \ldots, x_n$ such that

$$f(x_1, \ldots, x_n) = m.$$

We call f *positive definite* if for all $x_1, \ldots, x_n \in \mathbb{Z}^n$, not all of which are zero, we have

$$f(x_1, \ldots, x_n) > 0.$$

The following proposition is central to our discussion:

Proposition 12.7. *Suppose f, g are quadratic forms, and $f \sim g$. Then*

1. *The quadratic forms f and g represent the exact same set of numbers.*
2. *The quadratic form f is positive definite if and only if the quadratic form g is.*

Proof. Following the notation of Equation (12.1) write

$$f(x_1, \ldots, x_n) = v^T A v, \quad g(x_1, \ldots, x_n) = v^T B v$$

with $v = \begin{pmatrix} x_1 \\ x_2 \\ \vdots \\ x_n \end{pmatrix}$ the column vector with entries $x_1, \ldots, x_n$. For simplicity we write

$f(v)$ and $g(v)$ instead of $f(x_1, \ldots, x_n)$ and $g(x_1, \ldots, x_n)$, respectively. The assumption on the f and g means there is a matrix $P \in \mathrm{SL}_n(\mathbb{Z})$ such that $B = P^T A P$. In terms of f and g this means that for all v, $g(v) = f(P \cdot v)$. As a result, $g(\mathbb{Z}^n) = f(P \cdot \mathbb{Z}^n)$. Once we show $P \cdot \mathbb{Z}^n = \mathbb{Z}^n$, the first assertion follows. Since P has integer entries, $P \cdot \mathbb{Z}^n \subset \mathbb{Z}^n$. Similarly, since $P \in \mathrm{SL}_n(\mathbb{Z})$, P^{-1}, too, has integer entries. Therefore, $P^{-1} \cdot \mathbb{Z}^n \subset \mathbb{Z}^n$. Multiplying by P gives $\mathbb{Z}^n \subset P \cdot \mathbb{Z}^n$. Putting the inclusions $P \cdot \mathbb{Z}^n \subset \mathbb{Z}^n$ and $\mathbb{Z}^n \subset P \cdot \mathbb{Z}^n$ together gives $P \cdot \mathbb{Z}^n = \mathbb{Z}^n$, and we are done with the first part. The second statement follows from the first statement, and the statement that for $v \in \mathbb{Z}^n$, $v = 0$ if and only if $Pv = 0$.   □

Lemma 12.8. *If for a quadratic form f, disc f is square-free, then f is primitive. Equivalence preserves primitivity.*

Proof. If $f = mg$, then disc $f = m^n \mathrm{disc}\, g$. This observation implies the first assertion. The second statement is obvious.   □

12.2   Binary forms

We now discuss the case where $n = 2$, the so-called binary forms, in detail. Here we do not address questions of representability of integers by binary forms. The wonderful book Cox [14], especially Chapter 1, provides an accessible introduction to this important topic.

Suppose we have a binary quadratic form f which is associated to the symmetric matrix

$$A = \begin{pmatrix} a & b \\ b & c \end{pmatrix}.$$

Then disc $f = \det A = ac - b^2$.

Lemma 12.9. *The form f is positive definite if and only if $a > 0$ and disc $f > 0$.*

Proof. Suppose f is positive definite. Since $f(1, 0) = a$ we immediately see $a > 0$. Next,

$$0 < f(-b, a) = ab^2 - 2b^2a + ca^2 = -b^2a + ca^2 = a(ac - b^2) = a\,\text{disc } f.$$

Since we have already established $a > 0$, $a\,\text{disc } f > 0$ implies disc $f > 0$.
    Now suppose $a > 0$ and disc $f > 0$. Then

$$af(x, y) = a^2x^2 + 2abxy + acy^2 = (a^2x^2 + 2abxy + b^2y^2) + (ac - b^2)y^2$$

$$= (ax + by)^2 + (\text{disc } f)y^2.$$

Since $a > 0$ and disc $f > 0$, the identity

$$af(x, y) = (ax + by)^2 + (\text{disc } f)y^2 \tag{12.5}$$

shows that $f(x, y) \geq 0$, and $f(x, y) = 0$ only if $(\text{disc } f)y^2 = 0$ and $(ax + by)^2 = 0$, which immediately implies $x = y = 0$. This means f is positive definite. $\square$

Theorem 12.10. *Every equivalence class of positive definite binary quadratic forms contains a form f whose associated matrix $A = \begin{pmatrix} a & b \\ b & c \end{pmatrix}$ satisfies*

$$2|b| \leq a \leq c.$$

Proof. Suppose we have a positive definite form g associated to $A_g = \begin{pmatrix} a_0 & b_0 \\ b_0 & c_0 \end{pmatrix}$. We wish to show that there is a form f with $f \sim g$ for which the inequalities of the theorem hold. Let a be the smallest positive number represented by g. There are integers r, t such that $g(r, t) = a$. We claim $\gcd(r, t) = 1$. Otherwise, if $p \mid r$ and $p \mid t$, then $p^2 \mid a$, and we would have $g(r/p, t/p) = a/p^2$, and that contradicts the choice of a. Since $\gcd(r, t) = 1$, there are integers s, u such that $ru - st = 1$. By Theorem 2.23 if we fix one solution s_0, u_0 every other solution is of the form

$$s(h) = s_0 + rh, \quad u(h) = u_0 + ht, \quad h \in \mathbb{Z}.$$

Now consider the functions $a(h)$, $b(h)$, and $c(h)$, for $h \in \mathbb{Z}$, defined by the following matrix identity

$$\begin{pmatrix} a(h) & b(h) \\ b(h) & c(h) \end{pmatrix} = \begin{pmatrix} r & s(h) \\ t & u(h) \end{pmatrix}^T \begin{pmatrix} a_0 & b_0 \\ b_0 & c_0 \end{pmatrix} \begin{pmatrix} r & s(h) \\ t & u(h) \end{pmatrix}.$$

Explicitly, we have

$$\begin{cases} a(h) = a_0 r^2 + 2b_0 rt + c_0 t^2 = a, \\ b(h) = s(h)(ra_0 + tb_0) + u(h)(rb_0 + tc_0), \\ c(h) = a_0 s(h)^2 + 2b_0 s(h)u(h) + c_0 u(h)^2. \end{cases}$$

Simplification gives

$$b(h) = s_0(a_0 r + b_0 t) + u_0(b_0 r + c_0 t) + (a_0 r^2 + 2b_0 rt + c_0 t^2)h$$

$$= s_0(a_0 r + b_0 t) + u_0(b_0 r + c_0 t) + ah.$$

Since the coefficient of h is $a > 0$, and h is arbitrary, we may choose an h_0 so that $b(h_0)$ satisfies $|b(h_0)| \le a/2$. The expression for $c(h)$ shows that

$$c(h_0) = g(s(h_0), u(h_0)),$$

and consequently $a \le c(h_0)$. It is clear that the quadratic form associated to the matrix

$$\begin{pmatrix} a(h_0) & b(h_0) \\ b(h_0) & c(h_0) \end{pmatrix}$$

satisfies the requirements.   □

Definition 12.11. A primitive binary form $f(x, y) = ax^2 + 2bxy + cy^2$ is called *reduced* if its coefficients satisfy the inequalities of Theorem 12.10.

For example, the forms $x^2 + y^2$ and $4x^2 + 2xy + 5y^2$ are reduced, and $5x^2 + 2xy + 4y^2$ is not.

Corollary 12.12. *Every positive definite binary quadratic form of discriminant 1 is equivalent to $x^2 + y^2$.*

Proof. By Theorem 12.10 and Proposition 12.5 every such quadratic form is equivalent to a quadratic form whose associated matrix $\begin{pmatrix} a & b \\ b & c \end{pmatrix}$ satisfies $2|b| \le a \le c$ and $ac - b^2 = 1$. Then we have

$$a^2 \le ac = b^2 + 1 \le \frac{a^2}{4} + 1.$$

Consequently, $a^2 \le 4/3$. From this inequality it follows that $a = 1$. Since $2|b| \le 1$, we see $b = 0$. Since $ac = b^2 + 1 = 1$, we see $c = 1$.   □

Let us now use this last result to give another proof for Theorem 5.7, namely that every prime of the form $4k + 1$ is a sum of two squares.

One more proof of Theorem 5.7. Suppose p is of the form $4k + 1$. We wish to show that p is represented by the binary quadratic form $x^2 + y^2$. Since by Corollary 12.12 every positive definite binary form of discriminant 1 is equivalent to $x^2 + y^2$, and by

Proposition 12.7 equivalent forms represent the same set of numbers, it suffices to find some positive definite binary form

$$ax^2 + 2bx + cy^2$$

with discriminant 1 which represents p. We will show that we may even take $a = p$, see Exercise 12.7. Clearly, the form

$$g(x, y) = px^2 + 2bxy + cy^2$$

represents p, as $g(1, 0) = p$. We just need to choose b, c so that disc $g = 1$. We have

$$\text{disc } g = pc - b^2.$$

As a result, the existence of b, c is equivalent to $b^2 \equiv -1 \bmod p$, or $(-1/p) = +1$. But for p of the form $4k + 1$ this is a consequence of Equation (6.3). □

12.3 Ternary forms

In this section we study quadratic forms in three variables. Our goal here is to prove the analogue of Corollary 12.12 in this setting. Namely, we will prove:

Theorem 12.13. *Every positive definite ternary quadratic form of discriminant 1 is equivalent to $x^2 + y^2 + z^2$.*

The proof of this theorem, though in principle similar to the proof of Corollary 12.12, is fairly complicated. The reader might want to skip the rest of this subsection in the first reading and go straight to §12.4 where the Three Square Theorem is proved.

Theorem 12.14. *Suppose f is a ternary quadratic form associated to the symmetric matrix*

$$A = \begin{pmatrix} a_{11} & a_{12} & a_{13} \\ a_{21} & a_{22} & a_{23} \\ a_{31} & a_{32} & a_{33} \end{pmatrix}.$$

Then f is positive definite if and only if

- $a_{11} > 0$;
- $\det \begin{pmatrix} a_{11} & a_{12} \\ a_{21} & a_{22} \end{pmatrix} > 0$;
- $\det A > 0$.

Before we prove the theorem, we need a lemma that is the analogue of Equation (12.5) for ternary forms:

Lemma 12.15. *With notations as above,*

$$a_{11} f(x, y, z) = (a_{11}x + a_{12}y + a_{13}z)^2 + K(y, z)$$

with $K(y, z)$ a binary quadratic form associated to the matrix

$$\begin{pmatrix} a_{11}a_{22} - a_{12}^2 & a_{11}a_{23} - a_{12}a_{13} \\ a_{11}a_{23} - a_{12}a_{13} & a_{11}a_{33} - a_{13}^2 \end{pmatrix}.$$

Furthermore, disc $K = a_{11}$disc f. Finally, if f is positive definite, K will be positive definite.

Proof. Every statement in the lemma, except for the last one, is a straightforward computation; see Exercise 12.10. The last statement follows from Lemma 12.9. $\square$

We can now prove the theorem:

Proof of Theorem 12.14. Since $a_{11} = f(1, 0, 0)$, we see that $a_{11} > 0$ if f is positive definite. So we will assume $a_{11} > 0$.

If f is positive definite, Lemma 12.15 implies that K is positive definite. Lemma 12.9, applied to K, implies that $a_{11}a_{22} - a_{12}^2 > 0$ and disc $K = a_{11}$disc $f > 0$. These are the conditions required by the theorem.

Conversely, suppose the inequalities of the theorem are satisfied. Then, as above, it follows that K is positive definite. Suppose, to achieve a contradiction, that f is not positive definite. Then for some $(x, y, z) \neq (0, 0, 0)$, $f(x, y, z) \leq 0$. Then we have

$$(a_{11}x + a_{12}y + a_{13}z)^2 + K(y, z) \leq 0.$$

Since K is positive definite, this equation implies $K(y, z) = 0$ and $a_{11}x + a_{12}y + a_{13}z = 0$. The first of these implies $y = z = 0$, and then we conclude $x = 0$ as well.
                                                                                    $\square$

Our next theorem is the analogue of Exercise 12.3 for ternary forms.

Theorem 12.16. *Every positive definite ternary quadratic form f of discriminant d is equivalent to some quadratic form g whose associated matrix $A = (a_{ij})$ satisfies*

$$a_{11} \leq \frac{4}{3}\sqrt[3]{d}, \quad 2|a_{12}| \leq a_{11}, \quad 2|a_{13}| \leq a_{11}.$$

Proof. Suppose f is associated to the matrix B, and let a_{11} be the smallest natural number represented by f. Then there are integers c_{11}, c_{21}, c_{31} such that

$$a_{11} = f(c_{11}, c_{21}, c_{31}).$$

As in the proof of Theorem 12.10 we have

$$\gcd(c_{11}, c_{21}, c_{31}) = 1.$$

Exercise 12.11 shows that there is a 3×3 matrix $C = (c_{ij})$ whose first column is the numbers c_{11}, c_{21}, c_{31} and whose determinant is 1. Let g be the quadratic form whose associated matrix is

$$D = C^T BC.$$

Now

$$g(1, 0, 0) = f(c_{11}, c_{21}, c_{31}) = a_{11}.$$

Next, consider a form h whose associated matrix is

$$E = \begin{pmatrix} 1 & r & s \\ 0 & t & u \\ 0 & v & w \end{pmatrix}^T D \begin{pmatrix} 1 & r & s \\ 0 & t & u \\ 0 & v & w \end{pmatrix}.$$

Here we assume $r, s, t, u, v, w \in \mathbb{Z}$, and $tw - uv = 1$, so that for every r, s the determinant of the transformation matrix is 1.

We write $D = (b_{kl})$ and $E = (a_{kl})$. If

$$\begin{pmatrix} x_1 \\ x_2 \\ x_3 \end{pmatrix} = \begin{pmatrix} 1 & r & s \\ 0 & t & u \\ 0 & v & w \end{pmatrix} \begin{pmatrix} y_1 \\ y_2 \\ y_3 \end{pmatrix},$$

then one can check

$$b_{11}x_1 + b_{12}x_2 + b_{13}x_3 = a_{11}y_1 + a_{12}y_2 + a_{13}y_3.$$

Now we apply Lemma 12.15 to obtain positive definite binary forms K and L such that

$$a_{11}g(x_1, x_2, x_3) = (b_{11}x_1 + b_{12}x_2 + b_{13}x_3)^2 + K(x_2, x_3)$$

and

$$a_{11}h(y_1, y_2, y_3) = (a_{11}y_1 + a_{12}y_2 + a_{13}y_3)^2 + L(y_2, y_3).$$

The form K is transformed to L via $\begin{pmatrix} t & u \\ v & w \end{pmatrix}$. The form L has discriminant $a_{11}\mathrm{disc}\, f$, and the coefficient of y_2^2 is $a_{11}a_{22} - a_{12}^2$. Consequently, by Exercise 12.3 we can choose u, v, w, t such that

$$a_{11}a_{22} - a_{12}^2 \leq \frac{2}{\sqrt{3}}\sqrt{a_{11}d}.$$

It is easy to see that

$$a_{12} = ra_{11} + tb_{12} + vb_{13}$$

and

$$a_{13} = sa_{11} + ub_{12} + wb_{13}.$$

Since r, s are arbitrary, we can choose them so that

$$|a_{12}| \leq a_{11}/2, \quad |a_{13}| \leq a_{11}/2.$$

Also, since $a_{22} = h(0, 1, 0)$, we must have $a_{22} \geq a_{11}$. Hence,

$$a_{11}^2 \leq a_{11}a_{22} = (a_{11}a_{22} - a_{12}^2) + a_{12}^2 \leq \frac{2}{\sqrt{3}}\sqrt{a_{11}d} + \frac{a_{11}^2}{4},$$

from which it immediately follows that

$$a_{11} \leq \frac{4}{3}\sqrt[3]{d}.$$ □

Now we proceed to prove the main theorem of this section:

Proof of Theorem 12.13. By Theorem 12.16 we know that our quadratic form is equivalent to a form whose associated matrix has the properties

$$a_{11} \leq 4/3, \quad 2|a_{12}| \leq a_{11}, \quad 2|a_{13}| \leq a_{11}.$$

Clearly, $a_{11} = 1$, $a_{12} = 0$, and $a_{13} = 0$. Consequently, our form is equivalent to a form

$$g = x_1^2 + K(x_2, x_3)$$

with K a positive definite binary quadratic form of discriminant 1. By Corollary 12.12 there is a transformation $\begin{pmatrix} t & u \\ v & w \end{pmatrix}$ that sends K to $x_2^2 + x_3^2$. Finally, $\begin{pmatrix} 1 & 0 & 0 \\ 0 & t & u \\ 0 & v & w \end{pmatrix}$ sends g to $x_1^2 + x_2^2 + x_3^2$, and we are done.   □

12.4   Three squares

In this section we give a proof of the most non-trivial part of Theorem 9.8. Namely, we will prove that if n is not of the form $4^a(8k + 7)$ then n is a sum of three squares. Clearly if $n = x^2 + y^2 + z^2$, then $4n = (2x)^2 + (2y)^2 + (2z)^2$, so we may factor out any factor 4^m from n and assume that either n is odd or it is twice an odd number. This means that we may assume

$$n \equiv 1, 2, 3, 5, 6 \mod 8.$$

Theorem 12.13 and Proposition 12.7 imply that it suffices to find a positive definite ternary form of discriminant 1 that represents n. This means we need to find a 3×3 matrix (a_{ij}) with integer entries and three integers x_1, x_2, x_3 such that

$$a_{11} > 0, \quad a_{11}a_{22} - a_{12}^2 > 0, \quad \det(a_{ij}) = 1,$$

and

$$n = \sum_{ij} a_{ij}x_i x_j.$$

We take

$$a_{13} = a_{31} = 1, \quad a_{23} = a_{32} = 0, \quad a_{33} = n, \quad x_1 = 0, \quad x_2 = 0, \quad x_3 = 1.$$

Then if we set $b = a_{11}a_{22} - a_{12}^2$, computing the determinant of (a_{ij}) using the bottom row gives

$$1 = \det(a_{ij}) = \det \begin{pmatrix} a_{11} & a_{12} & 1 \\ a_{21} & a_{22} & 0 \\ 1 & 0 & n \end{pmatrix} = -a_{22} + n \det \begin{pmatrix} a_{11} & a_{12} \\ a_{12} & a_{22} \end{pmatrix}$$

$$= -a_{22} + nb.$$

So we just need

- $a_{11} > 0$;

- $b = a_{11}a_{22} - a_{12}^2 > 0$;

- $a_{22} = bn - 1$.

If $n > 1$, then $a_{11} > 0$ is a consequence of the other statements. The reason for this is that

$$a_{22} = bn - 1 > b - 1 \geq 0,$$

and

$$a_{11}a_{22} = a_{12}^2 + b > 0.$$

The latter implies $a_{11} > 0$. So we need

- $b = a_{11}a_{22} - a_{12}^2 > 0$;

- $a_{22} = bn - 1$,

or, equivalently, we need to show that there is $b > 0$ such that the equation

$$X^2 \equiv -b \mod (bn - 1)$$

has a solution. We separate the cases where n is even or odd.

The even case: $n \equiv 2, 6 \mod 8$. Since $\gcd(4n, n - 1) = 1$, Dirichlet's Arithmetic Progression Theorem, Theorem 5.11, shows that there is a natural number v such that

$$p = 4nv + n - 1 = (4v + 1)n - 1$$

is prime. Note that $p \equiv 1 \mod 4$. Let $b = 4v + 1 > 0$. By Theorem 7.3 we have

$$\left(\frac{-b}{p}\right) = \left(\frac{-1}{p}\right)\left(\frac{b}{p}\right) = \left(\frac{b}{p}\right) = \left(\frac{p}{b}\right) = \left(\frac{bn-1}{b}\right) = \left(\frac{-1}{b}\right) = +1.$$

The odd case: $n \equiv 1, 3, 5 \mod 8$. First let us assume $n \equiv 3 \mod 8$. Then $(n-1)/2$ is odd, and consequently, $\gcd(4n, (n - 1)/2) = 1$. By Dirichlet's Arithmetic Progression Theorem, Theorem 5.11, there is an integer v such that

$$p = 4nv + \frac{n - 1}{2} = \frac{(8v + 1)n - 1}{2}$$

is prime, and $p \equiv 1 \mod 4$. Set $b = 8v + 1$. Then $b > 0$ and $2p = bn - 1$. Since $b \equiv 1 \mod 8$, by Theorem 7.3, $(-2/b) = 1$. Then

$$\left(\frac{-b}{p}\right) = \left(\frac{b}{p}\right) = (-1)^{\frac{b-1}{2} \cdot \frac{p-1}{2}} \left(\frac{p}{b}\right) = \left(\frac{p}{b}\right) = \left(\frac{p}{b}\right)\left(\frac{-2}{b}\right)$$

$$= \left(\frac{-2p}{b}\right) = \left(\frac{1 - nb}{b}\right) = \left(\frac{1}{b}\right) = 1.$$

If $n \equiv 1, 5 \bmod 8$, then we consider primes of the form $p = 4nv + \frac{3n-1}{2}$, and we let $b = 8v + 3$. The remainder of the argument is completely similar; see Exercise 12.13.

Exercises

12.1 Verify Equation (12.3).

12.2 This exercise uses the notations of the proof of Lemma 12.2. Suppose $A, A' \in M_n(R)$ for some ring R, and suppose for all j we have $A(j) = A'(j)$. Show that $A = A'$.

12.3 Show that every positive definite binary quadratic form of discriminant d is equivalent to a quadratic form whose associated matrix $\begin{pmatrix} a & b \\ b & c \end{pmatrix}$ satisfies

$$2|b| \leq a \leq \frac{2}{\sqrt{3}}\sqrt{d}.$$

12.4 Show that a reduced binary quadratic form cannot be equivalent to a different reduced binary quadratic form.

12.5 Show that for every natural number d there are only finitely many equivalence classes of positive definite binary quadratic forms of discriminant d.

12.6 Find representatives for equivalence classes of positive definite binary quadratic forms of discriminant d when

   a. $d = 2$;
   b. $d = 3$;
   c. $d = 5$.

12.7 We say that a binary form f represents m *properly* if there are $a, b \in \mathbb{Z}$ with $\gcd(a, b) = 1$ such that $f(a, b) = m$. Show that a binary quadratic form represents an integer m properly if and only if it is equivalent to a binary form $mx^2 + bxy + cy^2$ for some $b, c \in \mathbb{Z}$.

12.8 Find reduced forms that are equivalent to the following forms:

   a. $4x^2 + y^2$;
   b. $9x^2 + 2xy + y^2$;
   c. $126x^2 + 74xy + 13y^2$.

12.9 (✠) List all reduced primitive positive definite binary quadratic forms of discriminant bounded by 100. For each d, find the number of forms with that discriminant.

12.10 Prove Lemma 12.15.

12.11 Suppose $a, b, c \in \mathbb{Z}$ are such that $\gcd(a, b, c) = 1$. Then prove that there are integers d, e, f, g, h, i such that the matrix

$$\begin{pmatrix} a & b & c \\ d & e & f \\ g & h & i \end{pmatrix}$$

has determinant 1.

12.12 Prove that the Three Square Theorem implies the Four Square Theorem.

12.13 Finish the proof of the Three Square Theorem for $n \equiv 1, 5 \bmod 8$.

12.14 Show that if $p > 17$ is a prime number $p \equiv 5 \bmod 12$ then p is a sum of three distinct positive squares. Hint: Use the identity,

$$9(a^2 + b^2) = (2a - b)^2 + (2a + 2b)^2 + (2b - a)^2.$$

Notes

Gauss Composition

The easy identity

$$(x^2 + y^2)(z^2 + w^2) = (xz + yw)^2 + (xw - zy)^2 \tag{12.6}$$

has been known for hundreds of years. As we noted in the Notes to Chapter 3, the master Indian mathematician Brahmagupta discovered the more general identity

$$(x^2 + dy^2)(z^2 + dw^2) = (xz + dyw)^2 + d(xw - yz)^2 \tag{12.7}$$

at some point in the seventh century CE. Over a thousand years later, Lagrange discovered the identities

$$(2x^2 + 2xy + 3y^2)(2z^2 + 2zw + 3w^2) = (2xz + xw + yz + 3yw)^2 + 5(xw - yz)^2, \tag{12.8}$$

and

$$(3x^2 + 2xy + 5y^2)(3z^2 + 2zw + 5w^2) = (3x^2 + xw + yz + 5yw)^2 + 14(xw - yz)^2. \tag{12.9}$$

All of these identities are of the form

$$f(x, y)f(z, w) = g(B_1(x, y, z, w), B_2(x, y, z, w)); \tag{12.10}$$

with f and g positive definite binary quadratic forms of the same discriminant, and B_1, B_2 homogeneous quadratic forms in the four variables x, y, z, w. The binary quadratic forms in Equation (12.6) have discriminant 1, in Equation (12.7) they have discriminant d, in Equation (12.8) they have discriminant 5, and in Equation (12.9) they have discriminant 14. Gauss proved a truly impressive theorem that generalizes all such identities. In fact, he showed the following theorem: Let f_1, f_2 be positive definite binary quadratic forms of discriminant d. Then there are homogeneous polynomials B_1, B_2 of degree 2 in the variables x, y, z, t such that

$$f_1(x, y) f_2(z, w) = g(B_1(x, y, z, w), B_2(x, y, z, w));$$

for some positive definite binary quadratic form g of discriminant d. Gauss called the quadratic form g the *composition of f_1 and f_2*, and for that reason the theorem is called the *composition law*. The binary quadratic forms we studied in this chapter all had an even middle coefficient, i.e., they were of the form $ax^2 + 2bxy + cy^2$ with b an integer. Gauss considered the more general quadratic forms $ax^2 + bxy + cy^2$ with b integral. For such forms the discriminant as we defined it is not necessarily an integer, so the discriminant is generally defined to be $4ac - b^2 \in \mathbb{Z}$. Gauss illustrated his theory with the following example:

$$(4x^2 + 3xy + 5y^2)(3z^2 + zw + 6w^2)$$

$$= (xz - 3xw - 2yz - 3yw)^2 + (xz - 3xw - 2yz - 3yw)(xz + xw + yz - yw)$$

$$+ 9(xz + xw + yz - yw)^2.$$

Let us denote the composition of the forms f_1 and f_2 by $f_1 \circ f_2$. An important feature of Gauss's composition is that if f_1 is equivalent to a form f_1', then $f_1 \circ f_2 \sim f_1' \circ f_2$. This means that the composition provides a well-defined operation on the finite set of equivalence classes of binary quadratic forms of discriminant d, turning it into a finite abelian group, the *class group of binary forms*. It was Dirichlet who interpreted the composition of binary quadratic forms in terms of ideal multiplication, whereby connecting the class group of binary forms to the ideal class group of modern algebraic number theory. After about 200 years since the publication of [21], in a series of groundbreaking works, Manjul Bhargava generalized the Gauss composition laws and found numerous other composition laws. Gauss's proof of his composition law is *extremely* complicated; see [21, Ch. V]. Cox [14, §3] contains a motivated introduction to Gauss's theory of quadratic forms. We refer the reader to Andrew Granville's lecture at a summer school in 2014 for a review of Gauss's work and the works of other mathematicians that preceded it, as well as an introduction to Bhargava's works:

http://www.crm.umontreal.ca/sms/2014/pdf/granville1.pdf

Chapter 13
How many Pythagorean triples are there?

In this chapter we determine an asymptotic formula for the number of primitive right triangles with bounded hypotenuse, giving a proof of a theorem of Lehmer from 1900. We start by relating the quantity we are interested in, namely the number of elements of the set

$$S(B) = \{(a, b, c) \in \mathbb{Z}^3 \mid a^2 + b^2 = c^2, \gcd(a, b, c) = 1, |a|, |b|, |c| \leq B\},$$

using our solution to the Pythagorean Equation, to the number of pairs of coprime integers satisfying certain conditions. Determining the latter number requires two inputs: an analogue of Gauss's Circle Theorem (Theorem 9.4) and a tool to ensure the coprimality of the integers; the tool we use to sieve out the non-coprime pairs is the function μ whose basic properties are collected in Lemmas 13.2 and 13.3. In the course of the proof we need to determine a quantity $C_2 = \sum_{\delta \text{ odd}} \mu(\delta)/\delta^2$. In §13.2 we show that the value C_2 is related to the value of the Riemann zeta function at 2, $\zeta(2)$, and explicitly calculate it. The main theorem of the chapter is Theorem 13.5. In the Notes to this chapter, we give some references for a conjecture of Manin that puts Lehmer's Theorem in a conceptual, geometric framework. The next item in the Notes is a disambiguation of the three number theorists with the last name of Lehmer (Hint: They were related!). The last part of the Notes is concerned with the Riemann zeta function, its analytic continuation, and the Riemann Hypothesis.

13.1 The asymptotic formula

It is clear that there are infinitely many right triangles with integer sides, but it still makes sense to obtain finer quantitative information about the set of right triangles. How many triples of integers (a, b, c) are there such that $a^2 + b^2 = c^2$ and $|a|, |b|, |c|$ are bounded by a fixed number? What if we required that the numbers a, b, c be coprime? For a positive real number B, we define

© Springer Nature Switzerland AG 2018
R. Takloo-Bighash, *A Pythagorean Introduction to Number Theory*,
Undergraduate Texts in Mathematics, https://doi.org/10.1007/978-3-030-02604-2_13

$$S(B) = \{(a, b, c) \in \mathbb{Z}^3 \mid a^2 + b^2 = c^2, \gcd(a, b, c) = 1, |a|, |b|, |c| \le B\}.$$

and set $\mathscr{N}(B) = \#S(B)$. Can we find an exact formula for $\mathscr{N}(B)$? Or, in the absence of a useful explicit formula, can we study the behavior of the function, e.g., its asymptotic behavior as B goes to infinity? And a related question, how many primitive right triangles are there with side lengths bounded by B? It will become clear in a moment that these questions are fairly easily tractable, and that one can give a beautiful formula describing the asymptotic behavior of the function $\mathscr{N}(B)$.

We start with some preliminary observations. By the proof of Theorem 3.1, if $(a, b, c) \in S(B)$, with $c > 0$, there are odd coprime integers x, y such that

$$\begin{cases} a = \frac{x^2 - y^2}{2}; \\ b = xy; \\ c = \frac{x^2 + y^2}{2}, \end{cases}$$

if a is even, and

$$\begin{cases} a = xy; \\ b = \frac{x^2 - y^2}{2}; \\ c = \frac{x^2 + y^2}{2}, \end{cases}$$

if b is even. Also, since $|a|, |b|, |c| \le |c|$, this means that we just need to require $(x^2 + y^2)/2 \le B$. One needs to be careful about signs here. For examples, in these formulae $(x^2 + y^2)/2$ is always positive, whereas we wish to count *all* elements of $S(B)$. So, our first guess might be that $\mathscr{N}(B)$ is equal to

$$\mathscr{N}_1(B) = \#\{x, y \in \mathbb{Z} \mid x, y \text{ odd}, \gcd(x, y) = 1, x^2 + y^2 \le 2B\}.$$

But this is not the whole story. For one, we need to multiply $\mathscr{N}_1(B)$ by 2 to account for the sign of c. Also, we need to multiply it by another factor of 2 to account for the odd and evenness of a and b. But then we need to divide by 2, as changing (x, y) to $(-x, -y)$ does not change the triple (a, b, c). Consequently,

$$\mathscr{N}(B) = 2\mathscr{N}_1(B).$$

To study the function $\mathscr{N}_1(B)$ we introduce the related function

$$h(B) = \#\{(x, y) \ne (0, 0) \mid x, y \in \mathbb{Z}, \text{ odd}, \gcd(x, y) = 1, x^2 + y^2 \le B\}.$$

Then clearly, $\mathscr{N}_1(B) = h(2B)$ and $\mathscr{N}(B) = 2h(2B)$.

To get an asymptotic formula for $h(B)$, first we relax the coprimality condition and define

$$\tilde{h}(B) = \#\{(x, y) \ne (0, 0) \mid x, y \in \mathbb{Z}, \text{ odd}, x^2 + y^2 \le B\}.$$

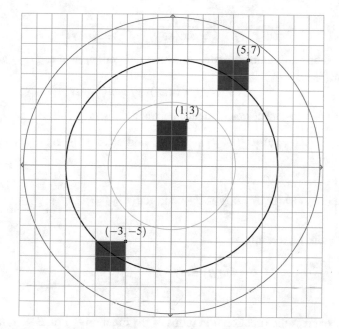

Fig. 13.1 The diagram for the proof of Lemma 13.1

Then we have the following lemma:

Lemma 13.1. *As $B \to \infty$,*

$$\tilde{h}(B) = \frac{1}{4}\pi B + O(\sqrt{B}).$$

Proof. Our proof of this lemma is modeled on the proof of Theorem 9.4. In this case, for every integral point (x, y) with x, y inside the circle, we draw a 2×2 square whose upper right corner is (x, y) as in Figure 13.1 with the point $(1, 3)$.

As in the proof of Theorem 9.4, not every square based on a point (x, y) inside the circle will be completely within the circle, e.g., the red square whose upper right corner is the point $(-3, -5)$ is not entirely within the circle of radius 7; and also some integral points outside the circle of radius 7 shown in the picture will have squares associated with them that intersect the circle, e.g., the blue square to the lower left of the point $(5, 7)$. Since the diameter of a 2×2 square is $2\sqrt{2}$ and its area is 4, by emulating the proof of Theorem 9.4, we have

$$\pi(\sqrt{B} - 2\sqrt{2})^2 \le 4\tilde{h}(B) \le \pi(\sqrt{B} - 2\sqrt{2})^2.$$

This proves the lemma.   □

We now relate the functions h and $\tilde{h}$. Suppose $(x, y) \neq (0, 0)$ is an integral point such that $x^2 + y^2 \le B$. Then we have

$$\left(\frac{x}{\gcd(x, y)}\right)^2 + \left(\frac{y}{\gcd(x, y)}\right)^2 \le \frac{B}{\gcd(x, y)^2}.$$

Clearly, if x, y are odd numbers, $\gcd(x, y)$ is odd, and $\gcd(x/\gcd(x, y), y/\gcd(x, y))$
$= 1$. The map

$$(x, y) \mapsto (x/\gcd(x, y), y/\gcd(x, y))$$

establishes a one-to-one correspondence between the sets

$$\{(x, y) \ne (0, 0) \mid x, y \in \mathbb{Z}, \ \text{odd}, \ x^2 + y^2 \le B\}$$

and

$$\bigsqcup_{\delta \le B} \left\{(x, y) \ne (0, 0) \mid x, y \in \mathbb{Z}, \ \text{odd}, \ \gcd(x, y) = 1, x^2 + y^2 \le \frac{B}{\delta^2}\right\},$$

a disjoint union. As a result,

$$\tilde{h}(B) = \sum_{\substack{\delta^2 \le B \\ \delta \text{ odd}}} h\left(\frac{B}{\delta^2}\right).$$

We now express the function h in terms of the function $\tilde{h}$. For $B < 1$, $h(B) = 0$.
If $1 \le B < 9$, then since $\delta^2 \le B$, with δ odd, means $\delta = 1$, we see that

$$h(B) = \tilde{h}(B)$$

for $1 \le B < 9$. Next, let $9 \le B < 25$. Then

$$\tilde{h}(B) = h(B) + h\left(\frac{B}{9}\right).$$

Now we note that for $9 \le B < 25$, $1 \le B/9 < 25/9 < 9$, and as a result $\tilde{h}(B/9) = h(B/9)$. Hence, for such B,

$$\tilde{h}(B) = h(B) - h\left(\frac{B}{9}\right).$$

We note that this formula is valid even if $B < 9$, as in that case $B/9 < 1$, and $h(B/9)$
$= 0$. Now let's suppose $25 \le B < 49$. Then as before,

$$\tilde{h}(B) = h(B) + h\left(\frac{B}{9}\right) + h\left(\frac{B}{25}\right).$$

Since $1 \le B/25 < 49/25 < 9$, we see that $\tilde{h}(\frac{B}{25}) = h(\frac{B}{25})$. Also, $1 \le B/9 < 16/9 <$
4, so again $\tilde{f}(\frac{B}{9}) = f(\frac{B}{9})$. Hence, for $9 \le B < 16$ we have

$$h(B) = \tilde{h}(B) - \tilde{h}\left(\frac{B}{9}\right) - \tilde{h}\left(\frac{B}{25}\right).$$

Again, this identity is valid for all $1 \leq B < 49$. Further experimentation with intervals of the form $k^2 \leq B < (k+1)^2$ suggests that there should exist a function $u : \mathbb{N} \to \{+1, -1\}$ such that

$$h(B) = \sum_{\substack{\delta^2 \leq B \\ \delta \text{ odd}}} \tilde{h}\left(\frac{B}{\delta^2}\right) u(\delta).$$

Suppose for a moment that this is indeed true. Then we would have

$$\tilde{h}(B) = \sum_{\substack{\delta^2 \leq B \\ \delta \text{ odd}}} h\left(\frac{B}{\delta^2}\right) = \sum_{\substack{\delta^2 \leq B \\ \delta \text{ odd}}} \sum_{\substack{\eta^2 \leq B/\delta^2 \\ \eta \text{ odd}}} \tilde{h}\left(\frac{B/\delta^2}{\eta^2}\right) u(\eta)$$

$$= \sum_{\substack{\delta^2 \leq B \\ \delta \text{ odd}}} \sum_{\substack{\eta^2 \delta^2 \leq B \\ \eta \text{ odd}}} \tilde{h}\left(\frac{B}{\delta^2 \eta^2}\right) u(\eta).$$

Now we switch the order of summation by letting $\delta\eta = n$. It is clear that n is odd and $n^2 \leq B$. Also, the η summation is over all divisors of n. So the above sum is equal to

$$\sum_{n^2 \leq B} \tilde{h}\left(\frac{B}{n^2}\right) \sum_{\eta|n} u(\eta).$$

So, in order for the latter to be equal to $\tilde{h}(B)$ for all $B \geq 1$, it would be sufficient to find a function $u : \mathbb{N} \to \{+1, -1\}$ such that

$$\sum_{\eta|n} u(\eta) = \begin{cases} 1 & n = 1; \\ 0 & n > 1. \end{cases}$$

(For the purposes of the problem we are discussing here it is sufficient to define the function u for odd numbers only, but this is a minor issue.) The interesting thing is that this last identity uniquely determines a function. In fact, it is clear that $u(1) = 1$. By setting $n = p$, a prime number, we see

$$u(1) + u(p) = 0$$

and, consequently, $u(p) = -1$. Next we try $n = pq$, with p, q distinct prime numbers. We have

$$u(1) + u(p) + u(q) + u(pq) = 0.$$

This gives, $u(pq) = +1$. Similarly, $u(pqr) = -1$ with p, q, r distinct primes. We can easily see using an easy inductive argument that if $p_1, \ldots, p_s$ are distinct prime numbers, then

$$u(p_1 \cdots p_s) = (-1)^s.$$

The function u above is called the Möbius function, and it is usually denoted by $\mu(n)$. This is a very important function in analytic number theory. See the exercises for the list of basic properties. In the sequel, we follow standard notation and use μ instead of u. We summarize this discussion as the following lemma:

Lemma 13.2.   *If we define a function μ by*

$$\mu(n) = \begin{cases} 1 & n = 1; \\ (-1)^s & n = p_1 \cdots p_s, \text{ with } p_i \text{ distinct primes}; \\ 0 & n \text{ not square-free}, \end{cases}$$

then for each natural number n

$$\sum_{d|n} \mu(d) = \begin{cases} 1 & n = 1; \\ 0 & n \neq 1. \end{cases}$$

Proof. Exercise 13.1. $\square$

Because of its importance we package the above discussion as the following lemma:

Lemma 13.3.   *Suppose F, G are functions defined on the set of positive real numbers. If for all $B > 0$,*

$$F(B) = \sum_{\substack{\delta \leq \sqrt{B} \\ \delta \text{ odd}}} G\left(\frac{B}{\delta^2}\right),$$

then

$$G(B) = \sum_{\substack{\delta \leq \sqrt{B} \\ \delta \text{ odd}}} F\left(\frac{B}{\delta^2}\right) \mu(\delta).$$

Remark 13.4.   This lemma is still valid if we remove the oddness condition.

Now that we know how to express h in terms of the function $\tilde{h}$, we can use Lemma 13.1 to find an asymptotic formula for the function h. By Lemma 13.3 and Lemma 13.1, we have

$$h(B) = \sum_{\substack{\delta^2 \leq B \\ \delta \text{ odd}}} \tilde{h}\left(\frac{B}{\delta^2}\right) \mu(\delta)$$

$$= \sum_{\substack{\delta^2 \leq B \\ \delta \text{ odd}}} \left(\frac{1}{4}\pi \frac{B}{\delta^2} + O(\sqrt{B/\delta^2})\right) \mu(\delta)$$

$$= \frac{1}{4}\pi B \sum_{\substack{\delta^2 \leq B \\ \delta \text{ odd}}} \frac{\mu(\delta)}{\delta^2} + O\left(\sqrt{B} \sum_{\substack{\delta^2 \leq B \\ \delta \text{ odd}}} \frac{1}{\delta}\right).$$

Note that we have replaced $O(\mu(\delta))$ by $O(1)$ in the last sum. We write the last sum as

$$= \frac{1}{4}\pi B \sum_{\substack{\delta=1 \\ \delta \text{ odd}}}^{\infty} \frac{\mu(\delta)}{\delta^2} - \frac{1}{4}\pi B \sum_{\substack{\delta^2 > B \\ \delta \text{ odd}}} \frac{\mu(\delta)}{\delta^2} + O\left(\sqrt{B} \sum_{\delta^2 \le B} \frac{1}{\delta}\right)$$

$$= \frac{1}{4}\pi B \sum_{\substack{\delta=1 \\ \delta \text{ odd}}}^{\infty} \frac{\mu(\delta)}{\delta^2} + O\left(B \sum_{\delta^2 > B} \frac{1}{\delta^2}\right) + O\left(\sqrt{B} \sum_{\delta^2 \le B} \frac{1}{\delta}\right).$$

By comparison with the convergent series $\sum_{\delta \ge 1} 1/\delta^2$ we see that the series $\sum_{\delta \text{ odd}} \mu(\delta)/\delta^2$ is convergent. Let's denote its value by C_2. We will calculate the exact value of C_2 in §13.2. Also,

$$\sum_{\delta^2 > B} \frac{1}{\delta^2} \le \int_{\sqrt{B}}^{\infty} \frac{dt}{t^2} \ll \frac{1}{\sqrt{B}},$$

and

$$\sum_{\delta^2 \le B} \frac{1}{\delta} \le \int_1^{\sqrt{B}} \frac{dt}{t} \ll \log B.$$

So we get

$$h(B) = \frac{1}{4}\pi C_2 B + O(\sqrt{B}) + O(\sqrt{B}\log B) = \frac{1}{4}\pi C_2 B + O(\sqrt{B}\log B). \quad (13.1)$$

We will show in §13.2 that $C_2 = 8/\pi^2$. Putting everything together, we get

Theorem 13.5. *As $B \to \infty$,*

$$\mathcal{N}(B) = \frac{4}{\pi}B + O(\sqrt{B}\log B)$$

Corollary 13.6 (Lehmer, 1900). *The number of primitive right triangles with hypotenuse bounded by B is*

$$\frac{1}{2\pi}B + O(\sqrt{B}\log B)$$

as $B \to \infty$.

13.2 The computation of C_2

In this section we will prove the following identity:

$$C_2 = \frac{8}{\pi^2}.$$

In fact, we will prove a more general result. For each natural number $k \geq 1$, let

$$C_{2k} = \sum_{\substack{n=1 \\ n \text{ odd}}}^{\infty} \frac{\mu(n)}{n^{2k}},$$

and

$$\zeta(2k) = \sum_{n=1}^{\infty} \frac{1}{n^{2k}},$$

The series $\zeta(2k)$ is convergent absolutely, and comparison implies that C_{2k} is absolutely convergent too.

Lemma 13.7. *For all natural numbers k,*

$$\left(1 - \frac{1}{2^{2k}}\right) C_{2k} \cdot \zeta(2k) = 1.$$

Proof. The first observation is that

$$\left(1 - \frac{1}{2^{2k}}\right) C_{2k} = \sum_{n=1}^{\infty} \frac{\mu(n)}{n^{2k}}.$$

Next, since all of our series are absolutely convergent we have

$$\left(1 - \frac{1}{2^{2k}}\right) C_{2k} \cdot \zeta(2k) = \sum_{n=1}^{\infty} \frac{\mu(n)}{n^{2k}} \sum_{m=1}^{\infty} \frac{1}{m^{2k}} = \sum_{n=1}^{\infty} \sum_{m=1}^{\infty} \frac{\mu(n)}{n^{2k} m^{2k}}$$

$$= \sum_{\delta=1}^{\infty} \sum_{mn=\delta} \frac{\mu(n)}{n^{2k} m^{2k}} = \sum_{\delta=1}^{\infty} \frac{1}{\delta^{2k}} \sum_{mn=\delta} \mu(n) = \sum_{\delta=1}^{\infty} \frac{1}{\delta^{2k}} \sum_{n\mid\delta} \mu(n).$$

Now by Lemma 13.2 whenever $\delta \neq 1$, the expression $\sum_{n\mid\delta} \mu(n)$ is equal to zero. Consequently, the only term that survives is $\delta = 1$, and the corresponding term is equal to 1. $\square$

This means in order to compute C_{2k} it suffices to compute $\zeta(2k)$.

The problem of computing the constant $\zeta(2)$, known as the *Basel Problem*, has a long history. Euler solved this problem in 1735 proving $\zeta(2) = \pi^2/6$. There are many proofs of this fact available in literature; see [64, 99]. Here we offer two proofs for Euler's identity using a product formula for the sine function. We will also suggest another approach using Fourier series in Exercise 13.20.

The starting point of both arguments is the infinite product formula

$$\sin z = z \prod_{n=1}^{\infty} \left(1 - \frac{z^2}{n^2 \pi^2}\right) \tag{13.2}$$

for the function $\sin z$; see [1, Ch. 5, §2.3].

We now give the first proof. We write the Taylor expansion of $\sin z/z$ to obtain

$$\sum_{k=0}^{\infty} (-1)^k \frac{z^{2k}}{(2k+1)!} = \prod_{n=1}^{\infty} \left(1 - \frac{z^2}{n^2\pi^2}\right)$$

If we equate the coefficients of z^2 we obtain

$$-\frac{1}{6} = -\sum_{n=1}^{\infty} \frac{1}{n^2\pi^2}.$$

Consequently,

$$\zeta(2) = \sum_{n=1}^{\infty} \frac{1}{n^2} = \frac{\pi^2}{6}.$$

In the second proof we actually compute $\zeta(2k)$ for all $k \in \mathbb{N}$. Again we use the formula (13.2). Take the logarithm of both sides to obtain

$$\log \sin z = \log z + \sum_{n=1}^{\infty} \log \left(1 - \frac{z^2}{n^2\pi^2}\right).$$

Differentiating gives

$$\frac{\cos z}{\sin z} = \frac{1}{z} + \sum_{n=1}^{\infty} \frac{\frac{-2z}{n^2\pi^2}}{1 - \frac{z^2}{n^2\pi^2}}$$

$$= \frac{1}{z} + \sum_{n=1}^{\infty} \frac{-2z}{n^2\pi^2} \sum_{k=0}^{\infty} \frac{z^{2k}}{n^{2k}\pi^{2k}}$$

$$= \frac{1}{z} - 2\sum_{k=0}^{\infty} \frac{z^{2k+1}}{\pi^{2k+2}} \sum_{n=1}^{\infty} \frac{1}{n^{2k+2}}.$$

Consequently,

$$z\frac{\cos z}{\sin z} = 1 - 2\sum_{k=1}^{\infty} \frac{\zeta(2k)}{\pi^{2k}} z^{2k}. \tag{13.3}$$

On the other hand, by Theorem A.1

$$\cos z = \frac{e^{iz} + e^{-iz}}{2}$$

and

$$\sin z = \frac{e^{iz} - e^{iz}}{2i}.$$

So we have

$$z\frac{\cos z}{\sin z} = iz\frac{e^{iz} + e^{-iz}}{e^{iz} - e^{-iz}} = \frac{2iz}{e^{2iz} - 1} + iz. \tag{13.4}$$

The function $t/(e^t - 1)$ whose value at $2iz$ appears in the above expression has a particularly well-known Taylor expansion with a long history. We define the *Bernoulli numbers* B_m, for $m \geq 0$, by

$$\frac{t}{e^t - 1} = \sum_{m=0}^{\infty} B_m \frac{t^m}{m!}.$$

It is not hard to see that $B_1 = -1/2$, and that for odd $m > 1$, $B_m = 0$. The first few non-zero B_m's are $B_0 = 1$, $B_2 = 1/6$, $B_4 = -1/30$, $B_6 = 1/42$, ... Furthermore, for all m, B_m is rational. See the exercises for more properties.

Going back to (13.4) we find that

$$z \frac{\cos z}{\sin z} = iz + \sum_{m=0}^{\infty} B_m \frac{(2iz)^m}{m!} = 1 + \sum_{k=1}^{\infty} (-1)^k \frac{2^{2k} B_{2k}}{(2k)!} z^{2k}.$$

Comparing this last expression with (13.3) gives:

Theorem 13.8. *For all natural numbers k,*

$$\zeta(2k) = (-1)^{k-1} \frac{2^{2k-1} B_{2k}}{(2k)!} \pi^{2k}.$$

Lemma 13.7 implies

Corollary 13.9. *With C_2 as above,*

$$C_2 = \frac{8}{\pi^2}.$$

Exercises

13.1 Prove Lemma 13.2.

13.2 Prove Corollary 13.6.

13.3 An *arithmetic function* is a function $f : \mathbb{N} \to \mathbb{C}$. For arithmetic functions f, g, we define the arithmetic function $f * g$ by

$$(f * g)(n) = \sum_{d|n} f(d) g\left(\frac{n}{d}\right).$$

Show that for all arithmetic functions f, g, h we have the following properties:

a. $f * (g * h) = (f * g) * h$;

b. $f * g = g * f$;

c. If $e(n) = \delta_{n0}$, Kronecker's delta, then $f * e = e * f = f$. Note that

$$e(n) = \begin{cases} 1 & n = 1; \\ 0 & n \neq 1. \end{cases}$$

13.4 Prove the claim in Remark 13.4.

13.5 (✠) Investigate the error term in Lemma 13.1.

13.6 (✠) Numerically verify the assertion of Theorem 13.5 and Corollary 13.6. Investigate the error terms in these results.

13.7 Define a function $\mathbf{1}$ by $\mathbf{1}(n) = 1$ for all n. Show that $\mathbf{1} * \mu = e$. Prove the *Möbius Inversion Formula*: If $f(n) = \sum_{d|n} g(d)$, then $g(n) = \sum_{d|n} \mu(d) f(\frac{n}{d})$.

13.8 Show that $\sum_{d|n} \varphi(d) = n$. Use this relation to derive a formula for the φ-function.

13.9 An arithmetic function f is called *multiplicative* if for every m, n with $\gcd(m, n) = 1$ we have $f(mn) = f(m)f(n)$. Show that if f, g are multiplicative, then so is $f * g$.

13.10 For a natural number n set $\sigma(n) = \sum_{d|n} d$. Find a formula for $\sigma(n)$ in terms of the prime factorization of n.

13.11 Show that for all $a, b \in \mathbb{N}$ with $a, b > 1$ we have

$$\frac{\sigma(a)}{a} < \frac{\sigma(ab)}{ab} < \frac{\sigma(a)\sigma(b)}{ab}.$$

13.12 Show that for $a, b > 1$,

$$\sigma(ab) > 2\sigma(a)^{1/2}\sigma(b)^{1/2}.$$

13.13 Show that for all $a, b \in \mathbb{N}$,

$$\sigma(a)\sigma(b) = \sum_{d|\gcd(a,b)} d\sigma\left(\frac{ab}{d^2}\right).$$

In particular, σ is a multiplicative function.

13.14 Find an asymptotic formula for

$$\sum_{\substack{a,b \leq X \\ \gcd(a,b)=1}} ab$$

as $X \to \infty$.

13.15 Find an asymptotic formula for

$$\sum_{n \leq X} \varphi(n)$$

as $X \to \infty$.

13.16 Prove the following statement: Let $(c_n)_n$ be a sequence of complex numbers, and $f : [1, \infty) \to \mathbb{C}$ a function with continuous derivative. Then

$$\sum_{n \le x} c_n f(n) = \left(\sum_{n \le x} c_n \right) f(x) - \int_1^x \left(\sum_{n \le t} c_n \right) f'(t) \, dt.$$

13.17 Show

$$\sum_{d \le x} \frac{1}{d} = \log x + O(1).$$

13.18 Recall the notion of *average order* from Definition 9.3.

a. Let $d(n)$ be the number of divisors of n. Show that

$$\sum_{k \le n} d(k) = \sum_{k \le n} \left[\frac{n}{k} \right].$$

Conclude that $d(n)$ has average order $\log x$;

b. Let $\phi(n)$ be the Euler totient function. Show that the average order of $\phi(n)$ is $\zeta(2)x$;

c. Let $\omega(n)$ be the number of distinct prime divisors of n. Show that the average order of $\omega(n)$ is $\log \log x$.

13.19 Find a multiplicative function f such that

$$\sum_{d \mid n} \frac{\mu(d) d^2 f(n/d)}{\phi(d)} = \sigma(n) f(n), \quad n \in \mathbb{N}.$$

13.20 Use Parseval's formula [41, Theorem 8.16] applied to the function $f(x) = x$ on the interval $[0, 1]$ to give another proof for Euler's identity, $\zeta(2) = \pi^2/6$.

13.21 Pick two natural numbers at random. What is the probably that they are coprime?

13.22 Prove that for each natural number r,

$$B_r = - \sum_{k=0}^{r-1} \binom{r}{k} \frac{B_k}{r - k + 1}.$$

Use this relation to find the first few Bernoulli numbers.

13.23 Show that all Bernoulli numbers are rational.

13.24 Show that for each natural number r, $B_{2r+1} = 0$.

13.25 Find an asymptotic formula for the number of primitive right triangles with perimeter bounded by X as $X \to \infty$.

Notes

Lehmer's theorem and Manin's conjecture

Lehmer [83] published a different proof of Corollary 13.6 in 1900. The argument we present here shows that any power saving improvement in the error term of Lemma 13.1 would improve the error terms in Theorem 13.5 and Corollary 13.6 to $O(\sqrt{B})$. The quantity considered in Corollary 13.6 appears in the *Online Encyclopedia of Integer Sequences*:

$$http://oeis.org/A156685$$

The question of counting integral solutions with bounded size to algebraic equations with infinitely many solutions is a very active area of research of current interest. Theorem 13.5 has now been greatly generalized. Yuri Manin has formulated several conjectures that connect the arithmetic features of some classes of equations where one expects a lot of solutions to the geometry of the resulting solution sets; see [104] for various questions and conjectures.

A family of number theorists

The Lehmer of Corollary 13.6 is Derrick Norman Lehmer (July 27, 1867–September 8, 1938). He was the father of Derrick Henry Lehmer (February 23, 1905–May 22, 1991) who was a mathematician credited with many contributions to number theory. D. H. Lehmer was married to Emma Markovna Lehmer (née Trotskaia) (November 6, 1906–May 7, 2007) who was a number theorist herself with over 50 publications to her name, [84]. There have been several other families of mathematicians in history, most notably the Bernoulli family. And here is a joke: What was the most influential mathematician family in history? Clearly Gauss's family, because it doesn't matter what the rest of his family did.

The Riemann zeta function

The complex function

$$\zeta(s) = \sum_{n=1}^{\infty} \frac{1}{n^s}$$

is called the *Riemann zeta function*. This series converges absolutely for $\Re s > 1$. Riemann was certainly not the first person to study this function. In fact, by the time of the publication of Riemann's work in 1859 various mathematicians, Euler in particular, had studied the values of the zeta function for integer values of s for at least two centuries; see [109] for a survey. The problem of computing $\zeta(2)$ which we discussed in this chapter was posed by Pietro Mengoli in 1650 and solved by Euler in 1735. Riemann, in a spectacular paper [93], proved the analytic continuation of the zeta function, proved the functional equation, discussed the connection to the distribution of prime numbers, and formulated a conjecture about prime numbers, nowadays known as the *Riemann Hypothesis*.

First a word about analytic continuation. Suppose we have a function $f(s)$ which is holomorphic on an open subset U of complex numbers, and suppose V is an open set in $\mathbb{C}$ containing U. We call a function g, holomorphic on V, the *analytic continuation* of f if the restriction of g to U is equal to f. It is not terribly hard to show that for $\Re s > 1$ we have

$$\zeta(s) = s \int_1^\infty \frac{[x]}{x^{s+1}} \, dx = \frac{s}{s-1} - s \int_1^\infty \frac{\{x\}}{x^{s+1}} \, dx.$$

The expression on the right-hand side is meromorphic on $\Re s > 0$ with a simple pole at $s = 1$, however, and this provides an analytic continuation for $\zeta(s)$ to a larger domain. But this is not where the analytic continuation stops. In fact, if we set

$$\xi(s) = s(s-1)\pi^{-s/2}\Gamma\left(\frac{s}{2}\right)\zeta(s),$$

then Riemann showed that $\xi(s)$ is holomorphic on $\Re s > 0$ and

$$\xi(1-s) = \xi(s). \tag{13.5}$$

Since $\xi(s)$ is holomorphic for $\Re s > 0$, and $\xi(1-s)$ is holomorphic for $\Re(1-s) > 0$, i.e., $\Re s < 1$, we obtain the holomorphy of $\xi(s)$ on the entire set of complex numbers. This further shows that $\zeta(s)$ has an analytic continuation to the entire complex plane to a meromorphic function with a unique simple pole at $s = 1$ with residue 1. Since we already have computed the value of $\zeta(s)$ for even positive integers $2k$, we can use the functional equation (13.5) to compute the values of the *analytic continuation* of $\zeta(s)$ for odd negative numbers. In fact, for $n \in \mathbb{N}$,

$$\zeta(1 - 2n) = -\frac{B_{2n}}{2n}.$$

For example, $\zeta(-1) = -1/12$. One can similarly compute the value of $\zeta(0)$ to be $-1/2$. Again, we should emphasize that these are the values of the analytically continued function, and they should not be taken to mean

$$1 + 1 + 1 + \cdots = -\frac{1}{2},$$

or

$$1 + 2 + 3 + \cdots = -\frac{1}{12}.$$

Let us illustrate what is happening here with an easy example. Suppose $U = \{s \in \mathbb{C} \mid |s| < 1\}$ and $f(s) = \sum_{k=0}^{\infty} s^k$. The series defining f is absolutely convergent on U and defines a holomorphic function there. By general properties of geometric series, for $|s| < 1$, we have

$$f(s) = \frac{1}{1-s}.$$

The function $g(s) = 1/(1-s)$ is holomorphic on the much larger domain $V = \{s \in \mathbb{C} \mid s \neq 1\}$. Note that outside the open set U the function $g(s)$ is not given by the original series defining $f(s)$. This important point is the source of many paradoxes in the theory of infinite series. For example, the value of the function $g(s)$ at $s = 2$ is equal to -1. If we set $s = 2$ in the formula for $f(s)$ we formally get

$$1 + 2 + 4 + 8 + 16 + 32 + 64 + \ldots$$

Does this then mean

$$1 + 2 + 4 + 8 + 16 + 32 + 64 + \cdots = -1?$$

Absolutely not! In fact the series defining $f(s)$ is not even defined for $s = 2$.

We now turn to the connections between the zeta function and the distribution of prime numbers. Euler observed the product formula that now bears his name: For $\Re s > 1$ we have

$$\zeta(s) = \prod_{p \text{ prime}} \frac{1}{1 - p^{-s}}.$$

If we use this formula to compute $(d/ds) \log \zeta(s)$ we obtain

$$-\frac{\zeta'(s)}{\zeta(s)} = \sum_{k \geq 1} \sum_{p \text{ prime}} \frac{\log p}{p^{ks}} = \sum_{n=1}^{\infty} \frac{\Lambda(n)}{n^s}, \qquad (13.6)$$

with $\Lambda(n)$ being the *von Mangoldt function* defined by

$$\Lambda(n) = \begin{cases} \log p & n = p^k, \ p \text{ prime}; \\ 0 & \text{otherwise}. \end{cases}$$

An idea that Riemann brought into this subject was *contour integration*. For a complex function $f(s)$ and a real number c let us define

$$\int_{(c)} f(s)\, ds = \lim_{R \to \infty} \int_{c-iR}^{c+iR} f(s)\, ds.$$

Fix a real number $c > 1$. A contour integration computation shows that for $x > 1$, non-integer,

$$\sum_{n < x} \Lambda(n) = \frac{1}{2\pi i} \int_{(c)} \left(-\frac{\zeta'(s)}{\zeta(s)} \right) \frac{x^s}{s}\, ds.$$

The function $-\zeta'(s)/\zeta(s)$ has a simple pole at $s = 1$ with residue 1. Suppose we can shift the contour back to (c'), for a number $c' < 1$. Then we would obtain

$$\sum_{n<x} \Lambda(n) = x + \frac{1}{2\pi i} \int_{(c')} \left(-\frac{\zeta'(s)}{\zeta(s)} \right) \frac{x^s}{s} \, ds. \tag{13.7}$$

Riemann's idea then was to prove that this last integral contributes less than x to the formula, and hence obtain

$$\sum_{n<x} \Lambda(n) \sim x, \quad x \to \infty. \tag{13.8}$$

Exercise 13.16 can now be used to prove

$$\#\{p \le x\} \sim \frac{x}{\log x} \tag{13.9}$$

which is the celebrated *Prime Number Theorem*, conjectured by Gauss. Also, knowing the specific value of c' would lead to error estimates for the Prime Number Theorem. So, the question that Riemann was faced with was to determine how far back the contour could be moved. In general, the logarithmic derivative of a meromorphic function has poles whenever the function has poles or zeros. In particular in order to know the poles of $\zeta'(s)/\zeta(s)$ we need to know where the function $\zeta(s)$ is zero. Riemann computed several zeros of the zeta function in the domain $\Re s > 0$ and observed that they are all on the line $\Re s = 1/2$, and conjectured that this would be the case for all zeros. If one assumes the Riemann Hypothesis, then it follows that

$$\#\{p \le x\} = \mathrm{Li}\, x + O(x^{1/2+\varepsilon})$$

for all $\varepsilon > 0$, with

$$\mathrm{Li}\, x = \int_2^x \frac{dt}{\log t}.$$

At present, the Riemann Hypothesis appears out of reach.

Titchmarsh's classic [52] is a much recommended, comprehensive introduction to the theory of the Riemann zeta function.

Chapter 14
How are rational points distributed, really?

In §3.2 we found a description of all the points with rational coordinates on the unit circle $x^2 + y^2 = 1$. In this chapter we examine some topological and analytic properties of these rational points. In particular, we will show that points with rational coordinates are *equidistributed* with a respect to a natural measure on the unit circle centered at the origin. The starting point of our investigation is the concept of *equidistribution* on the real line, and addressing the equidistribution properties of rational numbers according to a natural measure on the real line. This requires introducing an ordering of the set of rational numbers. The ordering we use is determined by the *height* of the rational number. The proof of Theorem 14.3, while in principle straightforward, is very complicated. We end the first section of this chapter with a strengthening of the latter theorem, Theorem 14.4. The proof of this theorem uses some technical tools from analysis. We prove the equidistribution of rational points on the unit circle in the second section of the chapter. In the Notes, we state a general theorem of Bohl, Sierpiński, and Weyl, proved independently of each other, about the distribution of a sequence of numbers in the interval $[0, 1]$. We also make some comments about the general question of the equidistribution of rational points on higher dimensional spheres.

14.1 The real line

It is a well-known fact that the set of rational numbers is dense in the set of real numbers. Our first goal here is to quantify this density statement.

Definition 14.1. Suppose $I = (\alpha, \beta)$ is an interval in $\mathbb{R}$, and ϑ a Riemann integrable function on I. We say a sequence $\{x_n\}_{n=1}^{\infty}$ of elements of I is ϑ-*equidistributed*, or *equidistributed with respect to the function* ϑ, if for each subinterval $J \subset I$ we have

$$\lim_{X \to \infty} \frac{\#\{n \leq X \mid x_n \in J\}}{X} = \int_J \vartheta(x)\, dx.$$

© Springer Nature Switzerland AG 2018
R. Takloo-Bighash, *A Pythagorean Introduction to Number Theory*,
Undergraduate Texts in Mathematics, https://doi.org/10.1007/978-3-030-02604-2_14

If $\vartheta(x) = 1/(\beta - \alpha)$ for all $x \in I$, we simply say the sequence $\{x_n\}$ is *equidistributed* in I.

We note that if a sequence $\{x_n\}$ is equidistributed in the interval I it will be dense in the interval, but not vice versa. In fact, it is possible to construct sequences $\{x_n\}_{n=1}^{\infty}$ and $\{y_n\}_{n=1}^{\infty}$ with the property that

$$\{x_n \mid n \in \mathbb{N}\} = \{y_n \mid n \in \mathbb{N}\}$$

with $\{x_n\}_n$ equidistributed, and $\{y_n\}_n$ not equidistributed; see Exercise 14.1. These examples also show that whether a sequence $\{x_n\}_n$ is equidistributed in an interval I depends strongly on the particular ordering of the elements of $\{x_n\}_n$.

We now turn our attention to the study of the distribution of rational numbers in real numbers. It is already an interesting problem to find a function ϑ such that the set of rational numbers is ϑ-equidistributed in the set of real numbers. As pointed out earlier, the function ϑ depends very much on the choice of the ordering of the set of rational numbers. Let us describe one such ordering which is particularly natural.

Definition 14.2. For a rational number $\gamma = r/s$ with $r, s \in \mathbb{Z}$ with $\gcd(r, s) = 1$, we define the *height* of γ by

$$H(\gamma) = \sqrt{r^2 + s^2}.$$

The motivation behind this definition is that we tend to think of the rational number

$$\frac{5000001473}{5000003010}$$

as a more *arithmetically complicated* rational number than $1.02 = 51/50$, even though both numbers are approximately 1. The height function quantifies this notion, in the sense that

$$H\left(\frac{5000001473}{5000003010}\right) = \sqrt{5000003010^2 + 5000001473^2}$$

is much bigger than

$$H(1.02) = \sqrt{51^2 + 50^2}.$$

An interesting property of our height function is that for all finite $B > 0$ the number of rational numbers γ with $H(\gamma) \leq B$ is finite. In fact, if $\gamma = r/s$ with $\gcd(r, s) = 1$, $H(\gamma) \leq B$ means $|r| \leq B$ and $|s| \leq B$. There are only finitely many such integers r and s. For example, the following rational numbers γ have the property $H(\gamma) \leq 4$:

$$0, \pm 1, \pm 2, \pm 1/2, \pm 3, \pm 1/3, \pm 2/3, \pm 3/2.$$

The proof of the following theorem occupies most of the remainder of this chapter:

Theorem 14.3. *Rational numbers ordered by their height are equidistributed in every interval (α, β), including unbounded intervals, with respect to*

$$\vartheta(t) = \frac{1}{\pi} \cdot \frac{1}{1+t^2}.$$

Proof. We need to compute the limit

$$S(\alpha, \beta) := \lim_{X \to \infty} \frac{\#\{\gamma \in \mathbb{Q} \cap (\alpha, \beta) \mid H(\gamma) \le X\}}{\#\{\gamma \in \mathbb{Q} \mid H(\gamma) \le X\}} \tag{14.1}$$

for each $\alpha < \beta$.

Two basic observations:

- For each $\alpha < \beta$, $S(\alpha, \beta) = S(-\beta, -\alpha)$;
- for each $\alpha < \beta < \delta$, we have $S(\alpha, \beta) + S(\beta, \delta) = S(\alpha, \delta)$.

These observations imply that it suffices to compute $S(\alpha, \beta)$ in the following three cases:

1. $0 < \alpha < \beta < 1$;
2. $1 < \alpha < \beta$.
3. $1 < \alpha$ and $\beta = +\infty$.

We will compute $S(\alpha, \beta)$ in each case.

First we find a formula for the denominator of the expression in Equation (14.1),

$$n(X) := \#\{\gamma \in \mathbb{Q} \mid H(\gamma) \le X\}.$$

For a non-zero rational number $\gamma = m/n$ with $\gcd(m, n) = 1$, we have

$$H(\gamma) = H(-\gamma) = H(\gamma^{-1}) = H(-\gamma^{-1}) = \sqrt{m^2 + n^2}.$$

It is now not hard to see (Exercise 14.7) that

$$n(X) = \frac{1}{2}f(X^2) + O(1). \tag{14.2}$$

with f defined by

$$f(B) = \#\{(x, y) \ne (0, 0) \mid x, y \in \mathbb{Z}, \gcd(x, y) = 1, x^2 + y^2 \le B\}.$$

By the computation of C_2 from §13.2 and Exercise 14.8, we have

$$n(X) = \frac{1}{2\zeta(2)}X^2 + O(X \log X). \tag{14.3}$$

Now we find an expression for

$$n_{\alpha, \beta}(X) := \#\{\gamma \in \mathbb{Q} \cap (\alpha, \beta) \mid H(\gamma) \le X\}. \tag{14.4}$$

Suppose $0 < \alpha < \beta < 1$, and that we have a reduced fraction $n/m \in (\alpha, \beta)$ with $H(n/m) \le X$. This means, $m, n \in \mathbb{N}$, $\gcd(m, n) = 1$, $m^2 + n^2 \le X^2$, and $\alpha m < n < \beta m$.

Our strategy is to write $n_{\alpha,\beta}$ as a sum of 1's over the defining conditions on m, n, and then use the function μ from Chapter 13 to handle the coprimality condition on m, n. Eventually we will use the geometric method of the proof of Theorem 9.4 to finish the computation.

By Lemma 13.2 applied to $\gcd(m, n)$ we have

$$\sum_{d\mid\gcd(m,n)} \mu(d) = \begin{cases} 1 & \text{if } m, n \text{ coprime;} \\ 0 & \text{otherwise.} \end{cases}$$

We have

$$n_{\alpha,\beta}(X) = \sum_{\substack{m,n\in\mathbb{N},\gcd(m,n)=1 \\ m^2+n^2\le X^2,\alpha m<n<\beta m}} 1 = \sum_{\substack{m,n\in\mathbb{N} \\ m^2+n^2\le X^2,\alpha m<n<\beta m}} \sum_{d\mid\gcd(m,n)} \mu(d)$$

$$= \sum_{d\le X} \mu(d) \sum_{\substack{m,n\in\mathbb{N},d\mid m,d\mid n \\ m^2+n^2\le X^2,\alpha m<n<\beta m}} 1 = \sum_{d\le X} \mu(d) \sum_{\substack{m,n\in\mathbb{N} \\ m^2+n^2\le X^2/d^2,\alpha m<n<\beta m}} 1.$$

Consequently, if we set

$$\tilde{n}_{\alpha,\beta}(X) = \sum_{\substack{m,n\in\mathbb{N} \\ m^2+n^2\le X^2,\alpha m<n<\beta m}} 1,$$

we have

$$n_{\alpha,\beta}(X) = \sum_{d\le X} \mu(d)\tilde{n}_{\alpha,\beta}\left(\frac{X}{d}\right).$$

Our immediate task is to find a formula for $\tilde{n}_{\alpha,\beta}(X)$. For simplicity we will assume that α, β are irrational numbers; also since eventually we will be letting $X \to \infty$, we will assume that $\alpha^{-1}(1 + \beta^2) < X$.

We start by writing

$$\tilde{n}_{\alpha,\beta}(X) = \sum_{\substack{m\le X \\ \max(\alpha m,1)\le n\le\min(\beta m,\sqrt{X^2-m^2})}} 1 = \sum_{\substack{m<\alpha^{-1} \\ 1\le n\le\min(\beta m,\sqrt{X^2-m^2})}} 1 + \sum_{\substack{\alpha^{-1}<m\le X \\ \alpha m\le n\le\min(\beta m,\sqrt{X^2-m^2})}} 1.$$

In the first sum, since $X > \alpha^{-1}\sqrt{1 + \beta^2}$, we have

$$\beta m < \sqrt{X^2 - m^2}.$$

Hence,

$$\sum_{\substack{m<\alpha^{-1} \\ 1\leq n\leq\min(\beta m,\sqrt{X^2-m^2})}} 1 = \sum_{\substack{m<\alpha^{-1} \\ 1\leq n\leq\beta m}} 1 = O(1)$$

as the whole sum can be bounded independent of X.

So we have shown

$$\sum_{\substack{m<\alpha^{-1} \\ 1\leq n\leq\min(\beta m,\sqrt{X^2-m^2})}} 1 = O(1). \tag{14.5}$$

Now we examine the second sum

$$\sum_{\substack{\alpha^{-1}<m<X \\ \alpha m\leq n\leq\min(\beta m,\sqrt{X^2-m^2})}} 1.$$

We note that if

$$\alpha^{-1} < m \le \frac{X}{\sqrt{1+\beta^2}},$$

then

$$\beta m \le \sqrt{X^2-m^2},$$

and if

$$\frac{X}{\sqrt{1+\beta^2}} < m \le X,$$

then

$$\sqrt{X^2-m^2} \le \beta m.$$

As a result the sum is equal to

$$\sum_{\substack{\alpha^{-1}<m<X \\ \alpha m\leq n\leq\min(\beta m,\sqrt{X^2-m^2})}} 1 = \sum_{\substack{\alpha^{-1}<m\leq\frac{X}{\sqrt{1+\beta^2}} \\ \alpha m\leq n\leq\beta m}} 1 + \sum_{\substack{\frac{X}{\sqrt{1+\beta^2}}<m\leq X \\ \alpha m\leq n\leq\sqrt{X^2-m^2}}} 1.$$

We analyze each piece separately.

We have

$$\sum_{\substack{\alpha^{-1}<m\leq\frac{X}{\sqrt{1+\beta^2}} \\ \alpha m\leq n\leq\beta m}} 1 = \sum_{\alpha^{-1}<m\leq\frac{X}{\sqrt{1+\beta^2}}} ([\beta m]-[\alpha m])$$

$$= \sum_{\alpha^{-1}<m\leq\frac{X}{\sqrt{1+\beta^2}}} (\beta-\alpha)m + O(1) = \frac{\beta-\alpha}{2(1+\beta^2)}X^2 + O(X),$$

after using Corollary A.6.

We have shown

$$\sum_{\substack{\alpha^{-1}<m\leq\frac{X}{\sqrt{1+\beta^2}}\\ \alpha m\leq n\leq\beta m}} 1 = \frac{\beta-\alpha}{2(1+\beta^2)}X^2 + O(X). \tag{14.6}$$

Next, we consider the sum

$$\sum_{\substack{\frac{X}{\sqrt{1+\beta^2}}<m\leq X\\ \alpha m\leq n\leq\sqrt{X^2-m^2}}} 1.$$

The important point to note is that for some values of m, $\alpha m > \sqrt{X^2-m^2}$, and for such values, the n-sum is empty. To have $\alpha m \leq \sqrt{X^2-m^2}$, we need to have $m \leq X(1+\alpha^2)^{-1/2}$ as an easy computation shows. Consequently,

$$\sum_{\substack{\frac{X}{\sqrt{1+\beta^2}}<m\leq X\\ \alpha m\leq n\leq\sqrt{X^2-m^2}}} 1 = \sum_{\substack{\frac{X}{\sqrt{1+\beta^2}}<m\leq\frac{X}{\sqrt{1+\alpha^2}}\\ \alpha m\leq n\leq\sqrt{X^2-m^2}}} 1 = \sum_{\frac{X}{\sqrt{1+\beta^2}}<m\leq\frac{X}{\sqrt{1+\alpha^2}}} ([\sqrt{X^2-m^2}]-[\alpha m])$$

$$= \sum_{\frac{X}{\sqrt{1+\beta^2}}<m\leq\frac{X}{\sqrt{1+\alpha^2}}} ([\sqrt{X^2-m^2}]-\alpha m + O(1))$$

$$= \sum_{\frac{X}{\sqrt{1+\beta^2}}<m\leq\frac{X}{\sqrt{1+\alpha^2}}} [\sqrt{X^2-m^2}]-\alpha \sum_{\frac{X}{\sqrt{1+\beta^2}}<m\leq\frac{X}{\sqrt{1+\alpha^2}}} m + O(X)$$

$$= \sum_{\frac{X}{\sqrt{1+\beta^2}}<m\leq\frac{X}{\sqrt{1+\alpha^2}}} [\sqrt{X^2-m^2}]-\alpha \left(\frac{X^2}{2(1+\alpha^2)} - \frac{X^2}{2(1+\beta^2)}\right) + O(X).$$

The sum

$$\sum_{\frac{X}{\sqrt{1+\beta^2}}<m\leq\frac{X}{\sqrt{1+\alpha^2}}} [\sqrt{X^2-m^2}]$$

is the number of integral points (m, n) within the disk $x^2 + y^2 \leq X^2$ with positive y-coordinates such that the x-coordinate is in the interval

$$\frac{X}{\sqrt{1+\beta^2}} < m \leq \frac{X}{\sqrt{1+\alpha^2}}.$$

These are the points with integral coordinates in the yellow region, including the boundary, in Figure 14.1.

By an argument similar to proof of Theorem 9.4 (Exercise 14.9), this number is equal to

Fig. 14.1 The integral
points (m, n) within the disk
$x^2 + y^2 \leq X^2$ with positive
y-coordinates such that the
x-coordinate is in the interval
$\frac{X}{\sqrt{1+\beta^2}} < m \leq \frac{X}{\sqrt{1+\alpha^2}}$

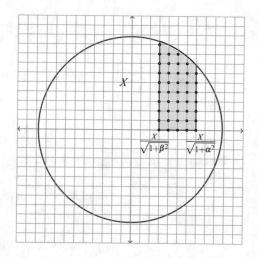

$$\int_{\frac{X}{\sqrt{1+\beta^2}}}^{\frac{X}{\sqrt{1+\alpha^2}}} \sqrt{X^2 - t^2}\, dt + O(X) = X^2 \int_{\frac{1}{\sqrt{1+\beta^2}}}^{\frac{1}{\sqrt{1+\alpha^2}}} \sqrt{1 - t^2}\, dt + O(X).$$

We have
$$\int \sqrt{1 - t^2}\, dt = \frac{1}{2} \sin^{-1} t + \frac{1}{2} t \sqrt{1 - t^2} + C.$$

Consequently,
$$\int_{\frac{1}{\sqrt{1+\beta^2}}}^{\frac{1}{\sqrt{1+\alpha^2}}} \sqrt{1 - t^2}\, dt = \frac{1}{2} \sin^{-1} \frac{1}{\sqrt{1+\alpha^2}} - \frac{1}{2} \sin^{-1} \frac{1}{\sqrt{1+\beta^2}} + \frac{\alpha}{2(1+\alpha^2)} - \frac{\beta}{2(1+\beta^2)}.$$

So we have proved
$$\sum_{\substack{\frac{X}{\sqrt{1+\beta^2}} < m \leq X \\ \alpha m \leq n \leq \sqrt{X^2 - m^2}}} 1 = X^2 \left(\frac{1}{2} \sin^{-1} \frac{1}{\sqrt{1+\alpha^2}} - \frac{1}{2} \sin^{-1} \frac{1}{\sqrt{1+\beta^2}} \right)$$

$$+ X^2 \left(\frac{\alpha}{2(1+\alpha^2)} - \frac{\beta}{2(1+\beta^2)} \right) - \alpha \left(\frac{X^2}{2(1+\alpha^2)} - \frac{X^2}{2(1+\beta^2)} \right) + O(X).$$

Putting everything together, we have
$$\tilde{n}_{\alpha,\beta}(X) = \frac{\beta - \alpha}{2(1+\beta^2)} X^2 + X^2 \left(\frac{1}{2} \sin^{-1} \frac{1}{\sqrt{1+\alpha^2}} - \frac{1}{2} \sin^{-1} \frac{1}{\sqrt{1+\beta^2}} \right)$$

$$+X^2\left(\frac{\alpha}{2(1+\alpha^2)} - \frac{\beta}{2(1+\beta^2)}\right) - \alpha\left(\frac{X^2}{2(1+\alpha^2)} - \frac{X^2}{2(1+\beta^2)}\right) + O(X)$$

$$= X^2\left(\frac{1}{2}\sin^{-1}\frac{1}{\sqrt{1+\alpha^2}} - \frac{1}{2}\sin^{-1}\frac{1}{\sqrt{1+\beta^2}}\right) + O(X).$$

We set

$$\eta(t) = \frac{1}{2}\sin^{-1}\frac{1}{\sqrt{1+t^2}}.$$

We can now analyze $n_{\alpha,\beta}(X)$. We have

$$n_{\alpha,\beta}(X) = \sum_{d\leq X}\mu(d)\tilde{n}_{\alpha,\beta}\left(\frac{X}{d}\right)$$

$$= \sum_{d\leq X}\mu(d)\left\{(\eta(\alpha)-\eta(\beta))\left(\frac{X}{d}\right)^2 + O\left(\frac{X}{d}\right)\right\}$$

$$= \frac{\eta(\alpha)-\eta(\beta)}{\zeta(2)}X^2 + O(X\log X).$$

Finally,

$$S(\alpha,\beta) = \lim_{X\to\infty}\frac{n_{\alpha,\beta}(X)}{n(X)} = \lim_{X\to\infty}\frac{\frac{\eta(\alpha)-\eta(\beta)}{\zeta(2)}X^2 + O(X\log X)}{\frac{\pi}{2\zeta(2)}X^2 + O(X\log X)} = \frac{\eta(\alpha)-\eta(\beta)}{\pi}.$$

Hence we have proved for $0 < \alpha < \beta < 1$,

$$S(\alpha,\beta) = \frac{1}{\pi}\left(\sin^{-1}\frac{1}{\sqrt{1+\alpha^2}} - \sin^{-1}\frac{1}{\sqrt{1+\beta^2}}\right) = \frac{1}{\pi}\int_\alpha^\beta\frac{dt}{1+t^2}.$$

Now we handle the case where $1 < \alpha < \beta$. In this case we have

$$n_{\alpha,\beta}(X) = \sum_{\substack{m,n\in\mathbb{N},\gcd(m,n)=1 \\ m^2+n^2\leq X^2,\alpha m<n<\beta m}} 1 = \sum_{\substack{m,n\in\mathbb{N},\gcd(m,n)=1 \\ m^2+n^2\leq X^2,\beta^{-1}n<m<\alpha^{-1}n}} 1 = n_{\beta^{-1},\alpha^{-1}}(X).$$

Consequently, if $1 < \alpha < \beta$, we have

$$S(\alpha,\beta) = \frac{1}{\pi}\int_{\beta^{-1}}^{\alpha^{-1}}\frac{dt}{1+t^2}.$$

It is easy to see (Exercise 14.10) that the latter integral is equal to

$$\frac{1}{\pi}\int_\alpha^\beta\frac{dt}{1+t^2}.$$

Now we treat the third case, where $\beta = +\infty$. The argument in this case is very similar to the first case, so we only sketch the proof. In this case we abbreviate $n_{\alpha,+\infty}(X)$ and $\tilde{n}_{\alpha,+\infty}(X)$ to $n_\alpha(X)$ and $\tilde{n}_\alpha(X)$, respectively. As before, we have

$$n_\alpha(X) = \sum_{d \leq X} \mu(d)\tilde{n}_\alpha\left(\frac{X}{d}\right).$$

We start by writing

$$\tilde{n}_\alpha(X) = \sum_{\substack{m \leq X \\ \alpha m \leq n \leq \sqrt{X^2 - m^2}}} 1.$$

Since we want $\alpha m \leq \sqrt{X^2 - m^2}$ we need to have

$$m \leq \frac{X}{\sqrt{1 + \alpha^2}}.$$

Thus,

$$\tilde{n}_\alpha(X) = \sum_{\substack{m \leq \frac{X}{\sqrt{1+\alpha^2}} \\ \alpha m \leq n \leq \sqrt{X^2 - m^2}}} 1 = \sum_{m \leq \frac{X}{\sqrt{1+\alpha^2}}} ([\sqrt{X^2 - m^2}] - [\alpha m])$$

$$= \sum_{m \leq \frac{X}{\sqrt{1+\alpha^2}}} ([\sqrt{X^2 - m^2}] - \alpha \sum_{m \leq \frac{X}{\sqrt{1+\alpha^2}}} m + O(X)$$

$$= X^2 \int_0^{\frac{1}{\sqrt{1+\alpha^2}}} \sqrt{1 - t^2}\, dt - \frac{\alpha}{2(1+\alpha^2)} X^2 + O(X)$$

$$= X^2 \left(\frac{1}{2} \sin^{-1} \frac{1}{\sqrt{1+\alpha^2}} + \frac{\alpha}{2(1+\alpha^2)}\right) - \frac{\alpha}{2(1+\alpha^2)} X^2 + O(X)$$

$$= X^2 \left(\frac{1}{2} \sin^{-1} \frac{1}{\sqrt{1+\alpha^2}}\right) + O(X).$$

Again if we set

$$\eta(t) = \frac{1}{2} \sin^{-1} \frac{1}{\sqrt{1+t^2}},$$

we have proved

$$\tilde{n}_\alpha(X) = \eta(\alpha)X^2 + O(X).$$

We can now analyze $n_\alpha(X)$. We have

$$n_\alpha(X) = \sum_{d \leq X} \mu(d)\tilde{n}_\alpha(\frac{X}{d}) = \sum_{d \leq X} \mu(d) \left\{\eta(\alpha)\left(\frac{X}{d}\right)^2 + O\left(\frac{X}{d}\right)\right\}$$

$$= \frac{\eta(\alpha)}{\zeta(2)} X^2 + O(X \log X).$$

Finally,

$$S(\alpha, +\infty) = \lim_{X \to \infty} \frac{n_\alpha(X)}{n(X)} = \lim_{X \to \infty} \frac{\frac{\eta(\alpha)}{\zeta(2)} X^2 + O(X \log X)}{\frac{\pi}{2\zeta(2)} X^2 + O(X \log X)} = \frac{2\eta(\alpha)}{\pi}.$$

Hence we have proved for $1 < \alpha$,

$$S(\alpha, +\infty) = \frac{1}{\pi} \sin^{-1} \frac{1}{\sqrt{1 + \alpha^2}} = \frac{1}{\pi} \int_\alpha^{+\infty} \frac{dt}{1 + t^2}.$$

$\square$

Theorem 14.3 has an interesting consequence. We call a real function f on $\mathbb{R}$ *locally Riemann integrable* if for each finite interval I, the restriction of f to I is Riemann integrable on I.

Theorem 14.4. *Let f be a bounded locally Riemann integrable function on $\mathbb{R}$. Then*

$$\lim_{X \to \infty} \frac{1}{\#\{\gamma \in \mathbb{Q} \mid H(\gamma) \leq X\}} \sum_{\gamma \in \mathbb{Q}, H(\gamma) \leq X} f(\gamma) = \frac{1}{\pi} \int_{-\infty}^{+\infty} \frac{f(t)}{1 + t^2} dt.$$

We note that for a bounded locally Riemann integrable function f as in the theorem, the integral

$$\int_{-\infty}^{+\infty} \frac{f(t)}{1 + t^2} dt$$

converges absolutely, Exercise 14.11.

Before we can start the proof of the theorem we need a general lemma:

Lemma 14.5. *A sequence $\{x_n\}$ is ϑ-equidistributed in a finite interval $I = (\alpha, \beta)$ if and only if for every Riemann integrable function f on I we have*

$$\lim_{N \to \infty} \frac{1}{N} \sum_{n=1}^N f(x_n) = \int_\alpha^\beta f(x) \vartheta(x) \, dx. \tag{14.7}$$

Proof. The definition of ϑ-equidistribution is equivalent to the validity of (14.7) for the characteristic function of each subinterval of I. This shows the sufficiency of the condition.

Now suppose the sequence $\{x_n\}$ is ϑ-equidistributed. Then Equation (14.7) is valid for all characteristic functions of subintervals of I. Since the two sides of (14.7) are linear in the function f, we conclude that (14.7) is also true for all linear combinations of characteristic functions of subintervals, i.e., step functions.

Now let f be a Riemann integrable function. Fix $\varepsilon > 0$. By [41, Theorem 6.6] there are step functions f_1, f_2 on I such that for all $x \in I$, $f_1(x) \leq f(x) \leq f_2(x)$, and

$$\int_\alpha^\beta f_2(x)\vartheta(x)\,dx - \int_\alpha^\beta f_1(x)\vartheta(x)\,dx < \varepsilon. \qquad (14.8)$$

Then

$$\int_\alpha^\beta f_1(x)\vartheta(x)\,dx = \lim_{N\to\infty} \frac{1}{N}\sum_{n=1}^N f_1(x_n) \le \liminf_{N\to\infty} \frac{1}{N}\sum_{n=1}^N f(x_n)$$

$$\le \limsup_{N\to\infty} \frac{1}{N}\sum_{n=1}^N f(x_n) \le \lim_{N\to\infty} \frac{1}{N}\sum_{n=1}^N f_2(x_n) = \int_\alpha^\beta f_2(x)\vartheta(x)\,dx.$$

Finally, (14.8) implies

$$\left| \limsup_{N\to\infty} \frac{1}{N}\sum_{n=1}^N f(x_n) - \liminf_{N\to\infty} \frac{1}{N}\sum_{n=1}^N f(x_n) \right|$$

$$\le \left| \int_\alpha^\beta f_2(x)\vartheta(x)\,dx - \int_\alpha^\beta f_1(x)\vartheta(x)\,dx \right| < \varepsilon.$$

Since $\varepsilon > 0$ is arbitrary, we have

$$\limsup_{N\to\infty} \frac{1}{N}\sum_{n=1}^N f(x_n) = \liminf_{N\to\infty} \frac{1}{N}\sum_{n=1}^N f(x_n).$$

Hence it makes sense to speak of the limit $\lim_N \sum_{n\le N} f(x_n)/N$. Revisiting the earlier inequalities gives

$$\int_\alpha^\beta f_1(x)\vartheta(x)\,dx \le \lim_{N\to\infty} \frac{1}{N}\sum_{n=1}^N f(x_n) \le \int_\alpha^\beta f_2(x)\vartheta(x)\,dx.$$

Since by definition

$$\int_\alpha^\beta f_1(x)\vartheta(x)\,dx \le \int_\alpha^\beta f(x)\,dx \le \int_\alpha^\beta f_2(x)\vartheta(x)\,dx,$$

we have

$$\left| \lim_{N\to\infty} \frac{1}{N}\sum_{n=1}^N f(x_n) - \int_\alpha^\beta f(x)\vartheta(x)\,dx \right|$$

$$\le \left| \int_\alpha^\beta f_2(x)\vartheta(x)\,dx - \int_\alpha^\beta f_1(x)\vartheta(x)\,dx \right| < \varepsilon.$$

Again, since $\varepsilon > 0$ is arbitrary, the theorem follows. $\square$

Now we can prove the theorem:

Proof of Theorem 14.4. Our first claim is that it suffices to prove the theorem for bounded locally Riemann integrable functions f which are nonnegative, i.e., $f(x) \ge 0$ for all $x \in \mathbb{R}$. In fact, for a function f, if we define the functions f_+, f_- by

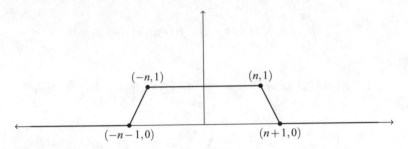

Fig. 14.2 The graph of u_k

$$f_+(x) = \max(f(x), 0), \quad f_-(x) = -\min(f(x), 0),$$

then by Exercise 14.12,

1. $f_+(x), f_-(x) \geq 0$ for all $x \in \mathbb{R}$;
2. f_+, f_- are locally Riemann integrable functions if f is.

It is clear that if we know the theorem for the nonnegative functions f_+, f_-, then we will know the result for the function f.

For reasons that will become clear in a moment, for a natural number k we define a function $u_k(x) : \mathbb{R} \to [0, 1]$ by

$$u_k(x) = \begin{cases} +1 & |x| < k; \\ k+1-|x| & k \leq |x| \leq k+1; \\ 0 & |x| > k+1. \end{cases}$$

The graph of the function u_k looks like the diagram in Figure 14.2.

Now fix a nonnegative bounded locally Riemann integrable function f on $\mathbb{R}$, and suppose for each $x \in \mathbb{R}$ we have $f(x) \leq C$ for some constant C. For each natural number n, define a function f_k by

$$f_k(x) = f(x)u_k(x).$$

Note that

- for $x \in \mathbb{R}$, $f_1(x) \leq f_2(x) \leq f_3(x) \leq \cdots$;
- $\lim_{k \to \infty} f_k(x) = f(x)$.
- for all k and all $x \in \mathbb{R}$, $f(x) - f_k(x) \leq C\chi_{[n,\infty]}(x)$.

For a function g and a real number X we set

$$S(g, X) = \frac{1}{\#\{\gamma \in \mathbb{Q} \mid H(\gamma) \leq X\}} \sum_{\gamma \in \mathbb{Q}, H(\gamma) \leq X} g(\gamma).$$

Note that $S(g, X)$ is linear and increasing in terms of g, meaning if $g_1(x) \leq g_2(x)$ for all x, then $S(g_1, X) \leq S(g_2, X)$.

By Lemma 14.5, for all k

$$\lim_{X \to \infty} S(f_k, X) = \frac{1}{\pi} \int_{-\infty}^{+\infty} \frac{f_k(t)}{1 + t^2} \, dt.$$

We have

$$S(f, X) = S(f_k, X) + S(f - f_k, X) \leq S(f_k, X) + CS(\chi_{[k,+\infty)}, X).$$

Hence for all k and all X,

$$S(f_k, X) \leq S(f, X) \leq S(f_k, X) + CS(\chi_{[k,+\infty)}, X). \tag{14.9}$$

By Theorem 14.3 we have

$$\lim_{X \to \infty} S(\chi_{[k,+\infty)}, X) = \lim_{X \to \infty} \frac{\#\{\gamma \in \mathbb{Q} \cap [k, \infty) \mid H(\gamma) \leq X\}}{\#\{\gamma \in \mathbb{Q} \mid H(\gamma) \leq X\}}$$

$$= \frac{1}{\pi} \int_k^{+\infty} \frac{1}{1 + t^2} \, dt < \frac{1}{\pi} \int_k^{+\infty} \frac{dt}{t^2} = \frac{1}{\pi k}.$$

Now in (14.9) we let $X \to \infty$ to obtain

$$\frac{1}{\pi} \int_{-\infty}^{+\infty} \frac{f_k(t)}{1 + t^2} \, dt = \lim_{X \to \infty} S(f_k, X) \leq \liminf_{X \to \infty} S(f, X)$$

$$\leq \limsup_{X \to \infty} S(f, X) \leq \lim_{X \to \infty} S(f_k, X) + \frac{C}{\pi k} = \frac{1}{\pi} \int_{-\infty}^{+\infty} \frac{f_k(t)}{1 + t^2} \, dt + \frac{C}{\pi k}.$$

Now we let $k \to \infty$. We obtain

$$\lim_{k \to \infty} \frac{1}{\pi} \int_{-\infty}^{+\infty} \frac{f_k(t)}{1 + t^2} \, dt \leq \liminf_{X \to \infty} S(f, X)$$

$$\leq \limsup_{X \to \infty} S(f, X) \leq \lim_{k \to \infty} \frac{1}{\pi} \int_{-\infty}^{+\infty} \frac{f_k(t)}{1 + t^2} \, dt.$$

At this point we can simply use the Monotone Convergence Theorem [41, Theorem 11.28] to conclude

$$\lim_{k \to \infty} \int_{-\infty}^{+\infty} \frac{f_k(t)}{1 + t^2} \, dt = \int_{-\infty}^{+\infty} \frac{f(t)}{1 + t^2} \, dt, \tag{14.10}$$

but we will prove this using an elementary argument to avoid relying on measure theory. By the remark after the statement of the theorem, the integrals

$$\int_{-\infty}^{+\infty} \frac{f(t)}{1+t^2}\, dt, \quad \int_{-\infty}^{+\infty} \frac{f_k(t)}{1+t^2}\, dt, \quad \int_{-\infty}^{+\infty} \frac{f(t)-f_k(t)}{1+t^2}\, dt,$$

are all absolutely convergent. Hence we can safely write

$$\int_{-\infty}^{+\infty} \frac{f(t)}{1+t^2}\, dt = \int_{-\infty}^{+\infty} \frac{f_k(t)}{1+t^2}\, dt + \int_{-\infty}^{+\infty} \frac{f(t)-f_k(t)}{1+t^2}\, dt.$$

The functions $f(x)$ and $f_k(t)$ are equal on the interval $[-k, k]$, and for $|x| > k$, $0 \le f(x) - f_k(x) \le C$. Hence,

$$0 < \int_{-\infty}^{+\infty} \frac{f(t)-f_k(t)}{1+t^2}\, dt$$

$$\le C \int_{|t|>k} \frac{dt}{1+t^2} < C \int_{|t|>k} \frac{dt}{t^2} = \frac{2C}{k} = O\!\left(\frac{1}{k}\right).$$

Consequently,

$$\int_{-\infty}^{+\infty} \frac{f(t)}{1+t^2}\, dt = \int_{-\infty}^{+\infty} \frac{f_k(t)}{1+t^2}\, dt + O\!\left(\frac{1}{k}\right).$$

Letting $k \to \infty$ establishes (14.10), and the theorem is proved. $\square$

14.2   The unit circle

We now turn our attention to rational points on the circle $S^1 : x^2 + y^2 = 1$. Our first statement is the following easy proposition:

Proposition 14.6. *The set of points with rational coordinates is dense in the unit circle.*

Proof. It is clear that it suffices to show that rational points are dense among points with positive y-coordinates. The points P and Q on the circle with positive y-coordinates are "close" to each other if and only if their x-coordinates are close to each other. Now suppose $P = (\alpha, \beta)$, with $\beta > 0$, is a point on the unit circle. Fix $\varepsilon > 0$. We will show that there is a point of the form

$$P_m := \left(\frac{1-m^2}{1+m^2}, \frac{2m}{1+m^2} \right)$$

with rational m such that the difference between the x-coordinates of P and P_m is less than ε. Without loss of generality assume $\alpha > 0$. We also assume that ε is much smaller than α and β. Note there is an $m \in \mathbb{Q}$ such that

$$\alpha - \varepsilon < \frac{1-m^2}{1+m^2} < \alpha + \varepsilon.$$

Indeed, in order for these inequalities to hold we need

$$\sqrt{\frac{1-(\alpha+\varepsilon)}{1+\alpha+\varepsilon}} < m < \sqrt{\frac{1-(\alpha-\varepsilon)}{1+\alpha-\varepsilon}}$$

and there is certainly a rational number m satisfying these inequalities. $\square$

Our purpose in the remainder of this chapter is to give a quantitative version of this density statement, and, as before, the concept that is central to our analysis is *equidistribution*.

In order to speak of equidistribution we need to have a notion of integral. In the case of the unit circle if we parametrize the circle as

$$(\cos\gamma, \sin\gamma), \quad 0 \le \gamma < 2\pi$$

then the natural integration will be relative to $d\gamma$, i.e., if f is a continuous function on the circle then we define

$$\int_{S^1} f := \frac{1}{2\pi} \int_0^{2\pi} f(\cos\gamma, \sin\gamma)\, d\gamma.$$

Definition 14.7. Suppose ϑ is a function on S^1. We say a sequence $\{x_n\}_{n=1}^\infty$ of elements of S^1 is ϑ-*equidistributed*, or *equidistributed with respect to the function* ϑ, if for every continuous function f on S^1 we have

$$\lim_{N\to\infty} \frac{1}{N} \sum_{n=1}^N f(x_n) = \int_{S^1} f\vartheta.$$

If $\vartheta(x, y) = 1$ for all $(x, y) \in S^1$, we simply say the sequence $\{x_n\}$ is *equidistributed* on S^1.

Recall from §3.2 that we have an explicit parametrization of the points on the circle S^1:

$$\eta(t) := \left(\frac{1-t^2}{1+t^2}, \frac{2t}{1+t^2}\right) \tag{14.11}$$

with $t \in \mathbb{R}$, plus the point $(-1, 0)$ which corresponds to t being equal to "infinity." Also, recall that if $\gamma \in \mathbb{Q}$, then $\eta(\gamma)$ is a point with rational coordinates on the circle, and that the set of points $\eta(\gamma)$ for $\gamma \in \mathbb{Q}$ with the point $(-1, 0)$ is equal to the set of points with rational coordinates on the circle.

In order to speak of equidistribution of rational points on the circle, we need a notion of ordering. A natural way to order rational points $\eta(\gamma)$, $\gamma \in \mathbb{Q}$, is according to the height of the rational numbers γ. As an example, earlier in this chapter we determined all rational numbers γ with $H(\gamma) \le 4$:

$$0, \pm 1, \pm 2, \pm 1/2, \pm 3, \pm 1/3, \pm 2/3, \pm 3/2.$$

Fig. 14.3 Points of the form
$\eta(\gamma)$ with $H(\gamma) \leq 4$

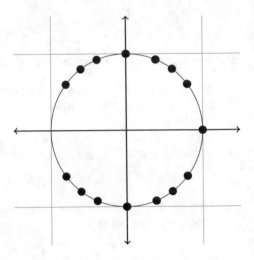

If we draw all points of the form $\eta(\gamma)$ for γ in the above list we obtain the following picture in Figure 14.3.

Theorem 14.8. *The rational points $h(\gamma)$, $\gamma \in \mathbb{Q}$, ordered according to the height of γ are equidistributed on the unit circle, i.e., for each arc ω with length t,*

$$\lim_{X \to \infty} \frac{\#\{\gamma \in \mathbb{Q} \mid H(\gamma) \leq X, \eta(\gamma) \in \omega\}}{\#\{\gamma \in \mathbb{Q} \mid H(\gamma) \leq X\}} = \frac{t}{2\pi}.$$

For a piecewise continuous function f on the unit circle,

$$\frac{1}{\#\{\gamma \in \mathbb{Q} \mid H(\gamma) \leq X\}} \sum_{H(\gamma) \leq X} f(\eta(\gamma)) \to \int_{S^1} f$$

as $X \to \infty$.

Proof. Since S^1 is a compact space and f is continuous, f is bounded. By Theorem 14.4, the limit is equal to

$$\frac{1}{2\pi} \int_{-\infty}^{+\infty} f(\frac{1 - t^2}{1 + t^2}, \frac{2t}{1 + t^2}) \frac{dt}{1 + t^2}.$$

A change of variable $t = \tan(\gamma/2)$ with $-\pi < \gamma < +\pi$ gives the result. The first statement of the theorem follows if we let f be the characteristic function of an arc. $\qquad \square$

Exercises

14.1 Construct examples of sequences $\{x_n\}_{n=1}^{\infty}$ and $\{y_n\}_{n=1}^{\infty}$ with the property that

$$\{x_n \mid n \in \mathbb{N}\} = \{y_n \mid n \in \mathbb{N}\}$$

with $\{x_n\}_n$ equidistributed, and $\{y_n\}_n$ not equidistributed.

14.2 Prove that for $\eta, \xi \in \mathbb{R}$, if $\xi < \eta$ then

$$\sum_{\xi < n \le \eta} 1 = [\eta] - [\xi]$$

14.3 Show that if $\xi \in \mathbb{R}$ and $\xi \ge 0$,

$$\sum_{0 \le n \le \xi} 1 = [\xi] + 1;$$

$$\sum_{-\xi \le n \le \xi} = 2[\xi] + 1.$$

14.4 Show that for all natural numbers n,

$$[\sqrt{n} + \sqrt{n+1}] = [\sqrt{n} + \sqrt{n+2}];$$

$$[\sqrt{n} + \sqrt{n+1}] = [\sqrt{4n+2}].$$

14.5 Let n be a natural number. Define a set D_n to be the collection of pairs $(x, y) \in \mathbb{Z}^2$ such that

$$0 < x \le n/2, \quad 0 < y \le n/2, \quad n/2 \le x + y < n.$$

Prove that

$$\#D_n = \begin{cases} \frac{(n-2)(n+8)}{8}, & 2 \mid n; \\ \frac{n^2-1}{8}, & 2 \nmid n. \end{cases}$$

14.6 Fix $n, r \in \mathbb{N}$. Find the number of solutions of

$$|x_1| + \cdots + |x_r| \le n$$

in integers $x_1, \ldots, x_r$.

14.7 Prove Equation (14.2).

14.8 Prove Equation (14.3).

14.9 Suppose $u > v > 1$. Prove that the number of integral points (m, n) within the disk $x^2 + y^2 \le X^2$ such that $n > 0$ and

$$\frac{X}{u} < m \le \frac{X}{v}$$

is equal to

$$\int_{\frac{x}{u}}^{\frac{x}{v}} \sqrt{X^2 - t^2}\, dt + O(X).$$

14.10 Show that for each $0 < \alpha < \beta$ we have

$$\int_{\beta^{-1}}^{\alpha^{-1}} \frac{dt}{1 + t^2} = \int_{\alpha}^{\beta} \frac{dt}{1 + t^2}.$$

14.11 Prove that for a bounded locally Riemann integrable function f on $\mathbb{R}$, the integral

$$\int_{-\infty}^{+\infty} \frac{f(t)}{1 + t^2}\, dt$$

converges absolutely.

14.12 For a function $f : \mathbb{R} \to \mathbb{R}$, we define the functions f_+, f_- by

$$f_+(x) = \max(f(x), 0), \quad f_-(x) = -\min(f(x), 0).$$

For all x, $f_+(x)$, $f_-(x) \geq 0$. Show that f is locally Riemann integrable if and only if f_+, f_- are locally Riemann integrable functions.

14.13 We can define another, and perhaps more natural, height function on the set of rational numbers as follows. For a rational number $\gamma = r/s$ with $r, s \in \mathbb{Z}$, $\gcd(r, s) = 1$, we set

$$H'(\gamma) := \max(|r|, |s|).$$

   a. List all rational numbers γ with $H'(\gamma) \leq 4$.
   b. Show that there is a real number $C > 1$ such that

$$C^{-1} H(\gamma) \leq H'(\gamma) \leq C H(\gamma)$$

   for all $\gamma \in \mathbb{Q}$.
   c. Find asymptotic formulae for

$$N'(X) := \#\{\gamma \in \mathbb{Q} \mid H'(\gamma) \leq X\}.$$

   and

$$N'_1(X) := \#\{\gamma \in \mathbb{Q} \cap [0, 1] \mid H'(\gamma) \leq X\}.$$

   d. Show that for a continuous function f on $[0, 1]$ we have

$$\lim_{X \to \infty} \frac{1}{N'_1(X)} \sum_{\gamma \in \mathbb{Q} \cap [0, 1]} f(\gamma) = \int_0^1 f(x)\, dx.$$

   Hint. Prove the statement for a function of the form $f(x) = x^k$, and then use the Stone–Weierstrass Theorem (in fact, Weierstrass's Theorem [41, Theorem 7.26] is sufficient).

   e. Find the function η with respect to which rational points listed according to their H' height are equidistributed.

    f. Find the function θ' on the circle S^1 which respect to which the points $\eta(\gamma)$ listed according to the H' of γ are equidistributed.

14.14 (✠) Draw a unit circle. Mark the points $\eta(t)$ with η as in Equation (14.11) and t ranging over rational number $\frac{a}{b}$ with $|a|, |b| < 1000$ and $\gcd(a, b) = 1$.

14.15 (✠) For each integral point $(x, y) \in \mathbb{Z}^2$ with $(x, y) \neq (0, 0)$, define a point $\sigma(x, y) \in \mathbb{R}^2$ with

$$\sigma(x, y) = \left(\frac{x}{\sqrt{x^2 + y^2}}, \frac{y}{\sqrt{x^2 + y^2}} \right).$$

Show that $\sigma(x, y) \in S^1$. Draw three unit circles and on each one mark one of the following collections of points:

a. $\sigma(x, y), (x, y) \in \mathbb{Z}^2, (x, y) \neq (0, 0), |x|, |y| \leq 1000$;
b. $\sigma(x, y), (x, y) \in \mathbb{Z}^2, (x, y) \neq (0, 0), |x| + |y| \leq 1000$;
c. $\sigma(x, y), (x, y) \in \mathbb{Z}^2, (x, y) \neq (0, 0), \sqrt{x^2 + y^2} \leq 1000$.

    Do you see any difference between the patterns you obtain?

14.16 (✠) Compare the patterns you obtain in the previous two exercises.

Notes

The theorem of Bohl, Sierpiński, and Weyl

Piers Bohl, Wacław Sierpiński, and Hermann Weyl proved the following important theorem around 1910 independently of each other: For each irrational α, the sequence $x_n = \{n\alpha\}, n \in \mathbb{N}$, is equidistributed in the interval $[0, 1]$, where here $\{n\alpha\}$ is the fractional part of the real number $n\alpha$. In 1916, Weyl proved the remarkable theorem that $\{n^2\alpha\}$, too, is equidistributed in $[0, 1]$, and that is how the theory of equidistribution started. Weyl also proved the following general criterion for the equidistribution of a sequence in the interval $[0, 1]$: Suppose $a_1, a_2, a_3, \ldots$ is a sequence of real numbers. Then the sequence $\{a_1\}, \{a_2\}, \{a_3\}, \ldots$ is equidistributed in the interval $[0, 1]$ if and only if for all non-zero integers m,

$$\lim_{N \to \infty} \frac{1}{N} \sum_{k=1}^{N} e^{2\pi i m a_k} = 0.$$

See [33, Ch. 12] or [22, Ch. 1] for comments on the proofs of these statements. The book [22] is a useful collection of articles exploring the various ways in which equidistribution makes an appearance in number theory.

Rational points on the sphere

In this chapter we proved the equidistribution of rational points on the unit circle. Proving the equidistribution of rational points on higher dimensional spheres, even the standard sphere in $\mathbb{R}^3$, is much more difficult. In fact, Duke [72] proved the equidistribution of rational points on the standard sphere in $\mathbb{R}^3$ only in 1998 (!). See [87] for a contemporary treatment of these problems.

Appendix A
Background

A.1   Sine, cosine, and exponentials

Theorem A.1. *For all complex numbers z,*

$$e^{iz} = \cos z + i \sin z.$$

Consequently,

$$\cos z = \frac{e^{iz} + e^{-iz}}{2}$$

and

$$\sin z = \frac{e^{iz} - e^{-iz}}{2i}.$$

Proof. It is well known that for a complex number z

$$e^z = \sum_{k=0}^{\infty} \frac{z^k}{k!};$$

$$\cos z = \sum_{k=0}^{\infty} (-1)^k \frac{z^{2k}}{(2k)!};$$

$$\sin z = \sum_{k=0}^{\infty} (-1)^k \frac{z^{2k+1}}{(2k+1)!}.$$

Once we observe $i^{4k+1} = i$, $i^{4k+2} = -1$, $i^{4k+3} = -i$, $i^{4k} = 1$, the theorem is an easy consequence of these Taylor expansions.   □

Theorem A.2. *There are n distinct complex numbers z such that $z^n = 1$. They can be expressed as*

$$e^{\frac{2\pi i k}{n}}, \quad k = 0, \ldots, n-1.$$

© Springer Nature Switzerland AG 2018
R. Takloo-Bighash, *A Pythagorean Introduction to Number Theory*,
Undergraduate Texts in Mathematics, https://doi.org/10.1007/978-3-030-02604-2

Proof. The equation $z^n = 1$ has at most n solutions. On the other hand, the above numbers, n distinct numbers, all satisfy the equation. $\square$

The following property of the exponential function is the basis of Fourier theory:

Theorem A.3. *Let k be an integer. Then*

$$\int_0^1 e^{2\pi i k x} \, dx = \begin{cases} 1 & k = 0; \\ 0 & k \neq 0. \end{cases}$$

Proof. See Exercise A.1.1. $\square$

A.2   The Binomial Theorem

For natural number n and k, with $0 \leq k \leq n$ we define

$$\binom{n}{k} = \frac{n!}{k!(n-k)!}.$$

The following theorem is fundamental:

Theorem A.4 (The Binomial Theorem). *If n is a natural number, then*

$$(x + y)^n = \sum_{k=0}^{n} \binom{n}{k} x^k y^{n-k}.$$

Proof. The proof is an easy induction and ultimately relies on the fact that

$$\binom{n}{k} = \binom{n-1}{k-1} + \binom{n-1}{k}.$$

$\square$

We now use the Binomial Theorem to prove the following theorem:

Theorem A.5. *For $k, y \in \mathbb{N}$ define*

$$\sigma_k(y) = \sum_{m=1}^{y} m^k.$$

Then there is a polynomial $f_k(x)$ with rational coefficients with leading term $x^{k+1}/(k+1)$ such that

$$\sigma_k(y) = f_k(y).$$

Proof. We will prove the theorem by induction. For $k = 1$ we have

$$\sum_{m=1}^{y} m = \frac{1}{2}y^2 + \frac{1}{2}y.$$

Now suppose we know the theorem for every $l < k$. By the Binomial Theorem

$$(m + 1)^{k+1} - m^{k+1} = \sum_{j=0}^{k-1} \binom{k}{j} m^j.$$

As a result

$$(y + 1)^{k+1} - 1 = \sum_{m=1}^{y} \{(m + 1)^k - m^k\} = \sum_{j=0}^{k} \binom{k + 1}{j} \sigma_j(y).$$

Consequently,

$$\binom{k + 1}{k} \sigma_k(y) = (y + 1)^{k+1} - 1 - \sum_{j=0}^{k-1} \binom{k + 1}{j} f_j(y).$$

By the induction hypothesis the right-hand side is a polynomial of degree $k + 1$ with leading term y^{k+1}. Once we observe

$$\binom{k + 1}{k} = k + 1$$

the theorem follows. □

Corollary A.6. *For all natural numbers k,*

$$\sigma_k(y) = \frac{y^{k+1}}{k + 1} + O(y^k).$$

A.3   The Pigeon-Hole Principle

The Pigeon-Hole Principle is the following intuitively obvious statement: If we distribute n balls among m boxes, with $n > m > 0$, then at least one box will end up with more than one ball. Stated differently, if we have n pigeon trying to get in m pigeon-holes, with $n > m > 0$, then at least one of the pigeon-holes will have two pigeons in it, hence the title *The Pigeon-Hole Principle*. The Pigeon-Hole Principle is also known as Dirichlet's Box Principle. Dirichlet (1834) used this principle to prove a theorem about rational approximation to irrational numbers. We present this theorem in Example A.11 below. The Pigeon-Hole Principle is an extremely useful statement with many applications. In this appendix we give a proof of this statement using mathematical induction. We then give several applications. The appendix ends with a few standard problems.

The Pigeon-Hole Principle should be thought of as a statement about functions. Let A be the set of pigeons and B the set of pigeon-holes. Then the process of sending pigeons to pigeon-holes is a function from $A \to B$. The technical statement of the Pigeon-Hole Principle is the following:

Theorem A.7. *Let A, B be finite sets with $\#A > \#B$. Then there are no injective maps $f : A \to B$.*

Proof. We will prove this by induction on $\#B$. If $\#B = 1$, and $\#A > 1$, it is clear that we cannot have an injective function $f : A \to B$ as there is only one option for the image of the function f. Now suppose $\#B = k \geq 2$ and that we know the theorem for every set of size $k - 1$. Suppose A is a set with $\#A > \#B$ and let $f : A \to B$ be an injective map. Pick an element $b \in B$. Since f is injective, $f^{-1}(b)$ consists of a single element $a \in A$. Then $\#(B - \{b\}) = k - 1$, and the restriction of f to $A - \{a\}$ gives a function $\tilde{f} : A - \{a\} \to B - \{b\}$. By the induction hypothesis this function $\tilde{f}$ is not injective, hence the original function f could not be injective. $\square$

Similarly one can show that if we have sets A, B with $\#A > k\#B$ for some natural number k, then there is at least one element $b \in B$ such that

$$\#f^{-1}(b) \geq k + 1.$$

We now give some examples.

Example A.8. Of every eight people, there are at least two who are born on the same day of the week. Of every fifteen people, there are at least three born on the same day of the week.

Example A.9. Of every $n + 1$ integers, there are at least two with difference divisible by n. In order to see this write $\mathbb{Z}$ as the disjoint union of the following n subsets $\mathbb{Z}_a$, $0 \leq a \leq n - 1$. For each a, let $\mathbb{Z}_a$ be the set of integers k such that $k \equiv a \bmod n$. Since we have $n + 1$ elements and n sets $\mathbb{Z}_a$, there is an a with the property that $\mathbb{Z}_a$ contains at least two elements x, y of the set. Since $x \equiv a$ and $y \equiv a$, it follows $x \equiv y \bmod n$ and consequently, $n \mid x - y$.

Example A.10. We will show that of every five distinct real numbers at least two of them satisfy

$$0 < \frac{a - b}{1 + ab} < 1.$$

Let the five numbers be $a_1, \ldots, a_5$. Since the map $\tan : (-\pi/2, \pi/2) \to \mathbb{R}$ is a bijection, there will be five angles $\theta_i \in (-\pi/2, \pi/2)$, $1 \leq i \leq 5$, such that $a_i = \tan \theta_i$. Now divide up the interval $(-\pi/2, \pi/2)$ to four subintervals $(-\pi/2, -\pi/4]$, $(-\pi/4, 0]$, $(0, \pi/4]$, and $(\pi/4, \pi/2)$. Since we have five θ_i's and four subintervals, by the Pigeon-Hole Principle at least two of them will be in the same subinterval. This means that there are indices i, j such that

$$0 < \theta_i - \theta_j < \pi/4.$$

Since $\tan$ is monotone increasing on the interval $(-\pi/2, \pi/2)$, we have

$$\tan 0 < \tan(\theta_i - \theta_j) < \tan(\pi/4).$$

Now we recall $\tan 0 = 0$, $\tan(\pi/4) = 1$, and that for angles α, β,

$$\tan(\alpha - \beta) = \frac{\tan \alpha - \tan \beta}{1 + \tan \alpha \cdot \tan \beta}.$$

We finally get

$$0 < \frac{a_i - a_j}{1 + a_i a_j} < 1$$

and we are done.

Example A.11 (Dirichlet). If α is an irrational number, then there are infinitely many rational numbers p/q, with $\gcd(p, q) = 1$, such that

$$\left| \alpha - \frac{p}{q} \right| < \frac{1}{q^2}.$$

Let n be a natural number. We will prove that there is a rational number p/q such that $1 \leq q \leq n$ with the property that

$$\left| \alpha - \frac{p}{q} \right| < \frac{1}{qn}. \tag{A.1}$$

It is not hard to see that the main claim of this example follows from this statement. Equation A.1 is equivalent to the existence of a pair of integers (p, q) with $1 \leq q \leq n$ such that

$$|q\alpha - p| < \frac{1}{n}.$$

Consider the fractional parts $\{\alpha\}, \{2\alpha\}, \ldots, \{n\alpha\}$. These are n numbers in the interval $(0, 1)$, and never a rational number, as otherwise α would be a rational number. In particular, each of them lands in the one of the following *pigeon-holes*: $(0, 1/n)$, $(1/n, 2/n), \ldots, (1 - 1/n, 1)$. If one of the $\{k\alpha\}$ falls in the first of these intervals $(0, 1/n)$, then we have $0 < \{k\alpha\} < 1/n$, which gives $0 < k\alpha - [k\alpha] < 1/n$. This verifies the assertion with $p = [k\alpha]$ and $q = k$. If none of the fractional parts falls in the first interval, then we have n fractional parts in $n - 1$ intervals. By the Pigeon-Hole Principle two of the fractional parts, $\{k\alpha\}$ and $\{l\alpha\}$ say, will be in the same interval. Without loss of generality assume $k > l$. Since the length of each of the intervals is $1/n$ we will have

$$|\{k\alpha\} - \{l\alpha\}| < 1/n.$$

The left-hand side of the inequality is equal to

$$|k\alpha - [k\alpha] - l\alpha + [l\alpha]| = |(k - l)\alpha - ([k\alpha] - [l\alpha])|.$$

The result follows with $q = (k - l) < n$ and $p = [k\alpha] - [l\alpha]$.

Exercises

A.1.1 Use Theorem A.1 or any other method to prove Theorem A.3.

A.1.2 Use Theorem A.1 to give a proof for the addition formula for sine and cosine:

$$\sin(\alpha + \beta) = \sin \alpha \cos \beta + \cos \alpha \sin \beta,$$

$$\cos(\alpha + \beta) = \cos \alpha \cos \beta - \sin \alpha \sin \beta.$$

A.1.3 Compute $\cos \frac{\pi}{7} \cdot \cos \frac{2\pi}{7} \cdot \cos \frac{3\pi}{7}$.

A.1.4 Compute the value of $\cos \frac{\pi}{7} - \cos \frac{2\pi}{7} + \cos \frac{3\pi}{7}$.

A.1.5 Let $\eta_1 = 1, \eta_2, \eta_3$ be the three third roots of 1 in $\mathbb{C}$. Find a formula for the value of $\eta_1^n + \eta_2^n + \eta_3^n$ for $n \in \mathbb{Z}$.

A.2.1 Show that for $n \in \mathbb{N}$,

$$\sum_{k=0}^{n} \binom{n}{k} = 2^n, \quad \sum_{k=0}^{n} (-1)^k \binom{n}{k} = 0.$$

A.2.2 Prove that for all natural numbers n,

$$\sum_{k=0}^{n} \binom{k}{r} = \binom{n+1}{r+1}.$$

A.2.3 Show that for all $n \in \mathbb{N}$,

$$\sum_{k=0}^{n} \binom{n}{k}^2 = \binom{2n}{n}.$$

A.2.4 Prove that for all natural n

$$\sum_{k=0}^{n} (-1)^k \binom{2n}{k} = \frac{-1}{n} \binom{2n}{n}.$$

A.2.5 Prove the identity

$$\sum_{k=1}^{n} k 2^{-2k} \binom{2k}{k} = \frac{n(n+1)}{3 \cdot 2^{2n+1}} \binom{2n+2}{n+1}.$$

A.2.6 Show that for all $n \in \mathbb{N}$, $n^2 \mid (n+1)^n - 1$.

A.2.7 Show that for all natural numbers n, k,

$$\frac{1}{k+1} n^{k+1} < \sum_{r=0}^{n} r^k < \left(1 + \frac{1}{n}\right)^{k+1} \frac{1}{k+1} n^{k+1}.$$

A.3.1 Show that if we have six numbers from the set $\{1, 2, \ldots, 10\}$ two of them add up to an odd number.

A.3.2 Show that if we have a subset $A \subset \{1, 2, \ldots, 100\}$ with ten elements, then the set A has disjoint subsets S, T whose elements have the same sum.

A.3.3 Show that if we choose a subset $S \subset \{1, 2, \ldots, 2n\}$ with $n + 1$ elements, then there are at least two integers $x, y \in S$ such that $x \mid y$.

A.3.4 Show that if we choose five points in a unit square, there are at least two of them that are at most $\sqrt{2}/2$ apart.

A.3.5 Show that of every group of n people there are two with an identical number of friends in the group.

A.3.6 Suppose we have an infinite array of natural numbers $(a_{ij})_{i,j \in \mathbb{N}}$ with the property that $a_{ij} \leq ij$. Show that for every natural number k, there is at least one natural number m which is repeated at least k times in the array.

Appendix B
Algebraic integers

Let $f \in \mathbb{Z}[x]$ be a polynomial with integer coefficients. We write

$$f(x) = \sum_{k=0}^{n} a_k x^k,$$

with $a_n \neq 0$. Then n is called the degree, and a_n the *leading coefficient*. If the leading coefficient of f is equal to 1, then f is called *monic*. For example, $3x^5 - 7x + 1$ is a polynomial of degree 5 with leading coefficient 3, and the polynomial $x^7 - 10^{487}x^2 + 57$ is monic.

Definition B.1. A complex number α is called an *algebraic integer* if there is a monic polynomial $f \in \mathbb{Z}[x]$ such that $f(\alpha) = 0$.

For example, it is clear that all integers are algebraic integers, and numbers like $1/5$ and $327/82$ are not. The complex number i is an algebraic integer as it satisfies $f(i) = 0$ with $f(x) = x^2 + 1$. More generally, every element of $\mathbb{Z}[i]$ is an algebraic integer. Every root of unity is an algebraic integer. The quadratic irrationality $-\sqrt{2}$ is an algebraic integer since it satisfies the equation $x^2 - 2 = 0$.

Lemma B.2. *If α is an algebraic integer, then there is a monic polynomial $f \in \mathbb{Z}[x]$ such that f is irreducible over $\mathbb{Q}$, and*

$$f(\alpha) = 0.$$

Proof. This is immediate from Gauss's Lemma (Corollary to Theorem 3.1, [25, Ch. 3]). □

The irreducible polynomial f in Lemma B.2 is called the *minimal polynomial* of α.

The following corollary is immediate from the lemma.

© Springer Nature Switzerland AG 2018
R. Takloo-Bighash, *A Pythagorean Introduction to Number Theory*,
Undergraduate Texts in Mathematics, https://doi.org/10.1007/978-3-030-02604-2

Corollary B.3. *If a rational number γ is an algebraic integer, then $\gamma \in \mathbb{Z}$.*

The following theorem is the main result of this section:

Theorem B.4. *If α, β are algebraic integers, then so are $\alpha + \beta$, $\alpha - \beta$, and $\alpha\beta$.*

The proof of the theorem requires a bit of preparation.

Definition B.5. A polynomial F in the n indeterminates $x_1, \ldots, x_n$ is called *symmetric* if for every $\sigma \in S_n$, the group of permutations of the set $\{1, \ldots, n\}$,

$$F(x_1, \ldots, x_n) = F(x_{\sigma(1)}, x_{\sigma(2)}, \ldots, x_{\sigma(n)}).$$

For example, the polynomial $x + y$ is a symmetric polynomial of the two variables x, y. The polynomial $x + y^2$ is not symmetric. The polynomial

$$x^2 + y^2 + z^2$$

is symmetric in the three variables x, y, z.

The simplest symmetric polynomials in the n indeterminates $x_1, \ldots, x_n$ are denoted by

$$s_1 = \sum_{1 \leq i \leq n} x_i;$$

$$s_2 = \sum_{1 \leq i < j \leq n} x_i x_j;$$

$$s_3 = \sum_{1 \leq i < j < k \leq n} x_i x_j x_k$$

$$\cdots$$

$$s_n = x_1 \cdots x_n.$$

These symmetric polynomials occur in nature as the coefficients of the polynomials with roots $x_1, \ldots, x_n$, i.e.,

$$(x - x_1) \cdots (x - x_n) = x^n - s_1 x^{n-1} + s_2 x^{n-2} + \cdots + (-1)^n s_n.$$

Not only are the s_i's the simplest symmetric polynomials, they are in fact the building blocks of all symmetric polynomials in the variables $x_1, \ldots, x_n$.

Theorem B.6. *Let $F \in \mathbb{Z}[x_1, \ldots, x_n]$ be a symmetric polynomial. Then there is a polynomial $G \in \mathbb{Z}[x_1, \ldots, x_n]$ such that*

$$F(x_1, \ldots, x_n) = G(s_1, s_2, \ldots, s_n).$$

Proof. Write F in the form

$$F(x_1, \ldots, x_n) = \sum_{r_1, r_2, \ldots, r_n \in \mathbb{N} \cup \{0\}} c(r_1, \ldots, r_n) x_1^{r_1} \cdots x_n^{r_n}$$

with $c(r_1, \ldots, r_n) \in \mathbb{Z}$. Pick the n-tuple $(r_1, \ldots, r_n)$ with the following three properties:

- $c(r_1, \ldots, r_n) \neq 0$;
- $r_1 \geq \cdots \geq r_n$;
- $w = nr_1 + (n-1)r_2 + \cdots + r_n$ is maximal. Call w the weight of F and denote it by $w(F)$.

Now consider the polynomial

$$F_1(x_1, \ldots, x_n) := F(x_1, \ldots, x_n) - c(r_1, \ldots, r_n)s_1^{r_1-r_2}s_2^{r_2-r_3}\cdots s_{n-1}^{r_{n-1}-r_n}s_n^{r_n}.$$

It is easy to see that F_1 has integral coefficients and that $w(F_1) < w(F)$. Apply the same procedure to F_1 to obtain a polynomial F_2 with $w(F_2) < w(F_1)$. By repeating this process we obtain a sequence of symmetric polynomials $F, F_1, F_2, \ldots$ such that $w(F) > w(F_1) > w(F_2) > \ldots$. For some k, we will have $w(F_k) = 0$, and that means F_k is a constant. This proves the theorem. $\square$

Now we can go back and prove our main theorem.

Proof of Theorem B.4. We will prove that $\alpha\beta$ is algebraic. The other cases are similar.

Suppose α satisfies the equation $f(\alpha) = 0$ with f a monic polynomial with integer coefficients. Write

$$f(x) = \prod_{i=1}^{k}(x - \alpha_i).$$

The algebraic integer α is one of the α_i's. As $f \in \mathbb{Z}[x]$, we see that

$$s_1 = \sum_i \alpha_i,$$

$$s_2 = \sum_{i<j} \alpha_i\alpha_j,$$

$$\vdots$$

$$s_k = \alpha_1 \cdots \alpha_k,$$

are integers.

Similarly, β satisfies an algebraic equation $g(x) = 0$ with $g \in \mathbb{Z}[x]$ a monic polynomial. Write

$$g(x) = \prod_{i=1}^{l}(x - \beta_i).$$

The algebraic integer β is one of the β_i's. Then, as before, the complex numbers

$$t_1 = \sum_i \beta_i,$$

$$t_2 = \sum_{i<j} \beta_i \beta_j,$$

$$\vdots$$

$$t_l = \beta_1 \cdots \beta_l$$

are integers.

Now consider the equation

$$h(x) = \prod_{i=1}^{k} \prod_{j=1}^{l} (x - \alpha_i \beta_j).$$

This expression has $\alpha\beta$ as a root. Also, it is symmetric in the variables α_i's and in the variables β_j's, separately. We want to show $h(x) \in \mathbb{Z}[x]$.

First write

$$h(x) = \sum_{r_1,\ldots,r_k,t \in \mathbb{N}\cup\{0\}} c(r_1,\ldots,r_k,t)\alpha_1^{r_1} \ldots \alpha_k^{r_k} x^t$$

with $c(r_1,\ldots,r_k,t)$ symmetric polynomials with integer coefficients in β_j's. By Theorem B.6 and the earlier remarks $c(r_1,\ldots,r_k,t) \in \mathbb{Z}$. Now we write

$$h(x) = \sum_t c_t x^t$$

with

$$c_t = \sum_{r_1,\ldots,r_k} c(r_1,\ldots,r_k,t)\alpha_1^{r_1} \ldots \alpha_k^{r_k}.$$

Again another application of Theorem B.6 shows that each c_t is an integer and we are done. $\square$

Remark B.7. There are several proofs for Theorem B.4. Here we briefly sketch two proofs of the theorem that rely on linear algebra methods. We encourage the reader to work out the details as an exercise.

The first proof uses the statement that a complex number α is an algebraic integer if and only if $\mathbb{Z}[\alpha]$ is $\mathbb{Z}$-module of finite rank. Now let α, β be algebraic integers. Then it is easy to see that $\mathbb{Z}[\alpha, \beta]$ is a $\mathbb{Z}$-module of finite rank, which, by the classification theorem of $\mathbb{Z}$-modules of finite rank, is free. Next, since $\alpha + \beta, \alpha\beta \in \mathbb{Z}[\alpha, \beta]$, it follows that $\mathbb{Z}[\alpha\beta]$ and $\mathbb{Z}[\alpha + \beta]$ are $\mathbb{Z}$-submodules of $\mathbb{Z}[\alpha, \beta]$, and consequently free of finite rank. This statement implies that $\alpha\beta$ and $\alpha + \beta$ are algebraic integers.

Another beautiful argument which we learned from Antoine Chambert-Loir uses the notion of the *companion matrix* of a polynomial. Let α be an algebraic integer and f_α be its minimal polynomial, and let n_α be the degree of f_α. We let C_α be the companion matrix of f_α. By definition, the characteristic polynomial of C_α is the polynomial f_α. The Cayley–Hamilton Theorem implies that C_α satisfies $f_\alpha(C_\alpha) = 0$, but since f_α is irreducible, this implies that f_α is the minimal polynomial of C_α. Then $C_\alpha : \mathbb{C}^{n_\alpha} \to \mathbb{C}^{n_\alpha}$ is a linear transformation with the roots of f_α as its eigenvalues. Similarly, for an algebraic integer β, we define f_β, n_β, and $C_\beta : \mathbb{C}^{n_\beta} \to \mathbb{C}^{n_\beta}$ as above. Then the fact that $\alpha + \beta$ is an algebraic integer follows from the following two statements:

- $\alpha + \beta$ is an eigenvalue of $C_\alpha \otimes I_{n_\beta} + I_{n_\alpha} \otimes C_\beta : \mathbb{C}^{n_\alpha} \otimes \mathbb{C}^{n_\beta} \to \mathbb{C}^{n_\alpha} \otimes \mathbb{C}^{n_\beta}$. Here for each n, $I_n : \mathbb{C}^n \to \mathbb{C}^n$ is the identity map.
- The characteristic polynomial of operator $C_\alpha \otimes I_{n_\beta} + I_{n_\alpha} \otimes C_\beta$ is monic with integer coefficients.

The proof for $\alpha\beta$ is similar, except that here one considers $C_\alpha \otimes C_\beta : \mathbb{C}^{n_\alpha} \otimes \mathbb{C}^{n_\beta} \to \mathbb{C}^{n_\alpha} \otimes \mathbb{C}^{n_\beta}$.

Exercises

B.1 Show that $\sqrt{2} + \sqrt[3]{5}$ is an algebraic integer by explicitly finding the algebraic equation that this number satisfies.

B.2 Write the following polynomials in the terms of the basic symmetric functions:

    a. $x^2 + y^2 + z^2$;
    b. $x^3 + y^3 + z^3$;
    c. $x^4 + y^4 + z^4$;
    d. $(x - y)^2(y - z)^2(z - x)^2$.

B.3 Let α, β, γ be the three roots of the polynomial $x^3 + 7x^2 - 8x + 3$. Find the polynomial with rational coefficients whose roots are the following numbers:

    a. $\alpha^2, \beta^2, \gamma^2$;
    b. $1/\alpha, 1/\beta, 1/\gamma$;
    c. $\alpha^3, \beta^3, \gamma^3$.

Appendix C
SageMath

SageMath is a free, open-source mathematical software which is a viable, powerful alternative to commercial computing packages such as Maple, or Mathematica. In this appendix we give a minimal introduction to SageMath. Bard's book [6], freely available online, is a good comprehensive introduction to the software with many examples. This book is our main reference for this appendix. Another useful reference for number theoretic applications of SageMath is Stein [49] where many numerical examples are worked out using SageMath.

SageMath is freely available for download from http://www.sagemath.org/. There are also two internet-based ways to use SageMath:

- SageMathCell is a web interface for SageMath, suitable for almost any everyday quick computation including all the computational exercises in this book. The website is https://sagecell.sagemath.org/
- CoCalc is a web service for online computation with the capability to support large volume computations, classroom support, etc., available at https://cocalc.com/

Here are some resources to get you started on SageMath. The online reference for SageMath is

www.sagemath.org/doc/reference

The online tutorial is available here

www.sagemath.org/doc/tutorial

A number of quick reference sheets containing very minimal lists of commands are available at

https://wiki.sagemath.org/quickref

To get acquainted with SageMath, the easiest way is to work within SageMathCell. This interface provides a window in which to type commands. There is also an Evaluate button to execute the commands (or one could press Shift and Enter at the same time).

© Springer Nature Switzerland AG 2018
R. Takloo-Bighash, *A Pythagorean Introduction to Number Theory*,
Undergraduate Texts in Mathematics, https://doi.org/10.1007/978-3-030-02604-2

C.1   Basic operations

To add numbers, one just types +, e.g., 2 + 3 gives 5. Multiplication is *, 2*3 will evaluate to 6, as it should. Power operation is written as 2^3, which will give 8. Division is more interesting: evaluating 4/5 gives 4/5. In order to get the decimal expansion, one needs to enter N(4/5), which returns 0.800000000000000. Square root is similar. Evaluating sqrt(8) produces 2 * sqrt(2). Typing N(sqrt(8)) and pressing Evaluate gives 2.82842712474619. For other roots, one can type in

```
N(3^(1/6)).
```

For the exponential function one can try exp(3) or e^3, or for the numerical value N(exp(3)). Logarithms are also easy: log(3) returns the natural log of 3, whereas log(3, 7) gives the logarithm of 3 in base 7. Entering sqrt(-4, all=true) gives [2*I , -2 *I], which means the list consisting of the complex numbers $2i$ and $-2i$. To try something a little more complicated one could try typing in

```
N(100*(1 + sqrt(2) + log(5, 62) )^5)
```

which immediately returns 17339.1704246701. For more precision, one could type

```
N(100*(1 + sqrt(2) + log(5, 62) )^5, prec=200)
```

or

```
numerical_approx(100*(1 + sqrt(2) + log(5, 62) )^5,
digits=200)
```

which returns 200 digits.

SageMath can, very easily, plot functions. For example, plot(3*exp(x+5)) plots the function $f(x) = 3e^{x+5}$ for $-1 < x < +1$. To get other ranges, e.g., $-3 < x < 5$, one types

```
plot(3*exp(x+5), -3, 5)
```

There are various other things one can do with plot, e.g., setting bounds in the y direction, superimposing graphs, etc., see [6, Ch. 3] for more details on plotting functions. One can also define functions. For example, one can define a function $f(x)$ by

```
f(x) = x^2 - 2
```

Next, evaluating `f(3)` returns 7. One could also plot the function by typing `plot(f(x))`.

C.2   Basic number theory

Here we review some of the most basic number theoretic operations that SageMath can do.

Prime numbers

The command

```
primes_first_n(55)
```

lists the first 55 prime numbers:

```
[2, 3, 5, 7, 11, 13, 17, 19, 23, 29, 31, 37, 41, 43,
47, 53, 59, 61, 67, 71, 73, 79, 83, 89, 97, 101,
103, 107, 109, 113, 127, 131,  137, 139, 149,
151, 157, 163, 167, 173, 179, 181, 191, 193,
197, 199, 211, 223, 227, 229, 233, 239, 241,
251, 257]
```

The command

```
is_prime(157)
```

checks the primality of 157, and returns `True`. Typing

```
next_prime(10057)
```

gives 10061 which is the next prime after 10057. There is also a similar command

```
previous_prime(10057)
```

The get the prime numbers in a certain range, e.g., 120 to 137, we use the command

```
prime_range(120, 137)
```

We get [127, 131] as the answer. If we need to find the 112th prime number, all we need to do is to type

```
nth_prime(112)
```

to see that that number is 613. Another useful command is

```
random_prime(10^20,10^30)
```

which returns a random prime number between 10^{20} and 10^{30}. Typing

```
prime_pi(x)
```

returns the number of prime numbers up to x.

Divisors

The command factor factorizes a number into a product of its prime factors, e.g., factor(12) gives

```
2^2 * 3
```

To get the list of divisors of a number we use the command divisors. For example divisors(325) gives the answer

```
[1, 5, 13, 25, 65, 325]
```

The function $\sigma_k(n) = \sum_{d|n} d^k$ is given by sigma(n, k). For example, sigma(325, 0) simply counts the number of divisors of 325 and returns 6. The command len(divisors(325)) would have done the same thing. The commands gcd and lcm compute gcd and lcm. For example, gcd(12, 18) returns 6, and lcm(12, 18) returns 36. The command xgcd(a,b) returns a triple (d, u, v) with $d = \gcd(a, b)$ and $au+bv = d$. For example, xgcd(12,15) gives (3, -1, 1).

Modular arithmetic

Suppose we divide a by b, and we write $a = bq + r$. To find the remainder r of a when divided by b, one can type a % b. For example

```
329 % 162
```

returns 5. We could have alternatively used the command `mod(329, 162)` to get the same answer. To find the integer quotient q, we write $a//b$. For example, `329 // 162` gives 2. To find the modular inverse of the number 3 modulo 2005 we enter

```
inverse_mod(3, 2005).
```

The answer is 1337. One can verify this by checking that

```
(1337*3)%2005
```

in fact returns 1.

SageMath has the capability to do modular arithmetic. Suppose we want to compute the order of 5 modulo 7. In order to do this, we type

```
R = Integers(7)
a = R(5)
multiplicative_order(a)
```

This will produce 6 as the answer, which means that 5 is a primitive root modulo 7. One can check this by entering

```
[c^i for i in range(6)]
```

This last command produces $[1, 5, 4, 6, 2, 3]$.

An alternative way to do modular arithmetic is to use the `Mod` operator. For example, if we want to compute 2^{75} mod 1000, we can simply type

```
Mod(2, 1000)^75
```

which very quickly returns 568. To compute the multiplicative inverse we can execute the command

```
Mod(3, 1000)^(-1)
```

which produces 667.

The Chinese Remainder Theorem

A useful command is the Chinese Remainder Theorem command `CRT`. Entering `CRT(a, b, m, n)` finds an integer x such that

$$\begin{cases} x \equiv a \mod m \\ x \equiv b \mod n. \end{cases}$$

For example, CRT (2, 1, 3, 5) returns 11. If we have more than two congruence equations, we have to use

```
CRT_list([a_1, a_2, \dots, a_m], [n_1, n_2, \dots, n_m])
```

For example,

```
CRT_list([1, 2, 3], [5, 7, 9])
```

returns 156.

The Euler totient function

To calculate the Euler totient function of a number, e.g., 10032 we type in

```
euler_phi(10032)
```

to obtain 2880. SageMath can also find primitive roots. Typing

```
primitive_root(25)
```

returns 2 which is a primitive root modulo 25—in fact, this command returns the smallest primitive root modulo 25. If one enters

```
primitive_root(36)
```

the output will be the message ValueError: no primitive root.

Quadratic residues

SageMath has built-in functions to handle quadratic residues and related functions. For example,

```
quadratic_residues(7)
```

produces [0, 1, 2, 4] which is the list of quadratic residues modulo 7 plus 0. Note that this is different from our convention in Chapter 6 where a quadratic residue was defined to be coprime to p. The command for the Legendre symbol is

```
legendre_symbol(a, p)
```

For example,

```
legendre_symbol(3, 7)
```

gives -1. The command for the Jacobi symbol is

```
jacobi_symbol(a, n)
```

which works similar to the Legendre symbol.

Sums of squares

The command

```
two_squares(5)
```

returns [1, 2], and $5 = 1^2 + 2^2$. The command

```
three_squares(6)
```

gives [1, 1, 2]. The command

```
four_squares(8)
```

produces [0, 0, 2, 2].

C.3   Polynomial operations

Here we briefly explain how to work with polynomials in SageMath.

Polynomials over the real or complex numbers

Let us define the polynomials $a(x)$ and $b(x)$ by setting

```
a(x) = x^3 - 1
b(x) = x^2 - x - 2
```

Evaluating a(2) gives 7. The command a(x) + b(x) returns

```
x^3 + x^2 - x - 3
```

Typing `a(x)*b(x)` gives

```
(x^3 - 1)*(x^2 - x - 2)
```

To do the multiplication one needs to enter `expand(a(x)*b(x))` which returns

```
x^5 - x^4 - 2*x^3 - x^2 + x + 2
```

The command `factor(a(x))` returns

```
(x^2 + x + 1)*(x - 1)
```

One can also compute the gcd of the polynomials by entering `gcd(a(x), b(x))` to obtain 1. Typing in `factor(lcm(a(x),b(x)))` gives

```
(x^2 + x + 1)*(x + 1)*(x - 1)*(x - 2)
```

To solve the equation `a(x)=0` one simply types `solve(a(x), x)`. The outcome is

```
[x == 1/2*I*sqrt(3) - 1/2, x == -1/2*I*sqrt(3) - 1/2,
x == 1]
```

The `solve` operator that we just introduced is a useful, versatile device that can be used in a variety of settings. For example, entering

```
var('z')
solve([a(x)-z==0, b(x)-2*z^2==5], x, z)
```

solves the system

$$\begin{cases} a(x) - z = 0, \\ b(x) - 2z^2 = 5. \end{cases}$$

The answer is

```
[[x == (1.214514354475611 + 0.4405103357723433*I),
z == (0.0844362836387264 + 1.863837112673745*I)],
[x == (1.214514354475611 - 0.4405103357723433*I),
z == (0.08443628363872642 - 1.863837112673745*I)],
 [x == (-0.9751234960329906 + 0.7411666213498296*I),
 z == (-0.3202238106249589 + 1.707106500754547*I)],
[x == (-0.9751234960329906 - 0.7411666213498296*I),
z == (-0.320223810624959 - 1.707106500754547*I)],
[x == (-0.2393908584426201 + 1.319030559283378*I),
z == (0.2357875269862346 - 2.068131317220872*I)],
[x == (-0.2393908584426201 - 1.319030559283378*I),
z == (0.2357875269862422 + 2.068131317220871*I)]]
```

Note that we did not have to declare the variable x as it is the default variable.

We refer the reader to the first chapter of [6] for other operations involving polynomials.

Polynomials modulo integers

We can specify the polynomial ring we work in using the command

```
R.<x> = PolynomialRing(Integers(7))
```

Then if we type

```
expand((3*x^2+5)*(2*x^3+3))
```

we obtain

```
6*x^5 + 3*x^3 + 2*x^2 + 1
```

If we type in

```
(x^3+1).roots()
```

we receive [(6, 1), (5, 1), (3, 1)] which lists the roots of $x^3 + 1$ in mod 7 numbers and their multiplicities. If we type

```
(3*x^2+5).roots()
```

we get [ ] in response which means the empty set, i.e., the polynomial $3x^2 + 5$ has no roots in mod 7 numbers.

Elliptic curves

In the Notes to Chapter 3 we defined a group law on the set of rational points on an elliptic curve $y^2 = x^3 + ax + b$ with $a, b \in \mathbb{Q}$. The command

```
E = EllipticCurve([0, 17])
```

defines the elliptic curve $y^2 = x^3 + 0 \cdot x + 17$, and typing the command

```
E
```

returns

```
Elliptic Curve defined by y^2 = x^3 + 17 over Rational Field
```

We can also add points on elliptic curves:

```
A=E([-1, 4])
B=E([2,5])
A+B
```

will produce

```
(-8/9 : -109/27 : 1)
```

or

```
A+A
```

will give

```
(137/64 : -2651/512 : 1)
```

Note that the answers are always produced as triples $(a : b : c)$ considered in the *projective space* with $c = 0$ or 1. If $c = 0$, then the resulting point is the identity point of the elliptic curve group law, i.e., the point at infinity. SageMath can compute elliptic curve invariants such as *torsion subgroup* and *rank* but since we are not using those quantities in this book, we will not review them in this brief appendix.

SageMath is incredibly diverse, and this brief appendix is far from a satisfactory introduction. As mentioned at the beginning of this appendix, there are a variety of resources available on the web which one can use to look up commands. The wonderful thing about SageMath is that it is an open-source Python-based software, and one can do actual Python programming within the software. Also, SageMath is constantly growing thanks to a large group of individuals who have devoted many, many hours developing the code to perform various mathematical tasks. And if anyone realizes that there is something that SageMath is missing, they can get involved in the effort.

References

1. Ahlfors, Lars V. *Complex analysis: An introduction of the theory of analytic functions of one complex variable*. Second edition McGraw-Hill Book Co., New York-Toronto-London 1966 xiii+317 pp.
2. Apostol, Tom M. *Introduction to analytic number theory*. Undergraduate Texts in Mathematics. Springer, New York-Heidelberg, 1976. xii+338 pp.
3. Aristotle, *Prior Analytics*. Book I. Translated with an introduction and commentary by Gisela Striker. Oxford Univdersity Press. 2009.
4. Artin, Emil, *The gamma function*. Translated by Michael Butler. Athena Series: Selected Topics in Mathematics Holt, Rinehart and Winston, New York-Toronto-London 1964 vii+39 pp.
5. Artmann, Benno. *Euclid—the creation of mathematics*. Springer, New York, 1999. xvi+343 pp.
6. G. Bard, *Sage for Undergraduates*, American Mathematical Society, available for download at http://bookstore.ams.org/mbk-87/
7. Berndt, Bruce C.; Evans, Ronald J.; Williams, Kenneth S. Gauss and Jacobi sums. Canadian Mathematical Society Series of Monographs and Advanced Texts. A Wiley-Interscience Publication. John Wiley and Sons, Inc., New York, 1998. xii+583 pp.
8. Borevich, A. I.; Shafarevich, I. R. *Number theory*. Translated from the Russian by Newcomb Greenleaf. Pure and Applied Mathematics, Vol. 20 Academic Press, New York-London 1966 x+435 pp.
9. Burton, David M. *The history of mathematics. An introduction*. Second edition. W. C. Brown Publishers, Dubuque, IA, 1991. xii+678 pp.
10. Carmichael, Robert, *Diophantine Analysis*, First edition. John Wiley and Sons. 1915
11. Cassels, J. W. S. *An introduction to the geometry of numbers*. Corrected reprint of the 1971 edition. Classics in Mathematics. Springer, Berlin, 1997. viii+344 pp.
12. Cassels, J. W. S. *Rational quadratic forms*. London Mathematical Society Monographs, 13. Academic Press, Inc. [Harcourt Brace Jovanovich, Publishers], London-New York, 1978. xvi+413 pp.
13. Conway, John H.; Smith, Derek A. On quaternions and octonions: their geometry, arithmetic, and symmetry. A K Peters, Ltd., Natick, MA, 2003. xii+159 pp.
14. Cox, David A., *Primes of the form $x^2 + ny^2$*. Fermat, class field theory, and complex multiplication. Second edition. Pure and Applied Mathematics (Hoboken). John Wiley & Sons, Inc., Hoboken, NJ, 2013. xviii+356 pp.
15. Dickson, Leonard Eugene. *History of the theory of numbers. Vol. I: Divisibility and Primality*, Carnegie Institute of Washington, 1919.

© Springer Nature Switzerland AG 2018
R. Takloo-Bighash, *A Pythagorean Introduction to Number Theory*,
Undergraduate Texts in Mathematics, https://doi.org/10.1007/978-3-030-02604-2

16. Dickson, Leonard Eugene. *History of the theory of numbers. Vol. II: Diophantine analysis.* Chelsea Publishing Co., New York 1966 xxv+803 pp.

17. Dumbaugh, Della; Schwermer, Joachim. *Emil Artin and beyond—class field theory and L-functions.* With contributions by James Cogdell and Robert Langlands. Heritage of European Mathematics. European Mathematical Society (EMS), Zrich, 2015. xiv+231 pp.

18. Ebbinghaus, H.-D.; Hermes, H.; Hirzebruch, F.; Koecher, M.; Mainzer, K.; Neukirch, J.; Prestel, A.; Remmert, R. *Numbers.* With an introduction by K. Lamotke. Translated from the second 1988 German edition by H. L. S. Orde. Translation edited and with a preface by J. H. Ewing. Graduate Texts in Mathematics, 123. Readings in Mathematics. Springer, New York, 1991. xviii+395 pp.

19. Edwards, Harold M. *Fermat's last theorem. A genetic introduction to algebraic number theory.* Corrected reprint of the 1977 original. Graduate Texts in Mathematics, 50. Springer, New York, 1996. xvi+410 pp.

20. Euclid. *Elements. All thirteen books complete in one volume.* The Thomas L. Heath translation. Edited by Dana Densmore. Green Lion Press, Santa Fe, NM, 2002. xxx+499 pp.

21. Gauss, Carl Friedrich. *Disquisitiones arithmeticae.* Translated and with a preface by Arthur A. Clarke. Revised by William C. Waterhouse, Cornelius Greither and A. W. Grootendorst and with a preface by Waterhouse. Springer, New York, 1986. xx+472 pp.

22. Equidistribution in number theory, an introduction. Proceedings of the NATO Advanced Study Institute (the 44th Sminaire de Mathmatiques Suprieures (SMS)) held at the Universit de Montral, Montral, QC, July 11–22, 2005. Edited by Andrew Granville and Zev Rudnick. NATO Science Series II: Mathematics, Physics and Chemistry, 237. Springer, Dordrecht, 2007. xvi+345 pp.

23. Guy, Richard K. *Unsolved problems in number theory.* Third edition. Problem Books in Mathematics. Springer, New York, 2004. xviii+437 pp.

24. Hardy, G. H.; Wright, E. M. *An introduction to the theory of numbers.* Sixth edition. Revised by D. R. Heath-Brown and J. H. Silverman. With a foreword by Andrew Wiles. Oxford University Press, Oxford, 2008. xxii+621 pp.

25. Herstein, I. N. *Topics in algebra.* Second edition. Xerox College Publishing, Lexington, Mass.-Toronto, Ont., 1975. xi+388 pp.

26. Hilbert, David. *Foundations of geometry.* Second edition. Translated from the tenth German edition by Leo Unger Open Court, LaSalle, Ill. 1971 ix+226 pp.

27. Jacobson, Michael J., Jr.; Williams, Hugh C. *Solving the Pell equation.* CMS Books in Mathematics/Ouvrages de Mathmatiques de la SMC. Springer, New York, 2009. xx+495 pp.

28. Joseph, G. G., *Crest of the Peacock: Non-European Roots of Mathematics*, Third Edition, Princeton University Press, 2011.

29. Kline, Morris, *Mathematical Thought from Ancient to Modern*, Vol 1, Oxford University Press, 1990.

30. Koblitz, Neal. *Introduction to elliptic curves and modular forms.* Second edition. Graduate Texts in Mathematics, 97. Springer, New York, 1993. x+248 pp.

31. Landau, Edmund. *Elementary number theory.* Translated by J. E. Goodman. Chelsea Publishing Co., New York, N.Y., 1958. 256 pp.

32. Lemmermeyer, Franz, *Reciprocity laws. From Euler to Eisenstein.* Springer Monographs in Mathematics. Springer, Berlin, 2000. xx+487 pp.

33. Miller, Steven J.; Takloo-Bighash, Ramin. *An invitation to modern number theory.* With a foreword by Peter Sarnak. Princeton University Press, Princeton, NJ, 2006. xx+503 pp.

34. G. H. Mossaheb, *Elementary Theory of Numbers* (in Persian), Vol 2, Soroush, Tehran. 1979. 1803 pp.

35. Murty, M. Ram. *Problems in analytic number theory.* Second edition. Graduate Texts in Mathematics, 206. Readings in Mathematics. Springer, New York, 2008. xxii+502 pp.

36. Mozzochi, C. J. *The Fermat diary.* American Mathematical Society, Providence, RI, 2000. xii+196 pp.

37. Murty, M. Ram; Esmonde, Jody. *Problems in algebraic number theory.* Second edition. Graduate Texts in Mathematics, 190. Springer, New York, 2005. xvi+352 pp.

38. Jowell, B. *The Dialogues of Plato, with analyses and introductions*, Vol IV. Oxford University Press, 1892.
39. Plofker, Kim, *Mathematics in India*, Princeton University Press, 2009.
40. Rashed, R. *Encyclopedia of the History of Arabic Science*, Vol 2.
41. Rudin, Walter. *Principles of mathematical analysis*. Third edition. International Series in Pure and Applied Mathematics. McGraw-Hill Book Co., New York-Auckland-Düsseldorf, 1976. x+342 pp.
42. Russell, Bertrand. *A history of western philosophy, and its connection with political and social circumstances from the earliest times to the present day*. New York, Simon and Schuster, 1945. xxiii+895 pp.
43. Samuel, Pierre. *Algebraic theory of numbers*. Translated from the French by Allan J. Silberger Houghton Mifflin Co., Boston, Mass. 1970, 109 pp.
44. Serre, J.-P., *A course in arithmetic*. Translated from the French. Graduate Texts in Mathematics, No. 7. Springer, New York-Heidelberg, 1973. viii+115 pp.
45. Siegel, Carl Ludwig *Lectures on the geometry of numbers*. Notes by B. Friedman. Rewritten by Komaravolu Chandrasekharan with the assistance of Rudolf Suter. With a preface by Chandrasekharan. Springer, Berlin, 1989. x+160 pp.
46. Sierpiński, W. *Elementary theory of numbers*. Second edition. Edited and with a preface by Andrzej Schinzel. North-Holland Mathematical Library, 31. North-Holland Publishing Co., Amsterdam; PWN—Polish Scientific Publishers, Warsaw, 1988. xii+515 pp.
47. Silverman, Joseph H. *The arithmetic of elliptic curves*. Corrected reprint of the 1986 original. Graduate Texts in Mathematics, 106. Springer, New York, 1992. xii+400 pp.
48. Silverman, Joseph H.; Tate, John T. *Rational points on elliptic curves*. Second edition. Undergraduate Texts in Mathematics. Springer, Cham, 2015. xxii+332 pp.
49. Stein, William, *Elementary number theory: primes, congruences, and secrets*. A computational approach. Undergraduate Texts in Mathematics. Springer, New York, 2009. x+166 pp.
50. Stewart, Ian; Tall, David, *Algebraic number theory and Fermat's last theorem*. Fourth edition. CRC Press, Boca Raton, FL, 2016. xix+322 pp.
51. Thomas, I. *Selections Illustrating the history of Greek Mathematics*, Vol. I. From Thales to Euclid. xvi+505 pp. Vol. II. From Aristarchus to Pappus. x+683 pp. Harvard University Press, Cambridge, Mass.; William Heinemann, Ltd., London, 1951.
52. Titchmarsh, E. C. *The theory of the Riemann zeta-function*. Second edition. Edited and with a preface by D. R. Heath-Brown. The Clarendon Press, Oxford University Press, New York, 1986. x+412 pp.
53. Trappe, Wade; Washington, Lawrence C., *Introduction to cryptography with coding theory*. Second edition. Pearson Prentice Hall, Upper Saddle River, NJ, 2006. xiv+577 pp.
54. Vaughan, R. C. *The Hardy-Littlewood method*. Second edition. Cambridge Tracts in Mathematics, 125. Cambridge University Press, Cambridge, 1997. xiv+232 pp.
55. van der Waerden, B. L. *Geometry and Algebra in Ancient Civiliazations*, Springer, 1983.
56. Weil, André. *Basic number theory*. Reprint of the second (1973) edition. Classics in Mathematics. Springer, Berlin, 1995. xviii+315 pp.
57. Weil, André. *Number theory. An approach through history from Hammurapi to Legendre*. Reprint of the 1984 edition. Modern Birkhuser Classics. Birkhäuser Boston, Inc., Boston, MA, 2007. xxii+377 pp.
58. Agrawal, M., Kayal, N., and Saxena, N. *PRIMES is in P*. Annals of Mathematics 160(2), 2004, 781–793.
59. Alter, Ronald; Curtz, Thaddeus B.; Kubota, K. K. *Remarks and results on congruent numbers*. Proceedings of the Third Southeastern Conference on Combinatorics, Graph Theory and Computing (Florida Atlantic Univ., Boca Raton, Fla., 1972), pp. 27–35. Florida Atlantic Univ., Boca Raton, Fla., 1972.
60. Alter, Ronald; Curtz, Thaddeus B. *A note on congruent numbers*. Math. Comp. 28 (1974), 303-305.
61. Ankeny, N. C. *Sums of three squares*. Proc. Amer. Math. Soc. 8 (1957), 316-319.

62. Baez, John C. The octonions. Bull. Amer. Math. Soc. (N.S.) 39 (2002), no. 2, 145-205.
63. Baker, Alan. *Experiments on the abc-conjecture*, Publ. Math. Debrecen, 65 (2004), pp. 253–260.
64. Chapman, R., *Evaluating* $\zeta(2)$, http://empslocal.ex.ac.uk/people/staff/rjchapma/etc/zeta2.pdf.
65. Chen, J.R. *On the representation of a larger even integer as the sum of a prime and the product of at most two primes*. Sci. Sinica 16 (1973), 157–176.
66. Cilleruelo, J., *The distribution of the lattice points on circles*, Journal of Number Theory, 43, 198–202 (1993).
67. Cilleruelo, J.; Córdoba, A. *Trigonometric polynomials and lattice points*. Proc. Amer. Math. Soc. 115 (1992), no. 4, 899-905.
68. Cilleruelo, Javier; Granville, Andrew, *Lattice points on circles, squares in arithmetic progressions and sumsets of squares*. Additive combinatorics, 241–262, CRM Proc. Lecture Notes, 43, Amer. Math. Soc., Providence, RI, 2007.
69. Conrad, K., *The Gaussian Integers*, available at http://www.math.uconn.edu/~kconrad/blurbs/ugradnumthy/Zinotes.pdf
70. Conrad, K., *The Congruent Number Problem*, available at http://www.math.uconn.edu/~kconrad/blurbs/ugradnumthy/congnumber.pdf
71. Davenport, H. *The geometry of numbers*. Math. Gaz. 31, (1947). 206–210.
72. Duke, W. *Rational points on the sphere*. Rankin memorial issues. Ramanujan J. 7 (2003), no. 1-3, 235–239.
73. Erdös, P. *On sets of distances of n points in Euclidean space*. Magyar Tud. Akad. Mat. Kutató Int. Közl. 5 (1960) 165–169, available at http://www.renyi.hu/~p_erdos/1960-08.pdf
74. Estermann, T. *On the representations of a number as a sum of squares*, Prace Matematyczno-Fizyczne (1937) Volume: 45, Issue: 1, page 93–125.
75. Gelbart, Stephen, *An elementary introduction to the Langlands program*. Bull. Amer. Math. Soc. (N.S.) 10 (1984), no. 2, 177–219.
76. Goldston, Daniel A.; Pintz, János; Yıldırım, Cem Y. *Primes in tuples. I*. Ann. of Math. (2) 170 (2009), no. 2, 819–862.
77. Granville, Andrew; Tucker, Thomas J. *It's as easy as abc*. Notices Amer. Math. Soc. 49 (2002), no. 10, 1224–1231.
78. Gross, Benedict H. *The work of Manjul Bhargava*. Proceedings of the International Congress of Mathematicians–Seoul 2014. Vol. 1, 56–63, Kyung Moon Sa, Seoul, 2014.
79. Hardy, G. H. *On the representation of a number as the sum of any number of squares, and in particular of five*. Trans. Amer. Math. Soc. 21 (1920), no. 3, 255–284.
80. Hirschhorn, M. D. *A simple proof of Jacobi's four-square theorem*. Proc. Amer. Math. Soc. 101 (1987), no. 3, 436–438.
81. Hooley, C. *Artin's conjecture for primitive roots*, J. Reine Angew. Math. 225 (1967), 209–220.
82. Laishram, Shanta. *Baker's explicit abc-conjecture and Waring's problem*. Hardy-Ramanujan J. 38 (2015), 49–52.
83. Lehmer, Derrick Norman; Asymptotic Evaluation of Certain Totient Sums. Amer. J. Math. 22 (1900), no. 4, 293–335.
84. Brillhart, John. *Emma Lehmer 1906–2007*. Notices Amer. Math. Soc. 54 (2007), no. 11, 1500–1501.
85. Maynard, James. *Small gaps between primes*. Ann. of Math. (2) 181 (2015), no. 1, 383–413.
86. Mazur, B. *Number theory as gadfly*. Amer. Math. Monthly 98 (1991), no. 7, 593–610.
87. Michel, Philippe; Venkatesh, Akshay, *Equidistribution, L-functions and ergodic theory: on some problems of Yu. Linnik*. International Congress of Mathematicians. Vol. II, 421–457, Eur. Math. Soc., Zrich, 2006.
88. Moree, Pieter. *Artin's primitive root conjecture: a survey*. Integers 12 (2012), no. 6, 1305–1416.
89. Murty, M. Ram. *Artin's conjecture for primitive roots*. Math. Intelligencer 10 (1988), no. 4, 59–67.

90. Pieper, Herbert, *On Euler's contributions to the four-squares theorem*. Historia Math. 20 (1993), no. 1, 12–18.

91. Polymath, D. H. J. *Variants of the Selberg sieve, and bounded intervals containing many primes*. Res. Math. Sci. 1 (2014), Art. 12, 83 pp.

92. Rice, Adrian; Brown, Ezra. *Why ellipses are not elliptic curves*. Math. Mag. 85 (2012), no. 3, 163–176.

93. Riemann, B. *On the Number of Prime Numbers less than a Given Quantity*. Translated from German by David R. Wilkins. Available at http://www.claymath.org/sites/default/files/ezeta.pdf

94. Rousseau, G. *On the quadratic reciprocity law*. J. Austral. Math. Soc. Ser. A 51 (1991), no. 3, 423–425.

95. Smith, Alexander, *The congruence numbers have positive natural density*, preprint.

96. Smith, Alexander, 2^∞-*Selmer groups*, 2^∞-*class groups, and Goldfeld's conjecture*, preprint.

97. Stephens, N. M. *Congruence properties of congruent numbers*. Bull. London Math. Soc. 7 (1975), 182–184.

98. Soundararajan, K. *Small gaps between prime numbers: the work of Goldston-Pintz-Y?ld?r?m*. Bull. Amer. Math. Soc. (N.S.) 44 (2007), no. 1, 1–18.

99. Sullivan, W. R., *Numerous proofs of* $\zeta(2) = \frac{\pi^2}{6}$, http://math.cmu.edu/~bwsulliv/MathGradTalkZeta2.pdf.

100. Takloo-Bighash, Ramin. *Distribution of rational points: a survey*. Bull. Iranian Math. Soc. 35 (2009), no. 1, 1–30.

101. Tian, Ye. *Congruent numbers and Heegner points*. Camb. J. Math. 2 (2014), no. 1, 117–161.

102. Tian, Y., Yuan, X., and Zhang, S.-W. *Genus Periods, Genus Points and Congruent Number Problem*, To appear in Asia J. Math.

103. Trainin, J. *An elementary proof of Pick's theorem*, The Mathematical Gazette, Vol. 91, No. 522 (2007), pp. 536–540.

104. Tschinkel, Yuri, *Algebraic varieties with many rational points*. Arithmetic geometry, 243–334, Clay Math. Proc., 8, Amer. Math. Soc., Providence, RI, 2009.

105. Tunnell, J. B. *A classical Diophantine problem and modular forms of weight 3/2*. Invent. Math. 72 (1983), no. 2, 323–334.

106. Vaughan, R. C.; Wooley, T. D. *Waring's problem: a survey*. Number theory for the millennium, III (Urbana, IL, 2000), 301–340, A K Peters, Natick, MA, 2002.

107. Waldschmidt, Michel. *Lecture on the abc conjecture and some of its consequences*. Mathematics in the 21st century, 211–230, Springer Proc. Math. Stat., 98, Springer, Basel, 2015.

108. Weil, André. *Numbers of solutions of equations in finite fields*. Bull. Amer. Math. Soc. 55, (1949). 497–508.

109. Weil, André. *Prehistory of the zeta-function*. Number theory, trace formulas and discrete groups (Oslo, 1987), 1–9, Academic Press, Boston, MA, 1989.

110. Wiles, Andrew, *Modular elliptic curves and Fermat's last theorem*. Ann. of Math. (2) 141 (1995), no. 3, 443–551.

111. Wooley, T. D. *On Waring's problems for intermediate powers*, https://arxiv.org/pdf/1602.03221.pdf

112. Zhang, Yitang. *Bounded gaps between primes*. Ann. of Math. (?) 179 (2014), no. 3, 1121–1174.

113. http://mathoverflow.net/questions/217698/many-representations-as-a-sum-of-three-squares

114. NOVA, The proof, http://www.pbs.org/wgbh/nova/proof/

Index

© Springer Nature Switzerland AG 2018
R. Takloo-Bighash, *A Pythagorean Introduction to Number Theory*,
Undergraduate Texts in Mathematics, https://doi.org/10.1007/978-3-030-02604-2

Printed in the United States
by Baker & Taylor Publisher Services